2nd International Conference on Rheology and Modeling of Materials 2015 (IC-RMM2)

Journal of Physics: Conference Series Volume 790

Lillafured, Hungary
5 - 9 October 2015

ISBN: 978-1-5108-3678-5
ISSN: 1742-6588

Printed from e-media with permission by:

Curran Associates, Inc.
57 Morehouse Lane
Red Hook, NY 12571

CURRAN ASSOCIATES INC. proceedings.com

Some format issues inherent in the e-media version may also appear in this print version.

Printed by Curran Associates, Inc. (2017)

Institute of Physics
Dirac House, Temple Back
Bristol BS1 6BE UK

Phone: 44 1 17 929 7481
Fax: 44 1 17 920 0979

techtracking@iop.org

Additional copies of this publication are available from:

Curran Associates, Inc.
57 Morehouse Lane
Red Hook, NY 12571 USA
Phone: 845-758-0400
Fax: 845-758-2633
Email: curran@proceedings.com
Web: www.proceedings.com

Table of contents

Volume 790

2nd International Conference on Rheology and Modeling of Materials (IC-RMM2)5–9 October 2015, Lillafüred, Hungary

Accepted papers received: 7 December 2016
Published online: 16 February 2017

Preface

Papers

ic-rmm2

ic-rmm2

2ND INTERNATIONAL CONFERENCE ON RHEOLOGY AND MODELING OF MATERIALS

October 5-9, 2015/ Miskolc-Lillafüred, Hungary

Organized by the International Scientific Advisory Board (**ISAB**),

International Organizing Committee (**IOC**) and

Session Chairs

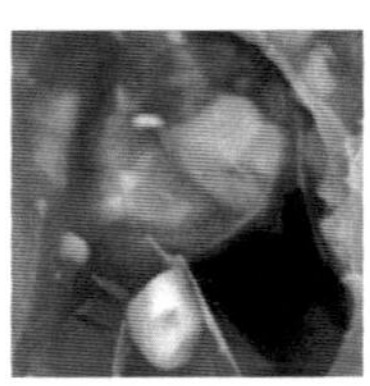

ic-rmm2

IOP Conf. Series: Journal of Physics: Conf. Series **790** (2017) 011001 doi:10.1088/1742-6596/790/1/011001

ic-rmm2

Abstract

Understanding the rheological properties of materials and their rheological behaviors during their manufacturing processes and in their applications in many cases can help to increase the efficiency and competitiveness not only of the finished goods and products but the organizations and societies also. The more scientific supported and prepared organizations develop more competitive products with better thermal, mechanical, physical, chemical and biological properties and the leading companies apply more competitive equipment and technology processes.
The aims of the 2nd **International Conference on Rheology and Modeling of Materials** (ic-rmm2) and the parallel organized symposiums of the 1st **International Symposium on Powder Injection Molding** (is-pim1) and the 1st **International Symposium on Rheology and Fracture of Solids** (is-rfs1) are the followings:

- Promote new methods and results of scientific research in the fields of modeling and measurements of rheological properties and behavior of materials under processing and applications;
- Change information between the theoretical and applied sciences as well as technical and technological implantations.
- Promote the communication and collaboration between the scientists, researchers and engineers of different disciplines, different nations, countries and continents.

The international conference ic-rmm2 and symposiums of is-pim1 and is-rfs1 provide a platform among the leading international scientists, researchers, PhD students and engineers for discussing recent achievements in measurement, modeling and application of rheology in materials technology and materials science of liquids, melts, solids, crystals and amorphous structures. Among thr major fields of interest are the influence of materials structures, mechanical stresses, temperatures, deformation speeds and shear rates on rheological and physical properties, phase transformation of foams, foods, polymers, plastics and other competitive materials like ceramics, nanomaterials, medical- and biomaterials, cosmetics, coatings, light metals, alloys, glasses, films, composites, hetero-modulus, hetero-viscous, hetero-plastic complex materials, petrochemicals and hybrid materials.

Multidisciplinary applications of rheology and rheological modeling in material science and technology encountered in sectors like alloys, ceramics, glasses, thin films, polymers, clays, construction materials, energy, aerospace, automotive and marine industry. Rheology in food, chemistry, medicine, biosciences and environmental sciences are of particular interests.

In accordance to the program of the conference ic-rmm2 and symposiums is-pim1 and is-rfs1 we have received more than **250** inquires and registrations from different organizations. Finally more than **240** abstracts were accepted for presentation. From them **12** were PLENARY lectures and **112** ORAL presentation. Researchers from **41** countries of Asia, Europe, Africa, North and South America arrived to Miskolc-Lillafüred (**Hungary**) and participated in the events of the conference. Including co-authors, the research work of more than **700** scientists were presented in the SESSIONS and SYMPOSIUMS of ic-rmm2 conference.

Prof. Dr. László A. Gömze
chair, ic-rmm2

IOP Conf. Series: Journal of Physics: Conf. Series **790** (2017) 011001 doi:10.1088/1742-6596/790/1/011001

ic-rmm2

THE INTERNATIONAL SCIENTIFIC ADVISORY BOARD (ISAB)

Prof. Dr. Igor EMRI, *President of International Committee of Rheology (**Slovenia**)*

Prof. Crispulo GALLEGOS *President of European Society of Rheology in 2010-2014*

Prof. Jozef L. KOKINI, *University of Illinois Urbana (**USA**)*

Prof. Jae Chun HYUN, *Korea University (**Korea**)*

Prof. Dr. László A. GÖMZE, *University of Miskolc, (**Hungary**)*

Prof. Dr. Sergei N. KULKOV, *Tomsk State University (**Russia**)*

THE INTERNATIONAL ORGANIZING COMMITTEE (IOC)

Dr. Fares BEAINY, *Volvo Construction Equipment (**USA**)*

Dr. Dmitry BORIN, *Technical Universizy of Dresden (**Germany**)*

Dr. Adorján BOROSNYÓI, *Technical University of Budapest (**Hungary**)*

Dr. Fatiha BOUDJEMAA, *Université de Khemis Miliana (**Algeria**)*

Dr. Suraj S. DESHMUKH, *Dow AgroSciences LLC (**India**)*

Dr. Ivica DURETEK, *Montanuniversitat Leoben (**Austria**)*

Dr. Nicky Eshtiaghi, *RMIT University (**Australia**)*

Dr. Ludmila N. GÖMZE, *IGREX Engineering Service Ltd. (**Hungary**)*

Prof. Dr. Maxim KHRAMCHENKOV, *Kazan Federal University (**Russia**)*

Prof. Dr. Sergey N. KULKOV, *Tomsk State University (**Russia**)*

Prof. Dr. Vasily LEBEDEV, *Petersburg Nuclear Physics Institute (**Russia**)*

Dr. Gizéla LECOMTE-NANA, *ENSCI Limognes University (**France**)*

Dr. Martin LEDERER, *University of Viena (**Austria**)*

Prof. Dr. André MOLENAAR, *Delft University of Tecnology (**The Netherlands**)*

Prof. Dr. Blaise NSOM, *Université de Bretagne Oc. (**France**)*

Prof. Dr. Saoshi OGATA, *Tokyo Metropoliten University (**Japan**)*

Prof. Dr. Tomasz SADOWSKI, *Lublin University of Technology (***Poiand***)*

Dr. Maria Angelas VARGAS, *University NAM (**Mexico**)*

Dr. Hao WANG, *Wuhan University of Science and Technology (**China**)*

Prof. Dr. Ichiro YONENAGA, *Tohoku University Sendai (**Japan**)*

CHAIR OF THE INTERNATIONAL ORGANIZING COMMITTEE

Prof. Dr. László A. GÖMZE, *University of Miskolc, (**Hungary**)*

The financial organizer of the conference: **IGREX Engineering Service Ltd** *(Hungary)*

IOP Conf. Series: Journal of Physics: Conf. Series **790** (2017) 011001 doi:10.1088/1742-6596/790/1/011001

ic-rmm2

SESSIONS AND SYMPOSIUMS

1. Rheology and Modeling of Clays, Minerals, Rocks and Soils
2. Rheology and Modeling of Polymers
3. Rheology of Foods and Dairies
4. Avalanche Dynamics in Rheology and Phase Transitions
5. Experimental Study and Modeling of Complex Flows
6. Rheology and Modeling of Petrochemical Materials and Asphalt Pavements
7. Multiscale Mechanics of Materials
8. Rheological Fluid Mechanics
9. Rheology and Modeling of Biological Systems
10. Rheology and Modeling of Construction Materials, Mortars and Concretes
11. Rheology of Emulsions, Suspensions and Foams
12. Rheology of Ferrofluids and Magneto-reological Materials
13. Rheology and Modeling of Fluid Interfaces
14. Rheology of Melted Metals and Glasses
15. Rheology of Nanofluids
16. Rheology of Drugs and Cosmetic Materials
17. Rheology Measureme Methods, Equipments and Errors
18. Rheology of Carbon Structures (RCS)
19. Rheology and Modeling of Hetero-modulus, Hetero-viscous Complex Materials and Composites

is-pim1 The 1st International Symposium on Powder Injection Molding
is-rfs1 The 1st International Symposium on Rheology and Fracture of Solids

Some of the presented works of the ic-rmm2 conference were selected by the Editorial Board of scientific journal **Építőanyag-JSBCM** and were published in it as the followings:

R. Bansil, J. Hardcastle and M.Constantino 2015 *Építőanyag-JSBCM* **67** 150 http://dx.doi.org/10.14382/epitoanyag-jsbcm.2015.25

Y. Wang, D.A. Lakho and D.Yao 2015 *Építőanyag-JSBCM* **67** 132 http://dx.doi.org/10.14382/epitoanyag-jsbcm.2015.21

S.O. Umerova, I.O.Dulina and A.V. Ragulya 2015 *Építőanyag-JSBCM* **67** 119 http://dx.doi.org/10.14382/epitoanyag-jsbcm.2015.19

A. Slobodov, A. Uspenskiy, R. Ralys and D. Kremnev 2015 *Építőnyag-JSBCM* **67** 159 http://dx.doi.org/10.14382/epitoanyag-jsbcm.2015.27

S.N. Kulkov, S.P. Buyakova, M. Chatzinikolaidou and I. Kocserha 2015 *Építőanyag-JSBCM* **67** 155 http://dx.doi.org/10.14382/epitoanyag-jsbcm.2015.26

A. Slobodov, A. Uspenskiy, S. Yavshits and A. Mischenko 2015 *Építőanyag-JSBCM* **67** 164 http://dx.doi.org/10.14382/epitoanyag-jsbcm.2015.28

L.A. Gömze, L.N. Gömze, S.N. Kulkov, I. Shabalin, I. Gotman, F. Pedraza, G. Lecomte, T. Mayorova, E. Kurovics and A. Hamza 2015 *Építőanyag-JSBCM* **67** 143 http://dx.doi.org/10.14382/epitoanyag-jsbcm.2015.24

H. Hornahad, E.A.B.Koenders, K. van Breugel 2015 *Építőanyag-JSBCM* **67** 135

IOP Conf. Series: Journal of Physics: Conf. Series **790** (2017) 011001 doi:10.1088/1742-6596/790/1/011001

ic-rmm2

ACKNOWLEDGEMENT

In the name of **ic-rmm Conference Boards** I would like acknowledge and say many thanks to our following sponsors for their support in press-campaign and contributions the news and information about **The 1st International Conference on Rheology and Modeling of Materials** between their members and in their media:

AIPEA – Internatonal Association for the Study of Clays
AIR – Associazione Italiana di REOLOGIA, (Italy)
biomat.net – The Biomaterials Network, (Portugal)
GFC – Groupe Francais de la Céramique, (France)
ICG – International Commission of Glass
ICR - International Committee of Rheology and its member organizations
IGREX Engineering Service Ltd, (Hungary)
Journals: Academic Journals
APPLIED RHEOLOGY
cfi - Ceramic Forum International
ÉPÍTŐANYAG, Journal of Silicate Based and Composite Materials
refractiries WORLDFORUM
Rheology Bulletin

ME – University of Miskolc (Hungary)
NRC – Nordic Rheology Society (Europe)
RAS – Institute of Strength Physics and Materials Science (Russia)
SOR – The Society of Rheology (USA)
TSU - Tomsk State University (Russia)

Many thanks to colleagues of **ISAB, IOB** and to **chairs** of the SESSIONS for their support and help in organization work and in successful transaction of conference ic-rmm2 the 2nd **International Conference on Rheology and Modeling of Materials**

I would like to say many thank personally to *Prof. Dr. Igor* ***Emri*** (the president of the International Committee of Rheology), *Prof. Dr. Sergey N.* ***Kulkov*** (Tomsk State University), *Dr. Giséle* ***Lecomte-Nana*** (ENSCI Limognes University), *Dr. Martin* ***Lederer*** (University of Viena) and *Prof. Dr. Blaise* **Nsom** (Université de Bretagne Oc.) for their excellent organization works in their countries and in continents of Africa, America, Asia, Australia and Europe. Especially many thanks to editors of journal **Applied Rheology** (Switzerland) **Építőanyag-JSBCM** (Hungary) and **Rheology Bulletin** (USA) for their strong support in scientific press and media.

Prof. Dr. László A. Gömze

chair of ic-rmm2 conference board

Peer review statement

All papers published in this volume of *Journal of Physics: Conference Series* have been peer reviewed through processes administered by the proceedings Editors. Reviews were conducted by expert referees to the professional and scientific standards expected of a proceedings journal published by IOP Publishing.

Rheo-mechanical model for self-healing asphalt pavement

Gömze, A L[1,2], Gömze, L N[2]

[1] University of Miskolc, Institute of Ceramics and Polymer Engineering, Miskolc, Hungary
[2] IGREX Engineering Service Ltd., Igrici, Hungary

E-mail: femgomze@uni-miskolc.hu and igrex2009@yandex.ru

Abstract. Examining the rheological properties of different asphalt mixtures at different temperatures, pressures and deformation conditions on the combined rheo-tribometers the authors have found that the generally used Burgers-model doesn't explain the deformation properties of asphalt mixtures and pavements under loading forces and loading pressures. To understand better the rheological and deformation properties of such complex materials like asphalt mixtures and pavements the authors used Malvern Mastersizer X laser granulometer, Bruker D8 Advance X-ray diffractometer, Hitachi TM 1000 Scanning Elektronmicroscope, Tristar 3000 specific surface tester and the combined rheo-tribometer developed and patented by the authors. After the complex investigation of different asphalt mixtures the authors have found a new, more complex rheological model for the asphalts including self-healing asphalt pavements.

1. Introduction

In our days rheology is playing very important role in examination of biological and polymer systems and materials science [1-4] also in development of ceramics and construction materials with required properties [5-8]. Knowledge of rheological properties is particularly important in processing and application of asphalt mixtures [9-13] as one of the most popular building materials in the world, thanking to the urbanization and increasing traffic by day to day. In spite of this popularity there are some shortages in explanation of the physical and mechanical properties of asphalt pavements made from different local bitumen and minerals using the traditional Burgers rheological model [14-19]. The reason is that using this rheological model the Maxwell element cannot explain the resistance of pavement to the mechanical loading forces especially in changing climatic conditions like winter and summer. To add this when construction engineers would like to prepare self-healing asphalt the wetting of surface of mineral raw material particles and fillers by binders like bitumen, is playing very important role [12, 20, 21].

The physical, chemical, mechanical and rheological properties of asphalt mixtures and consolidated asphalt pavements are very strong depend not only on working temperature, but on mineralogical composition, grain size distribution, filler containment and microstructure of used raw materials and the type and proportion of the used binders. It is also important the role of chemicals used for asphalt mixtures to increase wetting properties of bitumen binders [22-25]. The goals of this research are:

- understand the phenomena of rheo-mechanical properties of complex material systems like asphalt mixtures and pavements,
- understand the influence of temperatures, filler types and grain size distributions, and loading forces on the mechanical behavior and rheological properties,
- create new rheological model which can better support engineers in development more durable asphalt mixtures as well as self-healing asphalt pavements if possible.

2. Materials and experimental procedures

Asphalt is a mixture of coarse (graves and sands) and fine (fillers) aggregates with hydrocarbon binders like bitumen the structure of which is shown in Fig.1.

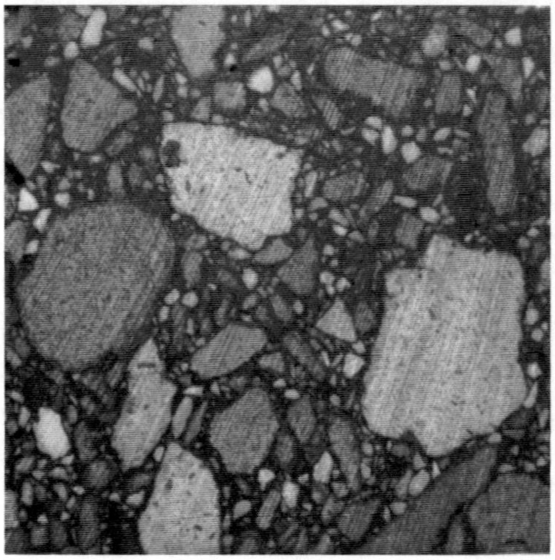

Figure 1. Structure of asphalt

Gezentsvei [26] was the first in 1950s who described the mechanical properties of this kind of material structure by rheological model of Burgers (Fig.2). He was also one of the first scientists who emphasized and underpinned the importance of temperature, granular structures on rheological properties of fresh asphalt mixtures and their behavior during compacting under the road roller [27].

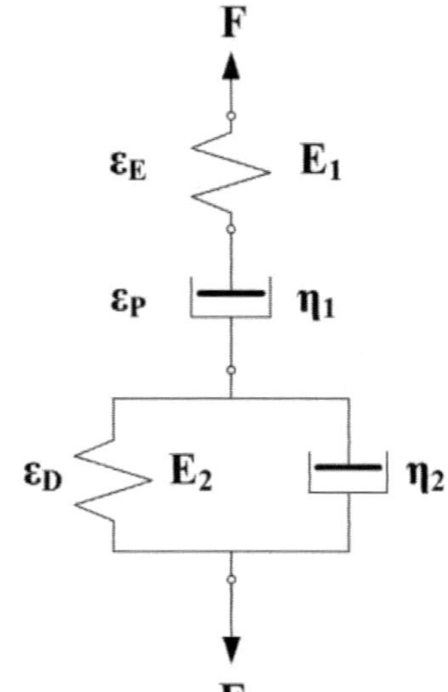

Figure 2. Rheological model of the Burgers-materials

It is obvious from the Burgers-model that these kinds of materials will have a residual and further permanent deformation in each case when an outstanding loading force will develop in it a mechanical stress:

$$\sigma > 0 \tag{1}$$

The deformation response function of this kind of materials was described by [28] as:

$$\varepsilon_0(t) = \varepsilon_l(t)\varepsilon_m(t) + \varepsilon_{k0}(t) \tag{2}$$

where:

$\varepsilon_1(t) = \sigma/E_1$ – the deformation response function of Hooke-element
$\varepsilon_1(t)$ -deformation response function of Newton-element
$\varepsilon_1(t)$ - deformation response function of Voigt-Kelvin-element

On the basis of Burgers-model the irreversible deformation of asphalt pavement is:

$$\varepsilon_m(t) = \sigma\tau/\eta_1 \tag{3}$$

and the value of which will permanently grow and never will reversible in time thanking to the new loading forces of car wheels on the road.

Generally there are three different asphalt pavements are used to prepare the roads. They are dense (closed), semi-dense (semi-closed) and light (open). All of them have different density depending on volumes of pores and gaps between the mineral particles. So the dense (closed) asphalt pavement can be characterized as:

$$\sum_{i=1}^{n} V_{ci} + \sum_{j=1}^{k} V_{pj} \leq 5\%V_0 \tag{4}$$

the semi-dense as:

$$5\%V_0 < \sum_{i=1}^{n} V_{ci} + \sum_{j=1}^{k} V_{pj} \leq 15\%V_0 \tag{5}$$

and the light as:

$$\sum_{i=1}^{n} V_{ci} + \sum_{j=1}^{k} V_{pj} > 15\%V_0 \tag{6}$$

where:

V_0 – the total volume of the examined asphalt pavement specimen (Marshall-probe)
V_{ci} – the volume of i-th elementary capillary
V_{pj} – the volume of j-th elementary pore
n – number of elementary capillaries in the examined asphalt pavement specimen
k –number of elementary pores in the examined Marshall-probe specimen

For examination the rheological properties of asphalt pavements at different temperatures under different loading forces the Marshall-probe specimens were prepared using the following five different fillers:

- Limestone filler 1
- Limestone filler 2
- Andesite + dolomite filler mixtures (50-50%)
- Andesite filler
- Basalt filler

The grain size distribution of dolomite filler is shown in Fig.3 and the specific surfaces of the fillers are shown in Table 1.

The typical microstructures of the filler components used for preparing Marshall-probe specimens of asphalt pavements are shown in Fig.4.

The rheological properties of Marshall-probes made from asphalt mixtures with different mineralogical, granular and bitumen compositions were examined on the new developed and patented [29] combined rheo-tribometer which is shown in Fig. 5. During the rheological tests the Marshall-specimens were put into the specimen holder (9), pressed with pneumatic cylinder (10) and heated through thermostat (14). The variable speed (v) of shear plate (6) was kept in a constant value depending on the experiment plan.

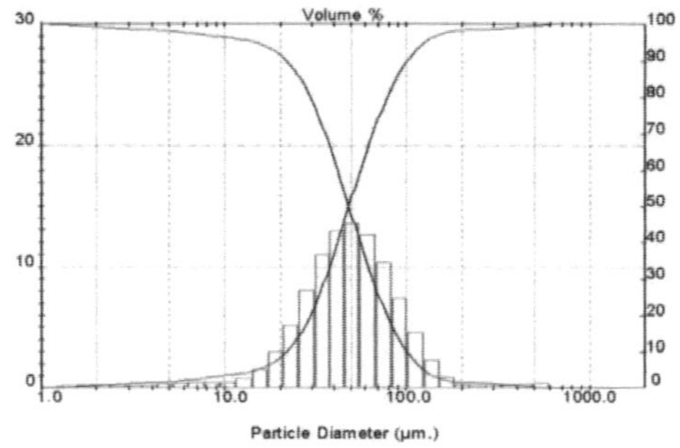

Figure 3. Grain size distribution of dolomite filler

Table 1. Specific surfaces of the used fillers

Name	pore volume	pore surface	outer surface	spec. surf. BET	spec. surf. Langmuir	spec. surf. single
	cm^3/g	m^2/g	m^2/g	m^2/g	m^2/g	m^2/g
1	0,000082	0,2185	2,0177	2,2362	3,0931	2,168
2	0,000095	0,2335	1,8329	2,0664	2,8455	2,0114
3	0,000906	2,0391	5,3887	7,4278	10,0959	7,339
4	0,000223	0,6263	7,0661	7,6924	10,5022	7,4422
5	0,000296	0,6826	1,7184	2,401	3,2578	2,3639

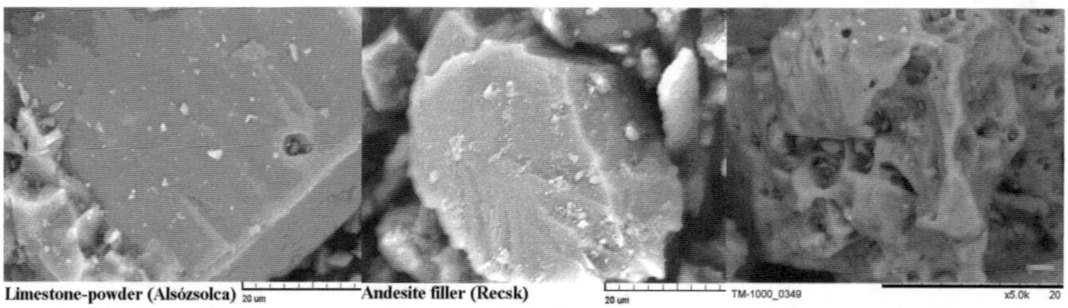

Limestone filler Andesit filler Dolomite filler

Figure 4. Microstructures of the filler components

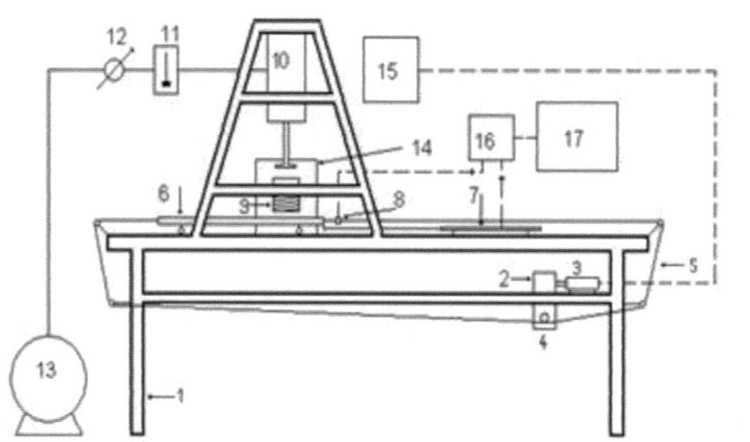

1. instrument table; 2. snail drive; 3. electric motor; 4. cable drum; 5. cableway; 6. batching car (with the shearing plate); 7. inductive displacement detector; 8. force-meter (spider); 9. heatable specimen holder; 10. pneumatic power cylinder; 11. magnetic valve; 12. pressure gauge; 13. compressor; 14. thermostat; 15. control unit; 16. data recorder (Spider 8); 17. computer (capturing and processing data)

Figure 5. Combined rheo-tribometer [29]

The operation scheme of the developed by author instrument is shown in Fig. 6.

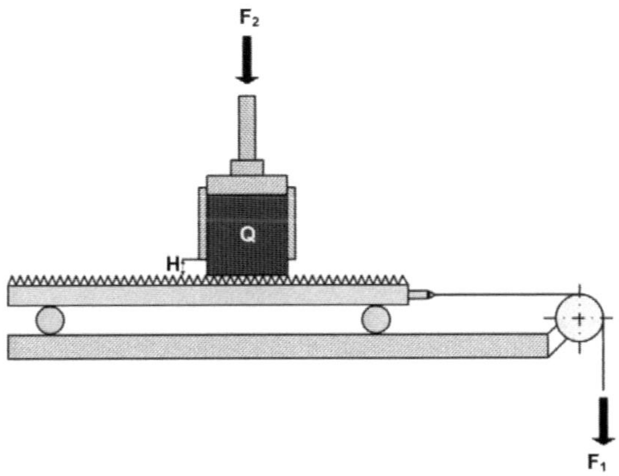

Figure 6. Operation scheme of the combined rheo-tribometer
$\mathbf{F_1}$ – moving force, $\mathbf{F_2}$ – loading force, **H** – working gap

Using this instrument we could measure the shear stress through the force meter (8) and fix it through the data recorder spider (16) by the computer (17). So we could measure and fix the developed shear stresses in Marshall-specimens as:

$$\tau = f(p, Q, T, v) \text{ [MPa]} \quad (7)$$

where:
p – the pressure effected on surface of Marshall-specimens [MPa];
Q – the material compositions of Marshall-specimens;
T – the temperature of Marshall-specimens during the shear tests [^{o}C];
v – the speeds of shear or deformations [m/s].

During the rheological tests of Marshall-specimens we were able to create the curves of specific deformation as function of time at different loading forces as it is shown in Fig. 7.

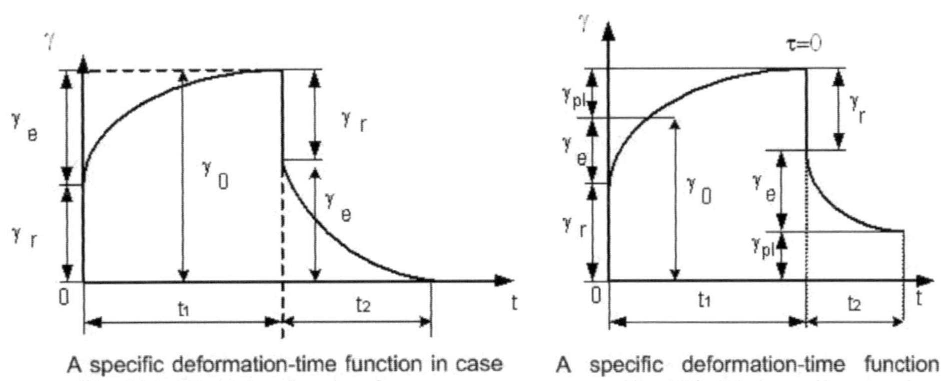

Figure 7. The specific deformation of asphalt pavements as function of time at different loading forces

3. Results and discussion

There are several papers about self-healing materials including also building materials in the last 8-10 years [30-35]. In accordance to the self-healing asphalts the scientist of The Netherland have achieved remarkable results under the supervision of professors A.A.A. Molenaar [36-40] and Erik Schlangen [41-44]. There are several methods to increase the self-healing capabilities of the asphalt pavements including application of fired nano clay particles, micro steel wools [45] and so on, but the most

IOP Conf. Series: Journal of Physics: Conf. Series **790** (2017) 012001 doi:10.1088/1742-6596/790/1/012001

important from them is the application of microcapsules containing oily rejuvenator [46-49] to enhance the self-healing ability of aged bitumen. Using only locally available aggregates in Latvia researchers could successfully develop high performance asphalt concrete. In order to achieve resistance against deformations at a high ambient temperature, a hard grade binder was used [50]. Workability, fatigue and thermal cracking resistance, as well as sufficient water resistance is achieved by low porosity (3-5%) and higher binder content compared to traditional asphalt mixtures To intensify the self-healing process generally necessary to add additional energy to asphalt pavement through increasing its temperature [36, 51] locally. All these additives and temperature will influence on the rheological parameters and properties of the asphalt pavements considerably, including also their rheo-mechanical model [28].

Using the developed by us combined rheo-tribometer (Fig. 5) we were able to measure the elastic modulus of Hooke's and Voigt-Kelvin's components of the asphalt pavements at different normal loading pressures at temperatures of 50, 60 and 80°C (Fig. 8).

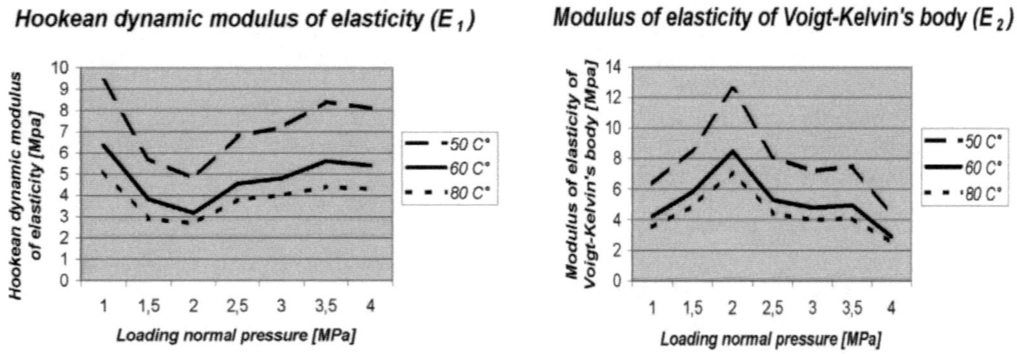

Figure 8. Hooke's ($\mathbf{E_1}$) and Voigt-Kelvin's modulus of elasticity of the tested asphalt specimens as function of normal pressure at different temperatures measured with the combined rheo-tribometer

Comparing the values of elastic modulus of Hooke's body with Voigt-Kelvin components of asphalt pavement we can underline that the Hooke element has minimal value at 2 MPa loading normal pressure meanwhile the Voigt-Kelvin element has maximum value at this pressure. Further increasing the loading pressure the Hooke's elastic modulus is growing meanwhile the values of Voigt-Kelvin's elastic modulus are decreasing. This interesting co-working of different elastic modulus of asphalt components are measured and determined first thanking to the new developed instrument. The combined rheo-tribometer gave us an opportunity to measure and determine the values of viscosities of destructed (Fig. 9a) and non-destructed Voigt-Kelvin (Fig. 9b) components of asphalt pavements.

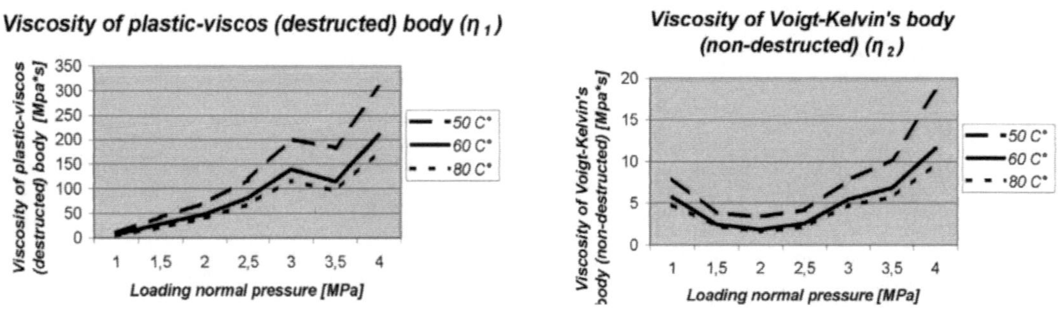

a b

Figure 9. Viscosity of destructed (a) and non-destructed (b) components of asphalt pavements as function of loading normal pressures at different temperatures

During the rheological test we found that viscosities of destructed components of self-healing asphalts are increasing continuously with increasing the values of loading normal pressures meanwhile the viscosities of non-destructed Voigt-Kelvin's components have local minimum at 1.5-2.5 MPa loading normal pressures. When the loading normal pressures are higher than 2.0-2.5 MPa the viscosities are increasing. These viscous and elastic properties of bitumen based asphalt mixtures together with the temperature very strong influence on the self-healing process of asphalt pavements.

The patented combined rheo-tribometer [29] gave us opportunity also to determine the static yield point of asphalt pavements (Marshall-specimen) as function of loading normal pressures at temperatures of 50, 60, 80°C (Fig. 10).

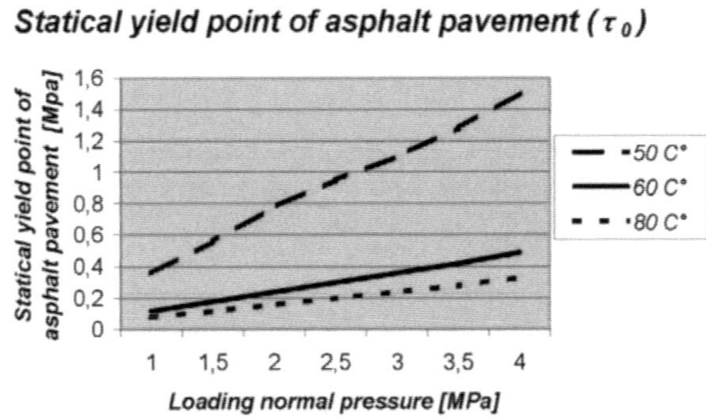

Figure 10. Static yield points of self-healing asphalt pavements as function of normal loading pressures at different temperatures

From Fig. 10 it is obvious that the self-healing asphalt mixtures – asphalt pavements – have excellent yield properties at normal climate conditions when the temperatures are less than 50°C, as increasing the loading normal pressures on them their static yield points are increasing considerably. So, their values of plasticity together with destructed and non-destructed viscosities and modulus of elasticity are increasing together with loading pressures. This means that as larger the mechanical load, as larger the mechanical resistance. The materials with these mechanical properties can be rheological modeling by Fig. 11, and their effective viscosity can be determined by eq. (8) which first time was published in Hungarian in 1983 [52]:

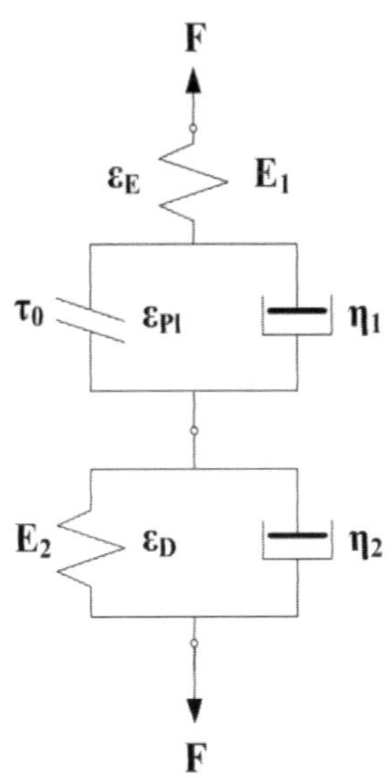

- E_1: Hookean dynamic modulus
- E_2: Elasticity modulus of the Voigt-Kelvin component
- ε_E: Actual elastic deformation
- ε_{Pl}: Plastic viscous deformation
- ε_D: Delayed elastic deformation
- η_1: Viscosity of the Newton component
- η_2: Viscosity of the Voigt-Kelvin component
- τ_0: static yield point of the plastic-viscous component

Figure 11. The rheological model of the self-healing asphalt mixtures and pavements

$$\eta_e = \frac{\tau_0+\eta_1\dot{\gamma}+\eta_1 n_\gamma \ddot{\gamma}}{\dot{\gamma}+n_\tau+n_\gamma\ddot{\gamma}+\dot{\gamma}\left[n_\tau+n_\gamma\left(1+\frac{\eta_1}{\eta_2}\right)\right]} \tag{8}$$

where

- η_1, η_2, τ_0 – as in Fig. 11
- n_γ – the delayed time of elastic deformation and can be determined as: $n_\gamma = \eta_2/E_2$
- n_τ – the stress relaxation time, and can be determined as: $n_\tau = \eta_1/E_1$
- $\dot{\gamma}$ - the total value of shear rate (the sum of shear rates of destructed and non-destructed components)
- $\ddot{\gamma}$ - derivative of total shear rate

The description of effective viscosity by equation (8) give us a better understanding of rheological properties – rheological requirements – of self-healing materials including self-healing asphalt mixtures and pavements.

4. Conclusion

The Burgers rheological model (Fig. 2) cannot explain the self-healing properties of bitumen bonded asphalt mixtures and pavements, but using the rheological model developed and described above (Fig. 11) can help us to better understand the self-healing process which very strong depends not only on mechanical pressures and temperatures but on stress relaxation times also (eq. 8).

Understanding the rheological properties of components can help us not only to prepare self-healing asphalt mixtures but keep better maintenance of asphalt pavements, roads and constructions.

References

[1] R. Bansil, J. Hardcastle and M.Constantino 2015 *Építőanyag-JSBCM* **67** 150 http://dx.doi.org/10.14382/epitoanyag-jsbcm.2015.25
[2] Y. Wang, D.A. Lakho and D.Yao 2015 *Építőanyag-JSBCM* **67** 132 http://dx.doi.org/10.14382/epitoanyag-jsbcm.2015.21
[3] S.O. Umerova, I.O.Dulina and A.V. Ragulya 2015 *Építőanyag-JSBCM* **67** 119 http://dx.doi.org/10.14382/epitoanyag-jsbcm.2015.19
[4] A. Slobodov, A. Uspenskiy, R. Ralys and D. Kremnev 2015 *Építőnyag-JSBCM* **67** 159 http://dx.doi.org/10.14382/epitoanyag-jsbcm.2015.27
[5] S.N. Kulkov, S.P. Buyakova, M. Chatzinikolaidou and I. Kocserha 2015 *Építőanyag-JSBCM* **67** 155 http://dx.doi.org/10.14382/epitoanyag-jsbcm.2015.26
[6] A. Slobodov, A. Uspenskiy, S. Yavshits and A. Mischenko 2015 *Építőanyag-JSBCM* **67** 164 http://dx.doi.org/10.14382/epitoanyag-jsbcm.2015.28
[7] L.A. Gömze, L.N. Gömze, S.N. Kulkov, I. Shabalin, I. Gotman, F. Pedraza, G. Lecomte, T. Mayorova, E. Kurovics and A. Hamza 2015 *Építőanyag-JSBCM* **67** 143 http://dx.doi.org/10.14382/epitoanyag-jsbcm.2015.24
[8] H. Hornahad, E.A.B.Koenders, K. van Breugel 2015 *Építőanyag-JSBCM* **67** 135 http://dx.doi.org/10.14382/epitoanyag-jsbcm.2015.22
[9] L.A. Gömze (2015) RHEOLOGY, Compilation of Scientific Papers I., *Published by IGREX Ltd, Hungary,* **ISBN 978-963-12-3088-8**
[10] K.Almássy and Cs. Tóth 2011 *Építőanyag-JSBCM* **63** 16, http://dx.doi.org/10.14382/epitoanyag-jsbcm.2011.3
[11] R. Géber 2007 *Építőanyag-JSBCM* **59** 73, http://dx.doi.org/10.14382/epitoanyag-jsbcm.2007.11
[12] L.A. Gömze (2016) APPLIED MATERIALS SCIENCE I., Compilation of Selected Scientific Papers I, *Published by IGREX Ltd, Hungary,* **ISBN 978-963-12-6600-9**
[13] Cs. Tóth and Z. Soós 2015 *Építőanyag-JSBCM* **67** 126, http://dx.doi.org/10.14382/epitoanyag-jsbcm.2015.20
[14] R. Géber and L.A Gömze 2013 *Materials Science Forum* **752** 209
[15] J.W. Romberg and R.N. Traxler 1947 *Journal of Colloid Science* **2** 33

[16] F. Martinez-Boza, P. Partal, F.J. Navarro and C. Gallegos 2001 *Rheologica Acta* **40** 135
[17] J. Mills-Beale, Z. You, E. Fini, B. Zada, C.H. Lee and Y.K. Yap 2014 *J. Mater Civ. Eng* **26** 358
[18] Y.R. Kim 2009 Modeling of Asphalt Concrete *McGraw-Hill Professional*
[19] J.M. Rojas, N.A. Hernández, O. Manero and J. Revilla 2009 *Jour. Applied Polymer Sci.* **115** 15
[20] R. Géber 2008 *Építőanyag-JSBCM* **60** 19, http://dx.doi.org/10.14382/epitoanyag-jsbcm.2008.5
[21] R. Géber and L.A Gömze 2010 *Materials Science Forum* **659** 471
[22] Y. Edwards, Y. Tasdemir and U. Isaccson 2007 *Construction and Building Materials* **21** 1899
[23] J.M. Krishnan and K.R. Rajagopal 2005 *Mechanics of Materials* **37** 1085
[24] H.S. Do, P.H. Mun and R.S. Keun 2007 *Waste Management* **3** 105
[25] M.A. Abdoli, A. Fathollahi, R. Babaei 2015 *Int. J.Environ. Res.* **9** 489
[26] L.B. Gezentsvei 1960 *Stroyizdat,* Moscow
[27] L.B. Gezencvej 1964 *KPM Publisher*, Budapest (in Hungarian)
[28] L.A. Gömze and Á. Kovács 2005 *Építőanyag-JSBCM* **57** 34 (in Hungarian) http://dx.doi.org/10.14382/epitoanyag-jsbcm.2005.7
[29] L.A. Gömze, I. Kocserha and G. Czél 2002 *Patent No.U0200079* Hungarian Patent Office, Budapest (in Hungarian)
[30] P. Fratzl and R. Weinkamer 2007 *Springer Series in Materials Science* **100** 323
[31] Y.D. Wu, S. Meure and D. Solomon 2008 *Progress in Polymer Science* **33** (5) 479
[32] P.R. Wool 2008 *Soft Matter* **4** (3) 400
[33] H. Fisher 2010 *Natural Science* **2** (8) 873 http://dx.doi.org/10.4236/ns.2010.28110
[34] B. Boh and B. Sumiga 2008 *RMZ-Materials and Geoenvironment* **55** (3) 329
[35] J.F. Su, J. Qiu, E. Schlangen and Y.Y. Wang 2015 Construction and Building Materials **74** 83 http://dx.doi.org/10.1016/j.conbuildmat.2014.10.018
[36] J. Qiu, M.F.C. van de Ven, S. Wu, J. Yu and A.A.A. Molenaar 2009 *Road Materials and Pavement Design* **10** 81
[37] Y. Xiao, M.F.C. van de Ven, A.A.A. Molenaar and S.P. Wu 2013 *Construction and Building Materials* **27** 24 http://dx.doi.org/10.1016/j.conbuildmat.2011.08.005
[38] J. Qiu, A.A.A. Molenaar, M.F.C. van de Ven, S. Wu and J.Yu 2012 *Materials and Structures* **45** (5) 777
[39] J. Qiu, M.F.C. van de Ven, S. Wu, J. Yu and A.A.A. Molenaar 2012 *Experimental Mechanics* **52** (8) 1163
[40] Y. Xiao, M.F.C. van de Ven, A.A.A. Molenaar and S.P. Wu 2013 *Construction and Building Materials* **48** 59 http://dx.doi.org/10.1016/j.conbuildmat.2013.06.080
[41] A. Garcia, E. Schlangen and M.F.C. van de Ven 2011 *Materials and Structures* **44** (2) 499
[42] J.F. Su, J. Qiu and E. Schlangen 2013 *Polymer Degradation and Stability* **98** 1205 http://dx.doi.org/10.1016/j.polymdegradstab.2013.03.008
[43] J.F. Su, J. Qiu, E. Schlangen and Y.Y. Wang 2014 *Materials and Structures* **48** 4067
[44] A. Tabakovic and E. Schlangen 2015 *Self-Healing Technology for Asphalt Pavements,* Springer Switzerland
[45] S.F. Brown, N.H. Thom and P.J. Sanders 2001 *Journal of the* Association *of* Asphalt Paving *Technologists,* **70**, 543
[46] A. Garcia, E. Schlangen, M.F.C. van de Ven and G. Sierra-Beltrán 2010 *J Hazard Materials* **184** 603
[47] J.F Su and E. Schlangen 2012 *Chemical Engineering Journal* **198-199** 289
[48] E.N. Brown, S.R. White and N. Sotton 2004 *Journal of Materials Science* **39** 1703
[49] E.N. Brown, S.R. White and N. Sotton 2005 *Composites Science and Technology* **65** 2474
[50] V. Haritonovs, J. Tihonovs and M. Zaumanis 2016 *IOP Conf. Ser.: Mater. Sci. Eng.* **123** 012055 http://dx.doi.org/10.1088/1757-899X/123/1/012055
[51] C. Spadaro, C.J.P. Plummer and J.A.E Manson 2011 *Journal of Materials Science* **46** 7449
[52] L.A. Gömze 1983 *Epitoanyag-JSBCM* **35** (5) 173 (in Hungarian)

Dynamical properties of the brain tissue under oscillatory shear stresses at large strain range

F Boudjema[1,2], **B Khelidj**[1], **M Lounis**[2]

[1] UDB-KM University, FIMA Laboratory, Khemis Miliana , Algeria
[2] USTO-MB University, LAAR Laboratory, Oran, Algeria

E-mail: fatiboudj27@yahoo.fr, b.khelidj@orange.fr, loumou2000@yahoo.fr

Abstract. In this experimental work, we study the viscoelastic behaviour of in vitro brain tissue, particularly the white matter, under oscillatory shear strain. The selective vulnerability of this tissue is the anisotropic mechanical properties of theirs different regions lead to a sensitivity to the angular shear rate and magnitude of strain. For this aim, shear storage modulus (G′) and loss modulus (G″) were measured over a range of frequencies (1 to 100 Hz), for different levels of strain (1 %, to 50 %). The mechanical responses of the brain matter samples showed a viscoelastic behaviour that depend on the correlated strain level and frequency range and old age sample. The samples have been showed evolution behaviour by increasing then decreasing the strain level. Also, the stiffness anisotropy of brain matter was showed between regions and species.

1. Introduction

To obtain insight into brain injury mechanisms, the internal mechanical response of the head under impact conditions has to be known. Obviously, it is not possible to study in experiments the influence of a violent shock on an *in vivo* human head, to reach the tolerance criteria of the head. We found in the literature that most studies on the mechanical behaviour of brain matter are digital simulations using finite elements FE; the other experimental studies often use either *in vitro* brain samples of material under dynamic conditions [1] or *in vivo* samples using image analysis techniques [2]. Most of the *in vitro* tests described in the literature were based on creep and relaxation in compression and shear tests, or on dynamic frequency or strain sweep experiments. To characterize viscoelastic brain behavior under small strain, most of the *in vitro* experiments have been performed dynamically, commonly using a dynamic frequency sweep. Thus, most of the dynamic frequency sweep protocols have investigated tissues in the linear strain domain. Although most of the dynamic frequency sweep tests have been carried out in shear, some dynamic frequency sweep experiments have been done in tensile/compression mode.

In 2008, Hrapko et al. published a literature review focusing on *in vitro* experimental protocols [1]. As well as Cheng et al. published literature reviews on comparing brain and spinal cord mechanicals properties [3]. In 2010, Di Iewa et al. presented a review using imaging applications for the mechanical characterization of brain tissue [4]. Also in 2010, Chatelin et al. complements these existing literature reviews on brain tissue mechanics [5] by presenting more recent results and by focusing on the comparison between *in vitro* and *in vivo* tests, greatly developed since 2005, especially those concerning Magnetic Resonance Elastography (MRE). Brain tissue consists of gray and white matters. Mechanically, the matters are relatively incompressible, nonlinear and viscoelastic, as concluded by [6-8]. The gray matter is typically considered to be isotropic and the mechanical response of white matter may depend on orientation [9]. Both quasi-static and dynamic brain matter analysis was dealt with in this study as they provide complementary information with more recent results [5]. In this light, we interest to study in this paper the dynamical response of brain tissue with time under the large strain range. As well as we tried to complete our contribution in understanding of the brain tissue mechanical properties in transient state [10], by the study of their properties under oscillatory strain solicitations. Dynamic strain and frequency sweep tests were used to test different old age samples in shear deformation.

2. Materials and Methods

In our study, we have used the dynamic shear test to determine the mechanical response of animal brain samples (lamb and bovine brain tissue) at large strain. These experiments were realized on samples harvested and preserved few hours to three days to study the degradation brain matter with time.

2.1. Samples Preparations

These experiments for this study were performed *in vitro* on lamb and cow brain aged between 6 and 12 months. Lamb brains were used because of their availability. The brains (one of bovine and four of lambs) were collected a few minutes to one day after the animals were killed at a local slaughterhouse. The brain tissue was preserved in physiological solution at 6°C, in icebox, to preserve the ionic balance during transport. The experiments reported in this paper were completed within 72h postmortem. Annulus sample of diameter 25 mm and thickness 0.8 to 3 mm were delicately excised from sections of gray matter (thalamus) and white matter (corona radiate and corpus callosum). Cylindrical plugs of the white matter (thickness = 4 mm) were then harvested manually from the corpus callosum and corona radiate using a punch (diameter =25 mm) and a surgical knife. The specimens were then bathed in the solution until testing. The experiments were performed at temperature (20±0.01°C).

2.2. Dynamics tests

Rheological tests were carried out on the brain samples at swelling equilibrium (25mm diameter disk, 2 mm thick) using a stress-controlled MCR302 rheometer (Physica Company, Germany) in parallel-plate geometry. The samples were tested in a dynamic regime at small and large amplitudes of deformation (0.1% to 50%). During the measurements, samples were kept in hydrated chamber at controlled temperature (20°C). The linear viscoelastic region was determined by strain sweep measurements which allowed us to set adequate stresses so that the resulting strains did not damage the brain sample just for confirmed our tests with literature. Then, the non-linear viscoelastic behaviour was study by dynamic frequency sweep under large strains. Dynamic controlled strains tests and dynamic frequency sweep tests were performed with frequencies ranging from 1 to 100 or 600 (rad/s) to determine the variation of storage and loss modulus (G′) and (G″), respectively), under various shear strains levels. To study the sample degeneracy the same tests were performed on samples preserved for one day, two days and three days.

3. Results and Discussions

The main objective of this work was to study the mechanical response of brain matter, obtained from different species (lambs and cow) and preserved from few hours to few days, to the large shear strains.

3.1. Species differences

To compare the response of the two different brains animals, dynamics frequency sweep measurement results obtained from same region of brain samples of two different species tested at *5%* strain level are shown in Figure 1 and 2. Numerous studies have demonstrated the non-linear viscoelastic behavior of adult brain tissue over a wide range of strains and strain rates [7], [11-15]. To predict tolerance criteria for non-linear, large deformation inertial brain injury, we were interesting to characterize the brain tissue properties at strains larger than 1%.

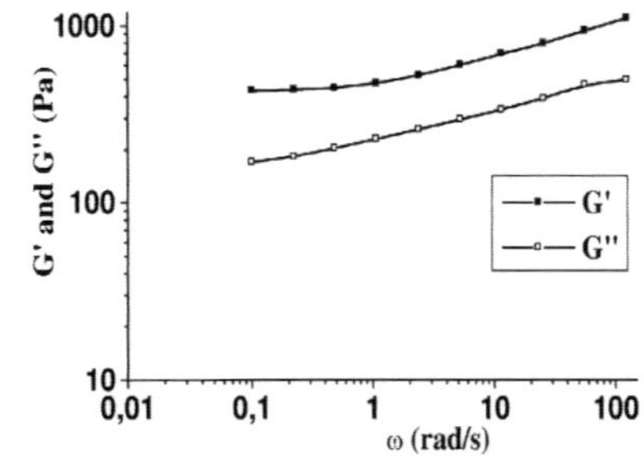

Figure 1. Storage modulus (G′) and loss modulus (G″) obtained in dynamic frequency sweep with 5 % of strain for corona radiate (CR) of the lamb brain tissue

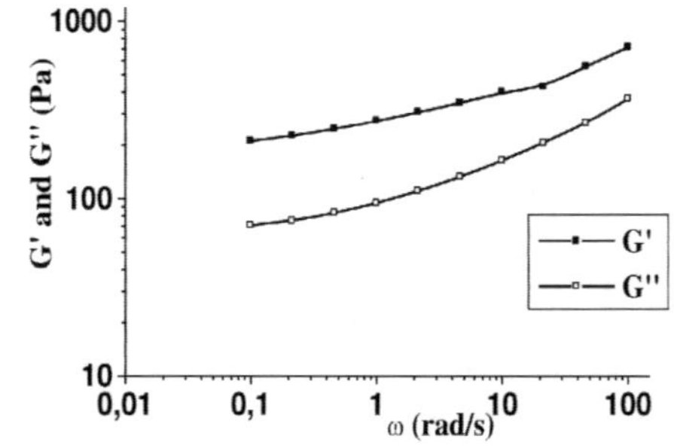

Figure 2. Storage modulus (G′) and loss modulus (G″) obtained in dynamic frequency sweep with a 5% of strain for corona radiate (CR) of bovine brain tissue

The storage (G′) and loss modulus (G″) varied with testing frequency indicating a viscoelastic response, (G′) is higher than (G″), with same evolution of the tow components as in literature results [1], [4], [11], but with differences in values level of each sample source.

3.2. Control strain

Measurements of controlled strain tests at different fixed frequencies were presented in Figure 3. From these experiments, the results showed that the storage modulus (G′) of the brain tissue started to decrease for an applied strain value about 1% of strain at low frequency.

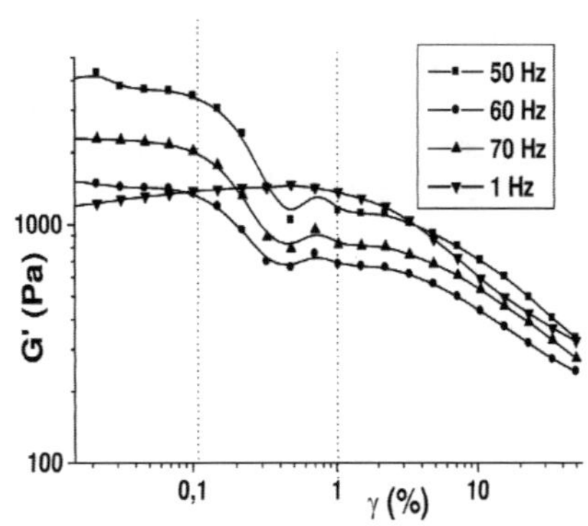

Figure 3. Mean amplitudes sweep response of (G′) of TH and CR samples at four different frequency values

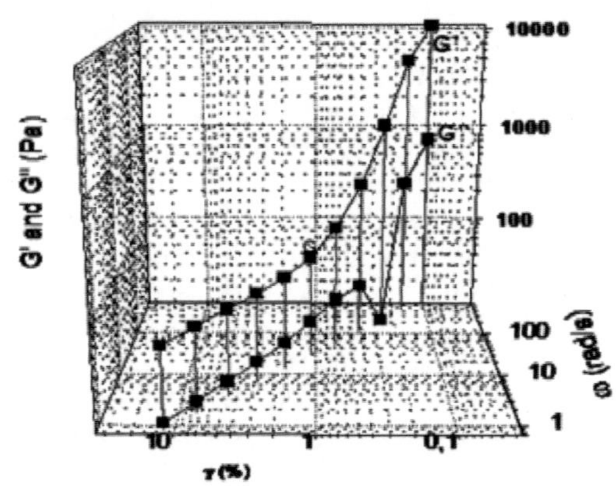

Figure 4. Storage modulus (G′) and loss modulus (G″) obtained in dynamic frequency sweep with a strain 5% of corona radiate (CR) from bovine brain tissue

Compared with similar work done on porcine cerebral tissue realized by Nicolle et al [16], the same limit was obtained. However, this linear viscoelastic limit decrease to 0.1% of strain for high frequency.

From another controlled strain test, with frequency sweep, results (figure 4) showed that:

- the storage modulus (G′) found to be larger than loss modulus (G″) (viscoelastic behaviour);
- storage modulus G′ and loss modulus G″ showed significantly frequency dependence;
- for high frequency we can't use large strain (bulck flow).

To study the strain level effect on the viscoelastic parameters (G′) and (G″), we have fixed the strain level in each test then we have varied the frequency.

The results of three large strain levels of CR from one day old bovine brain tissue were presented in Figure 5. In Figure 6, the results of CR from one day old lamb brain tissue were presented.

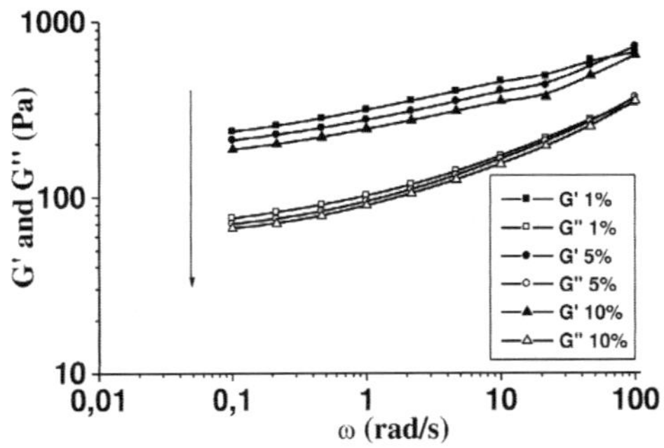

Figure 5. Mean frequencies sweep response of storage modulus, G' and loss modulus, G″ at three levels strain of CR from one day old bovine brain tissue

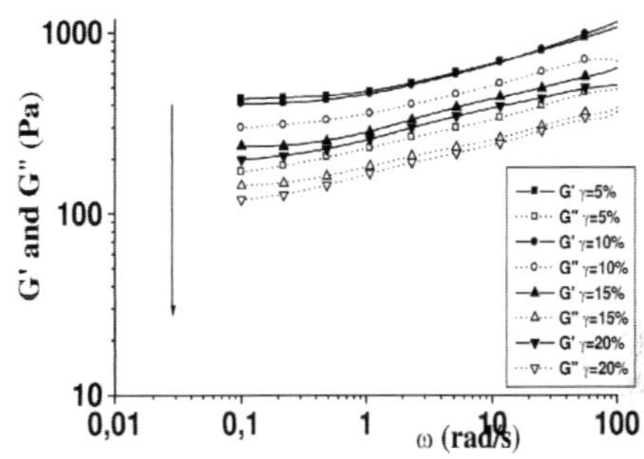

Figure 6. Mean frequencies sweep response of storage modulus, G' and loss modulus, G″ at four levels strain of CR from fresh lamb brain tissue

From these results, we can show that at large strain, storage modulus G′ and loss modulus G″ increase with increasing frequency, contrast they decrease with strain level. As well as, the difference between G′ and G″ become smaller with frequency.

To study the degradation effect, we have increased the strain level gradually in each sweep frequency test applied on Corpus callosum sample reserved for one day, and then we have decreased it for the second time as same. The results of storage modulus (G′) evolution are show in Figure 7.

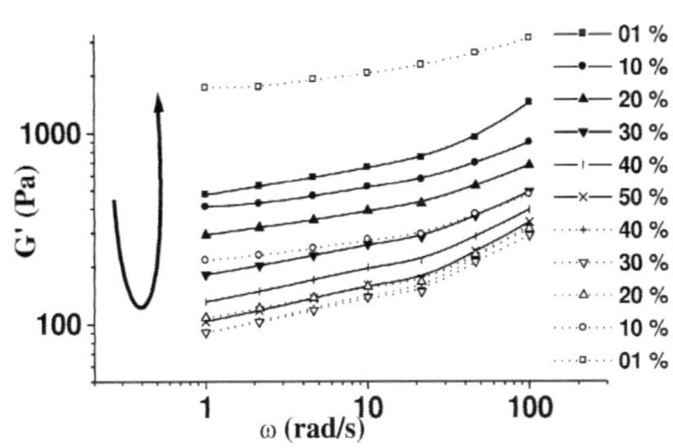

Figure 7. Series of frequency response of storage modulus, (G′), for the increasing then decreasing strain levels imposed at the same sample (CC) of one day old lamb brain

3.3. Syntheses results

The results of our study are compared with the published literature regarding animal and human brain tissue properties. A review of the mechanical properties of adult human and animal brain was realized by Chatelin et al. [7].

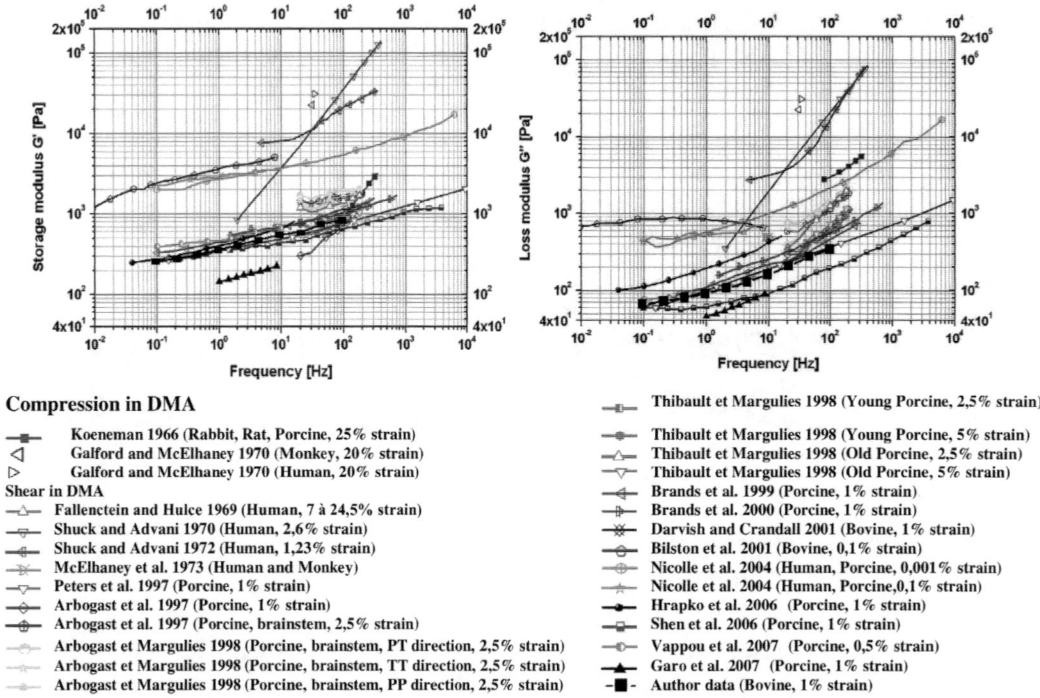

Figure 8. Comparison of (G′) and (G″) of in-vitro brain tissue reported in the literature in dynamic frequency sweep shear tests. In colour from the ref [2] with in black line and square symbol of author results (bovine, 1% of strain). In all studies the storage modulus is observed to increase with increasing frequency.

This review demonstrated that the reported properties of brain tissue from different species and testing conditions varied by as much as an order of magnitude summarize the storage modulus G′ (Figure 8) and loss modulus G″. However, in all results G′ and G″ increased with increasing frequency. Compared it with our results, taken a low strain range of 1%, we can show that these storage and loss modulus were superposed with Peters et al. results in which they tested porcine brain tissue under the same strain level.

4. Conclusion

This experimental study present oscillatory characteristic of *in vitro* animal brain tissue using large shear strain range. The measurements were carried out on samples taken from different brain species and different regions preserved from few hour to few days. The data demonstrated that the dynamical properties of brain tissue depend to various experimental conditions. It also proved that the viscoelastic parameters: storage modulus G′ and loss modulus G″ increase with increasing frequency which is confirmed with literature. These parameters decrease with strain level and time (fresh or not brain sample). From frequency sweep tests by varying the strain level applied on the same sample, we found that the brain tissue is not reproductive with increasing then decreasing strain level which indicate the degradation of sample.

5. References

[1] Hrapko M, van Dommelen JA, Peters GW and Wismans JSHM 2008 *J. Biomech. Eng.* **130** 031003.

[2] Macé E, Cohen I, Montaldo G, Miles R, Fink M and Tanter M 2011 *IEEE T Med Imaging* **30** 550 – 558.
[3] Cheng S, Clarke EC and Bilston LE 2008 *Med. Eng. Phys.* **30**, 1318–1337.
[4] Chatelin S, Constantinesco A and Willinger R 2010 *Biorheology*, **47** 255–276.
[5] Cheng S Clarke EC and Bilston 2008 LE *Med.Eng. Phys.* **30** 1318–1337.
[6] Estes and McElhaney 1970.
[7] Donnelly B, Medige J 1997 *J. Biomech. Eng.* **119** 423–432.
[8] Brandsetal.(2000)
[9] Nicolle S, Lounis M, Willinger R, 2004 *Stapp Car Crash Journal,* **48** 239.
[10] Boudjema F, Lounis M, Khelidj B and Bessai N 2015 *J. Phys. Conf. Ser.* **602** 012029.
[11] Thibault KL, Margulies SS, 1998 *J. Biomech.* **31** 1119–1126.
[12] McElhaney JH, Roberts VL, Hilyard JF (Eds.), 1976 Handbook of Human Tolerance. Jap. Auto. Res. Inst.
[13] Mendis K, Stalnaker R et al., 1995 *J. Biomech. Eng.* **117** 279–285.
[14] Miller K, Chunzei K, 1997 *J. Biomech.* **30** 1115–1121.
[15] Shuck, LZ, Advani, SH, 1972 *ASME J. Basic Eng.* **94** 905–911.
[16] Nicolle S, Lounis M, Willinger R and Palierne J-F, 2005 *Biorheology,* **42** 209–223.

Effect of crystallographic orientations of grains on the global mechanical properties of steel sheets by depth sensing indentation

P Burik[1], L Pesek[2], P Kejzlar[3], Z Andrsova[3] and P Zubko[2]

[1]Technical University of Liberec, Faculty of Mechanical Engineering, Department of Material Science, Liberec, Czech Republic
[2]Technical University of Kosice, Faculty of Metallurgy, Department of Materials Science, Kosice, Slovakia
[3]Technical University of Liberec, Institute for Nanomaterials, Advanced Technology and Innovation, Liberec, Czech Republic

E-mail: peter.burik@tul.cz

Abstract. The main idea of this work is using a physical model to prepare a virtual material with required properties. The model is based on the relationship between the microstructure and mechanical properties. The macroscopic (global) mechanical properties of steel are highly dependent upon microstructure, crystallographic orientation of grains, distribution of each phase present, etc... We need to know the local mechanical properties of each phase separately in multiphase materials. The grain size is a scale, where local mechanical properties are responsible for the behavior. Nanomechanical testing using depth sensing indentation (DSI) provides a straightforward solution for quantitatively characterizing each of phases in microstructure because it is very powerful technique for characterization of materials in small volumes.

The aim of this experimental investigation is: (i) to prove how the mixing rule works for local mechanical properties (indentation hardness H_{IT}) in microstructure scale using the DSI technique on steel sheets with different microstructure; (ii) to compare measured global properties with properties achieved by mixing rule; (iii) to analyze the effect of crystallographic orientations of grains on the mixing rule.

1. Introduction

Real materials such as multi-phase steels exhibit on the microscopic scale (microstructure) a heterogeneous material. Multiphase steels combine a high strength and a high formability, which is a condition for deep drawing parts in the automotive industry [1]. The overall behaviour of micro-heterogeneous steels depends strongly on the size, shape, spatial distribution and properties of the microstructural constituents and their respective interfaces [2]. As the properties of these materials are determined by the contributions from the constituent phases, the properties of separate phases and their volume fraction are the key factors for product design. Accordingly, there is a need for a reliable tool for phase analysis. Electron backscatter diffraction (EBSD), which is principally a microtexture determination technique, is a popular tool for microstructure characterization. EBSD is a technique based on the analysis of the Kikuchi pattern by the excitation of the electron beam on the surface of the sample in a scanning electron microscope (SEM) [3].

A micromechanical approach based on volume fraction components enables to describe the heterogeneities with the help of continuum mechanical quantities and by this to link mechanical properties at different scales. Thereby it is the goal to derive effective material properties for the component from microstructure quantities and to quantify the effect of microstructural features. The rule of mixture means - macroscopic properties of aggregates or composites consisting of two or more constituents can be obtained as the sum of the volume fraction of the components times their individual properties [1].

The composite property is estimated as the sum of the responses of the composite components weighted by the component volume fractions as shown in the equation below:

$$P = \sum_i^n V_i P_i \quad (1)$$

where P - some property and V - the volume fraction of the components, i = 1, 2, 3, ... n [4].

Mechanical properties of individual components can be determined by depth sensing indentation derived from the indentation load-displacement data used a micromechanical model developed by Oliver & Pharr (O&P) [5].

In the O&P analysis, the hardness, $H_{IT}^{O\&P}$, is determined from the load-displacement curve as:

$$H_{IT}^{O\&P} = \frac{P_{max}}{A_p^{O\&P}} \quad (2)$$

where P_{max} - the maximum indentation load and $A_p^{O\&P}$ - the projected contact area at P_{max}.
The projected contact area $A_p^{O\&P}$ is calculated by evaluating an empirically determined indenter shape function $A_p^{O\&P}$= f($h_c^{O\&P}$). For an ideal Berkovich indenter, it is given by:

$$A_p^{O\&P} = 24.56\, h_c^{2O\&P} \quad (3)$$

where $h_c^{O\&P}$- the contact depth between material and indenter at the P_{max} [6].

2. Experimental materials and methods

2.1Materials

The characteristic microstructure of the investigated material:

(i) XSG steel - interstitial free steel (ferrite microstructure), Figure 1a. The mean size of the ferrite grain is 19 μm.
(ii) HR 45 steel -microalloyed steel (ferrite-pearlite microstructure), Figure 1b. The mean size of the ferrite and pearlite grain is 7.9 ^m and 3.9 ^m, respectively. The volume fraction of pearlite is 14%.
(iii) DP 600 steel - dual phase steel (ferrite-martensite microstructure), Figure 1c. The mean size of the ferrite and martensite grain size is 12 ^m and 2.7 ^m, respectively. The volume fraction of martensite is 27 %.

Chemical composition, mechanical properties and thickness t of investigated steel sheets are in Table 1.

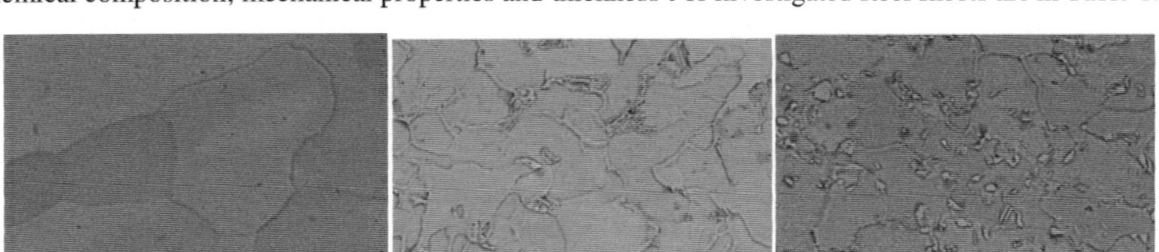

a) b) c)

Figure 1. Microstructure of investigated steels, a) XSG, b) HR 45, c) DP 600

Table 1. Chemical composition (wt. %), mechanical properties (Rp0.2 - yield stress, R_m - tensile strength, A_{80} - elongation, n - strain hardening exponent) and thickness t of investigated steel sheets

Steel	C	Mn	V	S	P	$R_p0.2$ (MPa)	R_m (MPa)	A_{80} (%)	n (-)	t (mm)
XSG	0.0013	0.082	0.002	0.0105	0.011	177	286	47.2	0.211	1.95
HR 45	0.156	0.654	0.002	0.004	0.013	360 R_e	449	27	0.139	1.80
DP 600	0.072	1.807	0.003	0.006	0.017	388	581	26.1	0.160	1.60

2.2 Methods

The nanoindentation tests were carried out at room temperature using a CSM instrument equipped with a Berkovich tip at a constant loading rate of 400 mN/min from 0 to the maximum force 5 mN and 500 mN, respectively, with 10 s hold period and constant unloading, the distances between the indentations were 10 μm (P_{max} = 5 mN) in the 11 x 10 matrix and 100 μm (P_{max} = 500 mN) in 3 x 1 matrix. The force-displacement curves were analyzed using the O&P analysis. After indentations, the indent location was observed via scanning electron microscope and the crystallographic orientation of grains with individual orientations was examined by means of EBSD method. High resolution ESBD images were prepared on an area ~ 120 x 110 μm using a step size of 300 nm and applying a tilt angle of 70°. The obtained EBSD data were analyzed by HKL Channel 5 software. Pairing of the measured mechanical properties to the corresponding crystallographic orientation of each indent was carried out so that the indents located inside the grains were taken into account, those which corresponded to the close grain boundary area of crystals (closer than the diameter of an indent) were neglected from the evaluation.

3. Results and discussion

Figure 2 shows inverse pole figure (IPF) map which describes the crystallographic orientation of detected grains of XSG steel. IPF map of the investigated area on which the nanoindentation tests were performed, practically completely covers the whole possible grain crystallographic orientations. The indentation hardness is dependent on the crystallographic orientation, Table 2. The indentation hardness is substantially lower in (001) plane than in both (101) and (111) planes in the all steels. XSG steel has the lowest difference between indentation hardness of (001) plane and indentation hardness of both (101) and (111) planes (about 7%). DP 600 steel has the greatest difference between indentation hardness of (001) plane and indentation hardness of both (101) and (111) planes (about 11%).

Effect of crystallographic orientation on the indentation hardness of pearlite component in the HR 45 steel and of martensite component in the DP 600 steel are not determined due to the fine grain microstructure of these components; indent size is equal to the grain size of the component and therefore it is not possible to accurately determine the hardness for the given crystallographic orientation of the grains.

The values of mechanical characteristics of ferrite in different crystallographic planes have a high standard deviation due to:

- High surface roughness: for example - HR 45 steel R_a = 0.053 ± 0.006 μm, the maximum depth in (001) plane h_{max} = 0.359 ± 0.017 μm; in (101) plane h_{max} = 0.346 ± 0.019 μm and in (111) plane h_{max} = 0.351 ± 0.013 μm.
- Different indentation distance from the grain boundary.
- Roughness of the soft phase and that of hard phase is not equal, slip of the indenter can occur.
- Depth of the indented area is unknown; it is likely that measured hardness value of hard component is affected by the soft component which is under indented hard grain and opposite the soft component may be affected by the hard one lying under that. Therefore the hardness value of hard component may be shifted to lower and that of soft component to higher value.

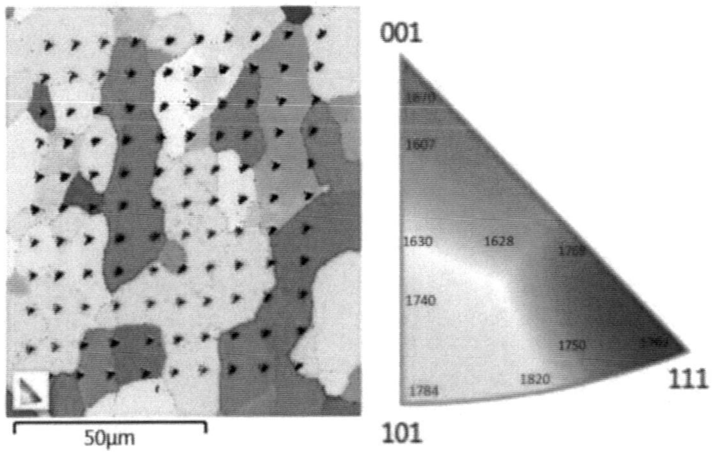

Figure 2. a) Indentation matrix in EBSD map of XSG steel, b) Inverse pole figure map of indentation hardness (MPa)

Table 2. Indentation hardness H_{IT} (MPa) of ferrite in various planes (001), (101), (111), between (001) and (101), between (101) and (111), between (001) and (111) (in direction normal to the sample surface)

Steel	(001)	(101)	(111)	(001)(101)	(101) (111)	(001)(111)
XSG HR	1660±115	1780±250	1760± 140	1540±120	1800±240	1680±250
45	1970±180	2040± 180	2050± 160	2050±130	2033 ±145	2111±45
DP 600	2045 ± 120	2290± 160	2260± 180	2290±280	2800±470	2800 ± 350

Table 3. The real volume fraction V_{real} (%) of the various planes (001), (101), (111), between (001) and (101), between (101) and (111), between (001) and (111) of ferrite in the investigated steels

Steel	(001)	(101)	(111)	(001)(101)	(101) (111)	(001)(111)
		a		b		
XSG HR	6.3	40.6	3	6.3	18.8	25
45	16	17	16	17	18	16
DP 600	10.4	23.9	19.5	10.4	12	23.8

Effect of crystallographic orientation on the global indentation hardness of ferrite was determined using various combinations of indentation hardness of individual planes to the volume fraction of individual planes according to the following models for indentation hardness of ferrite:

1. Based on relative ratio of basic planes volume fraction:

$$H_{IT}ferrite = H_{IT}(001)V_{(001)} + H_{IT}(101)V_{(101)} + H_{IT}(111)V_{(111)} \quad (4)$$

where the values of $V_{(001)}$, $V_{(101)}$, $V_{(111)}$ are calculated as:

$$V_{(001)} = \frac{100}{V_{real(001)}+V_{real(101)}+V_{real(111)}} + V_{real(001)} \quad (5)$$

$$V_{(101)} = \frac{100}{V_{real(001)}+V_{real(101)}+V_{real(111)}} + V_{real(101)} \quad (6)$$

$$V_{(111)} = \frac{100}{V_{real(001)}+V_{real(101)}+V_{real(111)}} + V_{real(111)} \quad (7)$$

2. Based on real ratio of the basic planes volume fraction and the arithmetic average of the other planes volume fraction:

$$H_{IT}\text{ferrite} = H_{IT}(001)V_{real(001)} + H_{IT}(101)V_{real(101)} + H_{IT}(111)V_{real(111)} + \frac{(H_{IT}(001)/(101)+H_{IT}(001)/(111)+H_{IT}(101)/(111)}{3}(V_{real(001)/(101)} + V_{real(001)/(111)} + V_{real(111)/(101)}) \quad (8)$$

3. Based on a real ratio of all planes volume fraction:

$$H_{IT}\text{ferrite} = H_{IT}(001)V_{real(001)} + H_{IT}(101)V_{real(101)} + H_{IT}(111)V_{real(111)} + H_{IT}(001)/(101)V_{real(001)/(101)} + H_{IT}(001)/(111)V_{real(001)/(111)} + H_{IT}(101)/(001)V_{real(101)/(001)} \quad (9)$$

where the values of $V_{real(001)}$, $V_{real(101)}$, $V_{real(111)}$, $V_{real(001)/(101))}$, $V_{real(001)/(111))}$, $V_{rea\acute{e}(111)/(101)}$ for model 2 and model 3 are from the Table 3.

Using a mixture rule we can obtain the model data. Indentation hardness of ferrite in mixture rule varies according to the individual model and the indentation hardness of other components (pearlite component in HR 45 steel, martensite component in DP 600 steel) in the mixture rule is constant.

Mixture rule can be rewritten as following equation:

XSG steel $$H_{IT_{global}} = H_{IT_i} \quad (10)$$

HR 45 steel $$H_{IT_{global}} = H_{IT_i} V_{ferrite} + H_{IT_{pearlite}} V_{pearlite} \quad (11)$$

DP 600 steel $$H_{IT_{global}} = H_{IT_i} V_{ferrite} + H_{IT_{martensite}} V_{martensite} \quad (12)$$

where HIT global - model for global indentation hardness, HITi - indentation hardness of ferrite calculated according to the model 1, 2, 3 (equations (4), (8), (9)), HIT indentation hardness of component (pearlite, martensite), Table 4, V - the volume fraction of the component.

Bulk indentation hardness - means data for series of large indents (P_{max} = 500 mN). Model for H_{IT} - means data calculated using mixture rule (equation (10) - (12)) with corresponding ferrite, ferrite / pearlite and ferrite / martensite content. Modeling parameter X = model for H_{IT} / bulk H_{IT}. The more the X is close to 1, the better works the mixture rule model.

Table 4. Indentation hardness H_{IT} (MPa) (P_{max}= 5 mN) for different structural components

Steel Model for H_m of ferrite (calculated)	XSG	HR 45	DP 600
1. model	1665	1980	2231
2. model	1670	2042	2372
3. model	1687	2042	2389
H_{IT} pearlite (P), martensite (M)) (measured)		2423 (P)	2582 (M)

Model 1, 2, 3 for indentation hardness H_m of ferrite calculated according to the model 1, 2, 3 (equation (4), (8), (9)), H_{IT} pearlite and martensite - indentation hardness H_{IT} of components of small indents (P_{max} = 5 mN) positioned into pearlite and martensite grains.

Global indentation hardness of ferrite is dependent on the individual model for indentation hardness of ferrite, Table 4. Models for global indentation hardness of ferrite contains "between planes" (model 2 and model 3) have greater indentation hardness than model based on relative ratio of basic planes volume fraction (model l).

Table 5 compares the measured and calculated (model) global mechanical properties. It can be observed the difference between the measured mechanical properties and calculated properties. Bulk indentation hardness gives greater value than model value for global indentation hardness because in bulk materials are included grain boundaries and precipitates which increase the indentation hardness. Model for global indentation hardness based on a real ratio of all planes volume fraction (model 3) is closer to bulk indentation hardness than model for global indentation hardness based on relative ratio of basic planes (model 1) volume fraction. XSG steel with coarse-grained microstructure of ferrite has the difference between the model based on real ratio of all plane volume fractions and the model based on relative ratio of basic planes volume fraction only 1.3%. DP 600 steel has the greatest difference between model based on real ratio of all plane volume fractions and model based on relative ratio of basic planes volume fraction (~ 4.1 %) because DP 600steel has fine grained microstructure of ferrite and DP 600 steel has the greater difference between individual planes between basic planes ((001), (101), (111)) and "between planes" ((001) (111) and (101) (111)).

Table 5. Bulk (measured) and model (calculated 1, 2, 3) global indentation hardness H_{IT} (MPa)

Steel	Bulk H_{IT} (Pmax= 500 mN)	Model for H_{IT} (Pmax = 5 mN).			X= Model for H_{IT} / Bulk H_{IT} (1)		
		1.	2	3.	1.	2.	3
XSG	1763	1665	1670	1687	0.944	0.947	0.957
HR 45	2425	1980	2042	2042	0.842	0.864	0.864
DP 600	2765	2231	2372	2389	0.841	0.878	0.882

4. Summary

Depth sensing indentation method is suitable for measuring the mechanical properties of each crystallographic orientation of ferrite grain in steel sheets with the aim to obtain the global mechanical properties from the particular mechanical properties using the mixture rule. The main limitation of the DSI method is measuring the mechanical properties of each crystallographic orientation of components with fine-grained microstructure.

Indentation hardness of ferrite in the steel sheets is dependent on the crystallographic orientation. Slip planes ((101), (111)) in body centred cubic crystal have higher hardness than the (001) plane. Investigated steel sheets contain approximately 50% of basic planes ((001), (101), (111)) oriented parallel to the surface.

Crystallographic orientations of ferrite have a significant effect on the global mechanical properties and therefore crystallographic orientation has to be included into the mixture rules in steel with ferrite. Model for indentation hardness calculated from the real ratio of crystallographic planes is closer to the bulk indentation hardness than model for indentation hardness calculated from the basic planes ((001), (101), (111)). Effect of crystallographic orientation of the ferrite on the mixing rules decreases with increasing size of the ferrite.

Acknowledgment

The paper was supported partly by Grant Agency of the Czech Republic thought the SGS project "Studium a hodnocení struktur a vlastností materiálú", partly by the project LO12O1 were obtained with co-funding from the Ministry of Education, Youth and Sports as part of targeted support from the "Národní program udrzitelnosti I" programme and partly by the VEGA project No. 1/O582/13 at the Technical University of Kosice, Slovakia.

References

[1] Prahl U Ramazani A Quade H and Twardowski R 2O12 *Forming Technology Forum* Damage modeling of multiphase steel using microstructure based RVE technique
[2] Fereiduni E and Ghasemi Banadkouki S S 2013 *Journal of Alloys and Compounds 577* Reliability / unreliability of mixture rule in a low alloy ferrite-martensite dual phase steel p 351-359
[3] Chen Z Yang Y and Jiao H 2012 *Scanning Electron Microscopy* Some applications of electron back scattering, diffraction (EBSD) in materials research
[4] Alabbasi F and Nemes J A 2003 *International Journal of Mechanical Sciences 45* Micromechanical modeling of dual phase steels p 1449-1465
[5] Hay J L 2000 *ASM Handbook 8, Mechanical testing and evaluation 236* Instrumented indentation testing p 232-244
[6] Zhou X Jiang Z Wang H and Yu R 2008 *Materials Science and Engineering A 488* Investigation on methods for dealing with pile-up errors in evaluating the mechanical properties of thin metal films at sub-micron scale on hard substrates by nanoindentation technique p 318-332

IOP Conf. Series: Journal of Physics: Conf. Series **790** (2017) 012004 doi:10.1088/1742-6596/790/1/012004

Effect of mechanical treatment on properties of zeolites with chabazite structure

A Y Buzimov[1,2], S N Kulkov[1,2], W Eckl[3], S Pappert[3], L A Gömze[4], E Kurovics[4], I. Kocserha[4], R. Géber[4]

[1] Institute of Strength Physics and Materials Science SB RAS, Tomsk, Russia
[2] National Research Tomsk State University, Tomsk, Russia
[3] Fraunhofer Institute for Chemical Technology ICT, Pfinztal, Germany
[4]Department of Ceramics and Silicate Engineering, University of Miskolc, Miskolc, Hungary

E-mail: buzimov92@gmail.com

Abstract. Zeolites are a valuable material having a wide variety of applications. We have examined the effect of mechanical activation on physical-chemical properties of commercial brands zeolite SAPO-34 and SCT-323. It has been shown that the amount of amorphous phase and the specific surface area depends on mechanical treatment. Specific surface area of the zeolites decreases strongly during the grinding process in mills. With the increase of milling time the particles size of zeolites decreased. An increase in amorphisation was observed. Specific surface area of zeolites after mechanical activation in a tumbling ball during 96 hours and annealing up to 800°C with an isothermal holding time of 1 hour does not lead to marked changes and decreases strongly after annealing at 1000°C/h. It has been shown that the milling time of ball milling is a powerful method to obtain the necessary specific surface.

1. Introduction

It is well known that zeolites are crystalline aluminosilicate materials having microporous (zeolite pore) in their structure. They are made of various connections of TO_4 (T = Si or Al) tetrahedral which result in the various zeolite pore sizes and structures [1]. Silicoaluminophosphate molecular sieves (SAPO) are an important class of molecular sieves, which play a unique role in the chemical industry because of their wide applications in catalysis, chemical separation, ion-exchange and membranes [2]. Small-pore SAPO with chabazite (CHA) structure has received great attention in recent years due to its high selectivity to the separation of light gases and methanol to olefin reaction [3].

The specific surface area is a dominant parameter for zeolites [4]. Obtain the necessary magnitude of specific surfaces of materials depends not only on their chemical and mineralogical composition, but also on physical, mechanical and rheological parameters of the used materials [5, 6]. It is known that mechanical processing in activator mills can change the structure, specific surface area, crystal composition of zeolite containing materials and occasionally improve their processing properties [7]. However, the existing published work on mechanical activation of the zeolites is limited. Therefore, the objective of this study was to investigate the effect of mechanical activation and annealing on physical-chemical properties of zeolites SAPO-34 and SCT-323.

2. Materials and Experimental Procedure

Commercial zeolite powders SAPO-34 and SCT-323 which are chabazite's structural analog from the silicoaluminophosphate group were used as objects of investigation. Mechanical activation of the zeolites was performed in a tumbling mill and a planetary ball mill for 5 minutes and for 100 hours, correspondently. Samples were annealed at temperatures up to 1000°C with an isothermal holding time of 1 hour at each temperatures. The crystal structure and phase composition of the zeolites before and after mechanical activation and annealing were determined by X-ray diffraction analysis (XRD) using diffractometer DRON-3 with Cu-Kα (λ= 1.5418 Å) radiation. The specific surface area of zeolites was determined on an SORBI instrument from low-temperature nitrogen adsorption and on a laboratory-scale apparatus from argon sorption using BET analysis. To study probable changes of morphology powders of zeolite after milling and annealing as well as to confirm the particle size results, all samples were subjected to Scanning Electron Microscopy (SEM).

3. Results and Discussion

3.1. Initial powders

Figure 1 and figure 2 represent the elemental analysis, SEM-pictures and X-ray patterns of SAPO-34 and SCT-323.

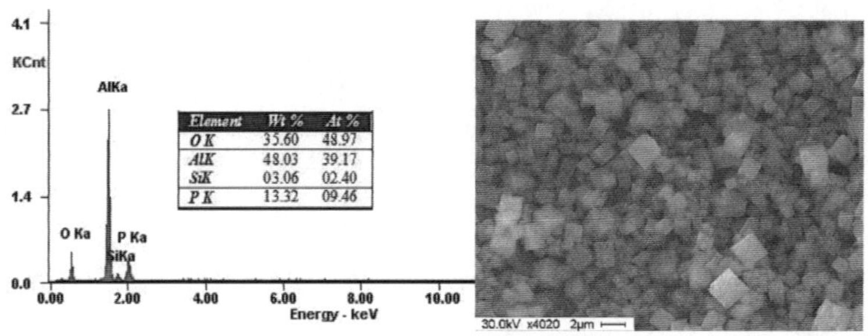

(a)

(b)

Figure 1. Elemental analysis, SEM-picture, for (a) SAPO-34, (b) for SCT-323

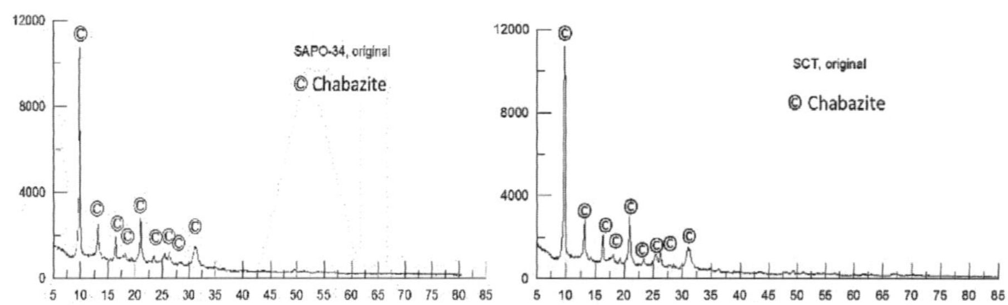

(a) (b)

Figure 2. X-ray pattern, for (a) SAPO-34, (b) for SCT-323

The results of elemental analysis showed that the zeolite powders in initial state consisted of Al, O, Si, P for SAPO-34; Zeolite SCT-323 containing titanium and relatively high content phosphorus is different from SAPO-34. The data of SEM showed that particles of zeolite powders are closely matched and have shapes close to cubic with an average particle size 1.5 μm for SAPO-34, 1.7 μm for SCT-323. Calculations of the X-ray patterns showed that the structure of the zeolites consisting Aluminum Silicate Phosphate (PDF-2 00-047-0630). XRD results showed presence of amorphous phase in zeolites, in an amount 83% for SAPO-34, 85% for SCT-323.

3.2. Mechanical activation

This series of zeolites have high specific surface area (561 m^2/g for SAPO-34, 573 m^2/g for SCT-323). Experimental results regarding the change in BET specific surface area of zeolite samples with grinding time in a tumbling ball mill are shown in figure 3. Apparently the specific surface area of the zeolites decreases during the grinding process. In the activation period of the first 12 hours, the specific surface area slightly decreases, and after this period rapidly decreases with prolonging the milling process, so it can be assumed that the minimum values of the specific surface area are achieved after 100 hours of grinding. BET analysis showed that the values of specific surface area of zeolite samples activated for 100 hours were 131 m^2/g for SAPO-34, 103 m^2/g for SCT-323.

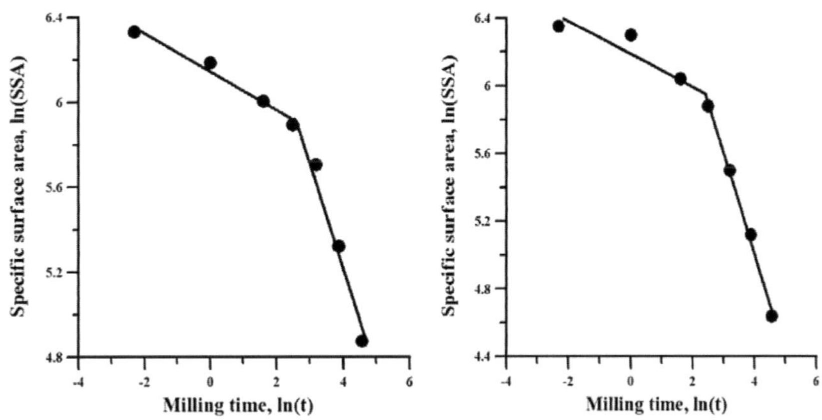

(a) (b)

Figure 3. The dependence of specific surface area values of Zeolites from milling times on logarithmic scale, (a) for SAPO-34, (b) SCT-323

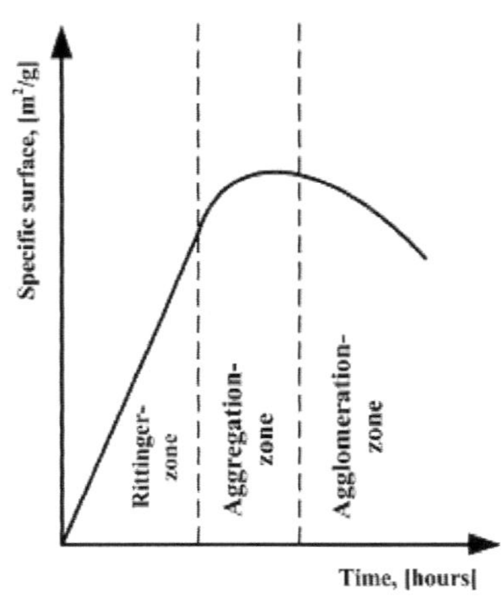

Figure 4. Theoretical curve of specific surface as function of grinding time

The theoretical curve of developed specific surface as function of grinding time is shown in Figure 4. Three well-shaped curves can be observed representing this theoretical curve. They are: Rittinger-zone: the increase of the specific surface is proportional to the grinding time; Aggregation-zone: the intensity with which the increase of the specific surface decreases with the grinding time;

Agglomeration-zone: the specific surface decreases with the comminution time [8, 9]. Because of the small particle size of synthetic zeolite SAPO-34 and SCT-323 with the increase in milling time takes place to agglomeration of particles only. Specific surface area of zeolites correspond to the agglomeration-zone on theoretical curve. Experimental results regarding the change in BET specific surface area of the milled zeolites samples with grinding time in a planetary ball mill are given in Table 1. The decrease in specific surface area is substantial at short grinding times, for instance, surface area reaches about 2.5 times of the initial value after 1 minute milling. BET analysis showed that the values of specific surface area of zeolite samples activated for 5 min were 83 m^2/g for SAPO-34 and 51 m^2/g for SCT-323.

Table 1. Specific surface area, average particle size (d) of zeolites with grinding time in a planetary ball mill

Milling time (min)	Specific surface area (m^2/g)		Grain size in the assumption of sphericity, d (nm)	
	SAPO-34	SCT-323	SAPO-34	SCT-323
0	561.15±13.5	572.99±13	5.3	5.2
1	241±3.61	278±6.93	12	10
3	146.74±2.57	127±2.34	20	23
5	83.63±2.02	51.34±1.61	35	58

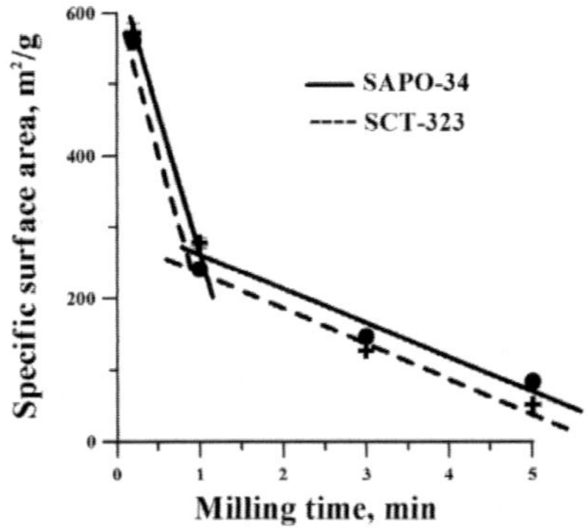

Figure 4. The dependence of specific surface area values of Zeolites from milling times in planetary ball mill

With the increase in milling time, the particles size decreased. The zeolite-average (value in microns) was 1.7 μm for SAPO-34, 1.5 μm for SCT-323. Then, after milling treatment at 24 hours the zeolite-average was 1.5 μm for SAPO-34 and 1.2 μm for SCT-323, after milling treatment at 96 hours the zeolite-average was 0.9 μm for SAPO-34 and 0.7 μm for SCT-323. Moreover most particles have lost their initial cubic shape and converted into spherical and also irregular shapes during the milling process on different milling times.

Selected XRD patterns of mechanically activated zeolite samples are shown in Figure 5. All of the diffraction peaks decreased and/or disappeared with the increase of mechanical activation time. In addition, amorphisation occurred and an increase in amorphisation was observed, from 83 to 90% for SAPO-34 and from 85% to 93% for SCT-323. The tendency of the peaks that belong to the zeolite phase Aluminum Silicate Phosphate to disappear and this tendency increased with the milling time. Koç et al. explained that mechanical activation results in the development of large numbers of dislocations and their associated strain fields, which might lead to an overall decrease in long range lattice periodicity [10].

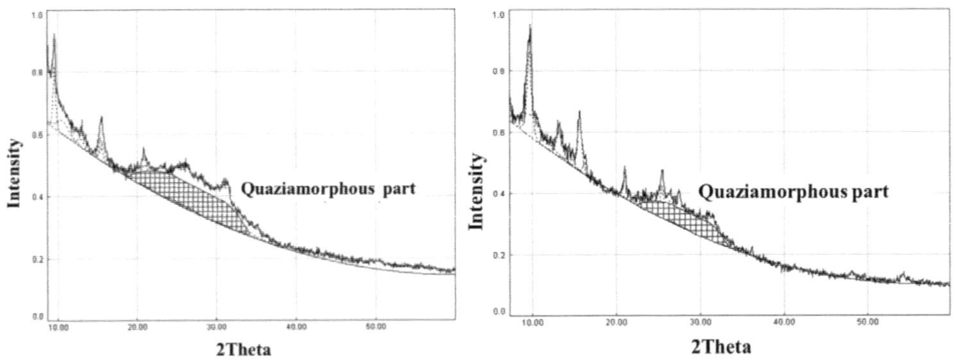

(a) (b)

Figure 5. X-ray diffraction patterns of zeolites during 96 hours milling time, for (a) SAPO-34, (b) for SCT-323

The properties of initial zeolites after annealing at 300°C with isothermal holding for 1 hour were studied. The specific surface area was determined at room temperature during a period of 1,000 hours. It has been shown that for the first 24 hours, the specific surface area after annealing was maximum, it decreased for the next 100 hours and after that it remained constant. One can observe the increase of the amount of amorphous phase from 83 to 90% for SAPO-34 and from 85% to 91% for SCT-323, during the 1000 hours period. Specific surface area of zeolites after mechanical activation in a tumbling ball mill during 96 hours and annealing up to 800°C with an isothermal holding time of 1 hour does not lead to marked changes and decreases strongly after annealing at 1000°C/h. Because of the high annealing temperature, sintering particles of zeolites to each other is observed.

4. Conclusion

Milling the particles during a period of 96 hours in a tumbling mill and 5 minutes in a planetary ball mill, the specific surface area is smaller than samples left for lesser periods of time. As for surface morphology, it was observed that the zeolites tended to be spherical in shape; but also irregular in shape after the milling treatment. It has been shown that the milling time of ball milling is a powerful method for obtaining the necessary specific surface due to changing the amount of amorphous phase in the system.

Acknowledgement

The authors would like to thank National research Tomsk State University for the financial support and to colleagues at Institute of Strength Physics and Materials Science of the Siberian Branch of the Russian Academy of Sciences, Fraunhofer Institute for Chemical Technology ICT

5. References

[1] Feng C, Khulbe K C, Matsuura T, Farnood R, Ismail A F 2015 *JMSR* **1** 49
[2] Hajfarajollah H, Askari S, Halladj R 2013 *Reac. Kinet. Mech. Cat.* **111** 723
[3] Peyman E, Shohreh F, Siamak A 2011 *Iran.J.Chem.Chem.Eng.* **30** 29
[4] Olson E 2012 *Journal of GxP Compliance* **16** 52
[5] Gömze L A, Gömze L N, Kulkov S N at all 2015 *Építőanyag-JSBCM* **67** 143 http://dx.doi.org/10.14382/epitoanyag-jsbcm.2015.38
[6] Gömze L A 2003 *Építőanyag-JSBMC* **55 (4)** 133 http://dx.doi.org/10.14382/epitoanyag-jsbcm.2003.23
[7] Nikashina V A, Streletskii A N, Meshkova I N, Kolbanev I V, Grinev V G, Serova I B 2011 *Inorganic Materials* **47** 1341
[8] Gömze L A 2010 *Mater. Sci. Forum* **659** 19
[9] Juhász Z A, Opoczky L 2003 *Építőanyag-JSBCM* **55** 86 http://dx.doi.org/10.14382/epitoanyag-jsbcm.2003.16
[10] Tunç T, Demirkiran A S 2014 *Powder Technology* **260** 7

Rheological characterization of geopolymer binder modified by organic resins

M Cekalová , T Kovárík , D Rieger

New Technologies - Research Centre, University of West Bohemia, Pilsen, Czech Republic

E-mail: cekalova@ntc.zcu.cz, toko@ntc.zcu.cz, davidpj@ntc.zcu.cz

Abstract. The purpose of this study is going to investigate properties of alkali-activated powder (calcined kaoilinitic clay and granulated blast furnace slag) prepared as a geopolymer paste and modified by various amount of organic resin. Hybrid organic-inorganic binders were prepared as a mix of organic resin and geopolymer inorganic paste under vacuum conditions. The process of solidification was investigated by measurements of storage (G') and loss modulus (G') in torsion. The measurement was conducted in oscillatory mode by constant strain of 0.01 %. This strain is set in linear visco-elastic region for minimization influence of paste structure. The effect of organic resin is presented and determined by changes of viscosity ('n*), modules in torsion and tangent of loss angle (tan 8). Results indicate that addition of organic resin significantly affects the initial viscosity and hardening kinetics.

1. Introduction

Geopolymers are a class of amorphous threedimensional aluminosilicate binder materials, synthesized by reaction of an aluminosilicate powder with a concentrated alkali metal silicate or hydroxide solution [1]. The exchange of silicon for aluminium in the polymer structure results in a net negative charge which is compensated by alkali cations. Geopolymers are generally prepared by mixing of an aluminosilicate raw material in form of a powder with an alkaline silicate solution followed by curing. The most common starting materials are metakaolinite and fly ash, although other aluminosilicate sources such as synthetic aluminosilicate powders and slags are also used [2].

Geopolymers have drawn attention because of their excellent mechanical properties, similar to those offered by the traditional Ordinary Portland Cement (OPC). Applications of geopolymer-based materials in the fields of new ceramics, binders, matrices for hazardous waste stabilization, fire-resistant materials, asbestos-free materials, and high-tech materials have been documented. These materials show excellent mechanical properties, low shrinkage, thermal stability, freeze-thaw, acid and fire resistance, long term durability and recyclability, so the application of geopolymer-based materials covers many fields. On the other hand geopolymers are ceramic materials, so they present a typical brittle mechanical behavior with the consequent low ductility and low fracture toughness. This behavior may represent a great limit in several structural applications. The development of hybrid materials represents a highly interesting and relevant research field for the potentially useful physical properties that could derive from the interfacial interactions of two chemically incompatible phases [3] [4] [5].

Rheology is the study of the fluidity and deformation of matter. The understanding of the rheological properties of cement pastes, mortars and concretes is requisite to ascertaining their consistency and workability and consequently their ease of casting or placement [6] [7] [8] [9].

The main aim of this study is to investigate the rheological properties of hybrid organic-inorganic binders using polyester and epoxy resin. The process of solidification was investigated by measurements of storage and loss modulus in torsion and complex viscosity.

2. Experimental

2.1. Materials

Materials was purchased as commercial product from Ceske lupkove zavody, a.s. (Czech Republic) as two component geopolymeric binder based on fired clays (SiO_2 - 47.4%, Al_2O_3 - 29.7%, CaO - 14.5%, MgO - 3.5%,TiO_2 - 1.8%, Fe_2O_3 - 0.8%, K_2O - 0.5%, Na_2O - 0.1%) activated with potassium silicate solution.

The epoxy (CHS-EPOXY 324) and polyester (Polyester 109) resin was purchased from Sincolor Company. Hybrid composites with epoxy resin (EPO) and polyester resin (PES) were prepared incorporating the resin into the geopolymeric matrix suspension under vacuum stirring, when both polymerization reactions have not been completed yet. Epoxy and polyester resins were obtained by mixing their components in weight ratio as specified in the technical data sheet supplied by the manufacturer.

2.2. Methods

Rheological analyses were performed by ARES G2 rheometer (TA Instrument). The measurement was conducted in oscillatory mode by constant strain of 0.01%. The conditions of measurement were as follows. Radial velocity 1 rad/sec and strain 0.01% were applied to ensure a minimal influence on the hardening process. The effect of organic resins is presented and determined by changes of complex viscosity (η^*), modules in torsion and tangent of loss angle (tan 5).

SEM analysis was carried out by means of a NeoScope JCM-5000 microscope (JEOL). Micrographs showed obtained from fracture surfaces of hybrid pastes EPO 0-15 % and PES 0-15 %.

3. Results and Discussion

3.1. GEO-EPO hybrid pastes

Figure 1. shows the complex viscosity versus time for the samples EPO 0-15%. The value of the initial complex viscosity pastes without resin were 1000 Pa ' s and increase of viscosity begins at 3000 s. The initial complex viscosity of pastes with epoxy resin were 100 Pa . s. The time when the viscosity began to increase was shortened with the amount of resin. Increase of viscosity began at 2000 s for pastes with 1% of epoxy resin and 300 s for pastes with a 15 % resin.

Figure 2. shows changes of loss angle with time. Table 1. shows these changes and it is evaluated as "transition point" - the time when there was a significant change in the properties examined pastes. This point can be considered as transition of inner structure from pseudoliquid solution to inorganic polymer network.

Figure 3. shows micrographs obtained from fracture surfaces of hybrid pastes EPO 0-15%.

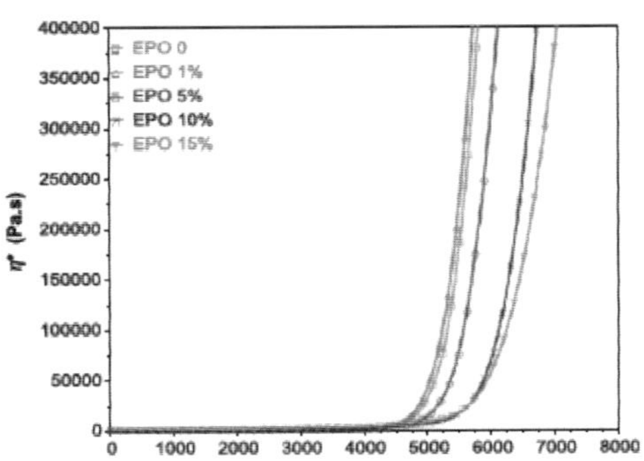

Figure 1. Complex viscosity versus time for samples EPO 0, 1, 5, 10 and 15 %

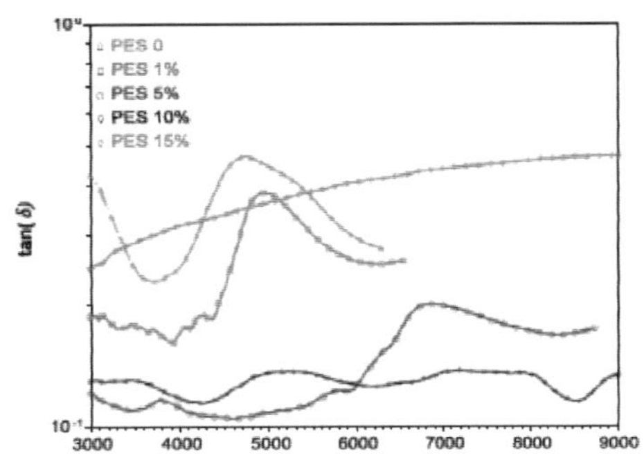

Figure 2. Changes of loss angle with time for samples EPO 0, 1, 5, 10 and 15%

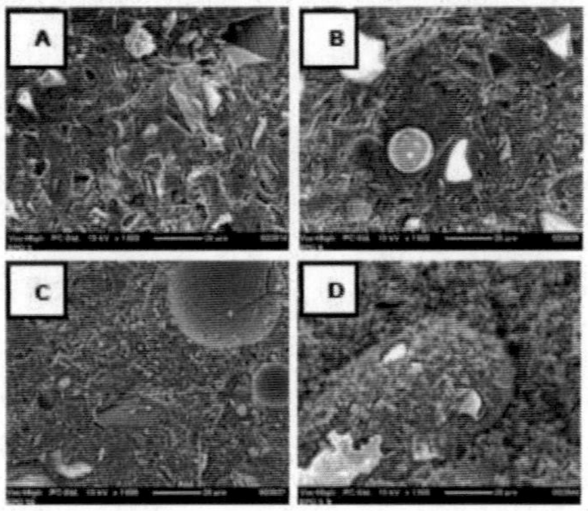

Figure 3. Scanning electron microscope (SEM) micrographs of (A, B, C, D) EPO 1-15 %

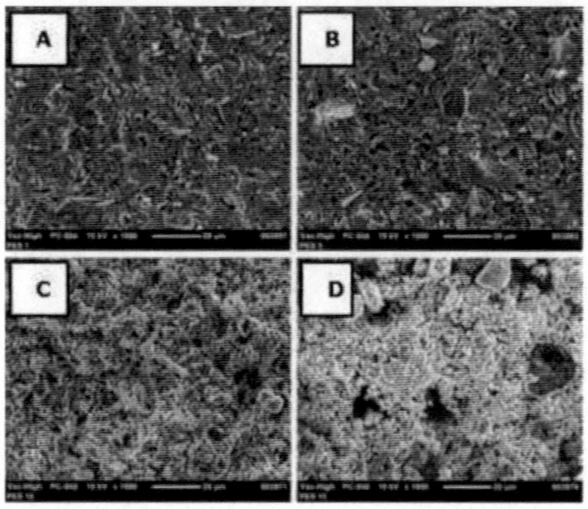

Figure 4. Scanning electron microscope (SEM) micrographs of (A, B, C, D) PES 1-15%

Table 1. Transition point - changes of loss angle with time

Sample	EPO %	0 EPO 1 %	EPO 5 %	EPO 10 %	EPO 15 %
Transition point (s) *	4676	4647	4981	5574	5630

* transition point of polycondensation reaction of geopolymer

3.2. GEO-PES hybrid pastes

Figure 5. shows the complex viscosity versus time for the samples PES 0-15%. The value of the initial complex viscosity pastes without resin were 1000 Pa. s and increase of viscosity begins at 3000 s. The initial complex viscosity of pastes with polyester resin were 100 Pa. s (PES 1%) and 2000 Pa ' s (PES 5-15%). The time when the viscosity began to increase was shortened with the amount of resin. Increase of viscosity began at 300 s for pastes with polyester resin (5-15%).

Figure 6. shows changes of loss angle with time. Table 2 shows these changes and it is evaluated as "transition point" - the time when there is a significant change in the properties examined pastes. This point can be considered as transition of inner structure from pseudoliquid solution to inorganic polymer network.

Figure 4. shows micrographs obtained from fracture surfaces of hybrid pastes PES 0-15%

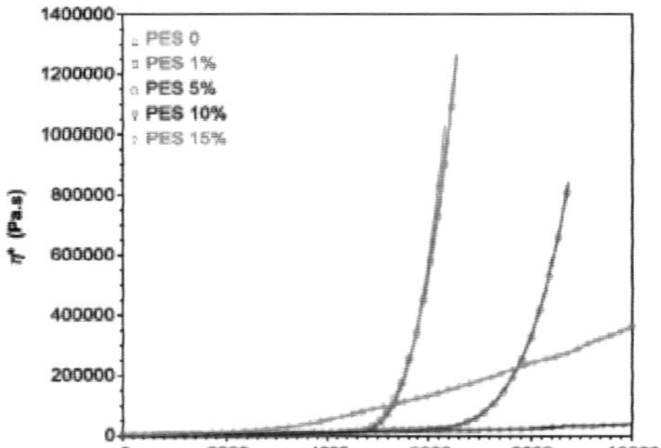

Figure 5. Complex viscosity versus time for samples PES 0, 1, 5, 10 and 15 %.

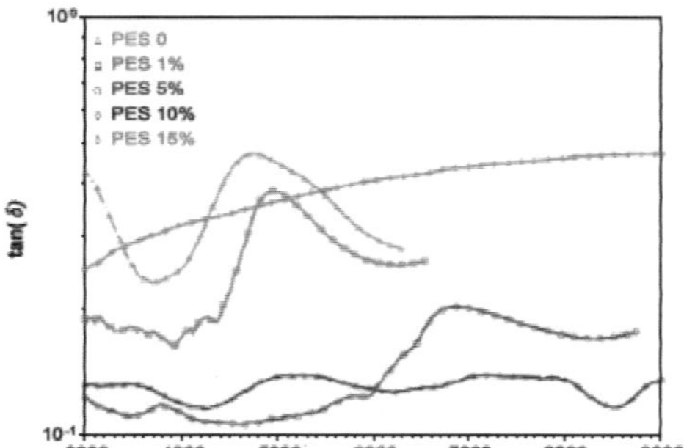

Figure 6. Changes of loss angle with time for samples PES 0, 1, 5, 10 and 15 %.

Table 2. Transition point - changes of loss angle with time

Sample	PES 0%	PES 1%	PES 5%	PES 10%	PES 15%
Transition point (s) *	4676	4958	6989	-	-

* transition point of polycondensation reaction of geopolymer

4. Conclusions

Series of alkali-activated geopolymer composites, incorporated with different amount of resin were synthesized. The effect of organic resin is presented and determined by changes of viscosity, modules in torsion and tangent of loss angle. The complex viscosity and tangent of loss angle appeared to be influenced by concentration of organic resin: when it increases the kinetic of geopolymerisation decreases. This influence is particularly apparent at higher concentrations of organic resins (10 and 15%).

The rheological properties of geopolymer hybrid pastes with polyester resin at concentrations of 10 and 15 % showed a significant deviation. These geopolymer hybrid pastes were poorly workable, they had significantly prolonged curing time (over 14 days) and poor cohesion. In this type of resin would seem appropriate initial compatibilization with the matrix in order to improve the connection of these two phases.

Acknowledgement

The result was developed within the CENTEM project, reg. no. CZ.1.05/2.1.00/03.0088, cofunded by the ERDF as part of the Ministry of Education, Youth and Sports OP RDI programme and, in the follow-up sustainability stage, supported through CENTEM PLUS (L01402) by financial means from the Ministry of Education, Youth and Sports under the "National Sustainability Programme I.

References

[1] J. Davidovits 1991 *J. of Therm. Anal.* **37** 1633-1656
[2] M.Romagnoli, C. Leonelli, E. Kamse, M. L. Gualtieri 2012 *Constr. Build. Mater.* **36** 251-258
[3] G.Roviello, L. Ricciotti, C. Ferone, F. Colangelo, R. Cioffi, O. Tarallo 2013 *Materials* **6** 3943-3962
[4] F. Colangelo, G. Roviello, L. Ricciotti, C. Ferone, R. Cioffi 2013 *Materials* **6** 2989-3006
[5] C. Ferone, G. Roviello, F. Colangelo, R. Cioffi, O. Tarallo 2013 *Appl. Clay Sci.* **73** 42-50
[6] F. Puertas, C. Varga, M.M. Alonso 2014 *Cement Concrete Comp.* **53** 279-288
[7] M.Romagnoli, C. Leonelli, E. Kamse, M. L. Gualtieri 2012 *Constr. Build. Mater.* **36** 251-258
[8] A.Poulesquen, F. Frizon, D. Lambertin 2011 *J. Non-Cryst Solids* **357** 3565-3571
[9] D.Rieger, T. Kovank, J. Riha, R. Medlin, P. Novotny, P. Belsky, J. Kadlec, P. Holba 2015 *Constr. Build. Mater.* **83** 26-33

Filling behaviour of wood plastic composites

I Duretek, T Lucyshyn, C Holzer

Department of Polymer Engineering and Science, Chair of Polymer Processing, Montanuniversitaet Leoben, Leoben, Austria

E-mail: Ivica.Duretek@unileoben.ac.at

Abstract. Wood plastic composites (WPC) are a young generation of composites with rapidly growing usage within the plastics industry. The advantages are the availability and low price of the wood particles, the possibility of partially substituting the polymer in the mixture and sustainable use of the earth's resources. The current WPC products on the market are to a large extent limited to extruded products. Nowadays there is a great interest in the market for consumer products in more use of WPC as an alternative to pure thermoplastics in injection moulding processes.
This work presents the results of numerical simulation and experimental visualisation of the mould filling process in injection moulding of WPC. The 3D injection moulding simulations were done with the commercial software package Autodesk® Moldflow® Insight 2016 (AMI). The mould filling experiments were conducted with a box-shaped test part. In contrast to unfilled polymers the WPC has reduced melt elasticity so that the fountain flow often does not develop. This results in irregular flow front shapes in the moulded part, especially at high filler content.

Keywords: wood plastic composites, injection moulding, simulation, process, mould

1. Introduction

Wood plastic composites (WPC) are highly filled thermoplastics; they are usually binary systems consisting of wood flour/spans/fibres and polymer matrix (Figure 1). These two main constituents are very different in origin, structure and performance. Polymers are high molecular weight materials whose performance is largely determined by its molecular architecture. The matrix polymers are typically low-cost commodity polymers that flow easily. The polymers tend to shrink and expand with temperature. Wood itself contains polymers such as lignin, cellulose, and various hemicelluloses but has very different properties from the synthetic polymers with which it is most often combined. Wood is less expensive, stiffer, and stronger than these synthetic polymers, making it a useful filler or reinforcement. Though wood does not shrink and swell much with temperature, it readily absorbs moisture. As with most natural materials, the anatomy of wood is complex. Wood is porous, fibrous, and anisotropic [16].

WPC are a very young segment of the polymer industry with a great potential. Not only due to its flexible ratio of mixture and as a consequence, a flexible setting of mechanical properties, but also because of numerous benefits compared with pure wood. WPC have a higher biological, UV-radiation and weathering resistance and lower water absorption than conventional wood products and are less likely to be harmed by fungi, which results in lower maintenance costs. Furthermore the WPC can be processed by commonly used plastics processing methods (e.g. injection molding, extrusion) which give a tremendous freedom of form and geometry. This gives a broad variety of possibilities regarding the shape of the product. Another benefit is the flexibility in visual appearance of WPC products achieved by post-manufactured brushing, embossing or shaping. [3].

Wood fibre has lately attracted considerable attention as a filler to reinforce plastics, which has been driven by the continuous increase of oil prices and concerns of recycling. Wood fibre has advantages regarding density, cost, mechanical properties and biodegradability compared to other fibres. However, the density of wood fibre is still higher than plastics, such as polyethylene (PE) and polypropylene (PP).

Figure 1. Example of WPC

The use of WPC in commercial products is limited today. The most frequent WPC products are to a large extent limited to extruded products. The use so far is a replacement of wood in outdoor railings and decking (see Figure 2) [1, 2]. In Europe WPC are used in a wide variety of applications, from decking and siding to sophisticated musical instruments, furniture, watches, pencils, tableware, toys, decoration and pallets [4].

Injection moulding is used more and more for the production of low-proportion natural fibre composites (NFC) – materials with up to 50% content of natural fibres by weight. A number of applications are currently undergoing commercial assessment, with the main focus on automotive applications (Figure 3), where WPC materials have to compete with the talcum and glass fibre-filled compounds currently in use, as well as compression-moulded NFC [4].

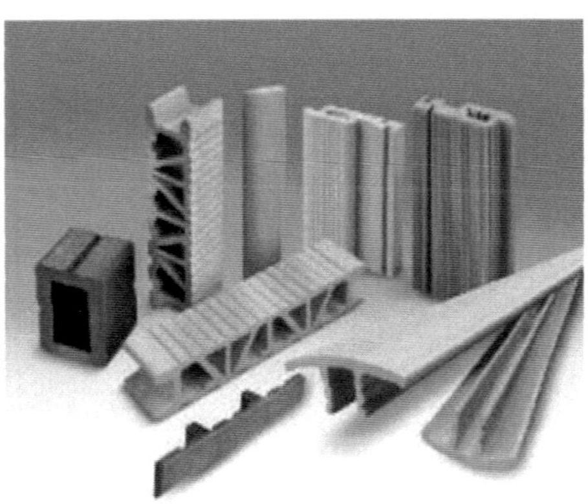

Figure 2. Typical extruded semi-finished WPC products [10]

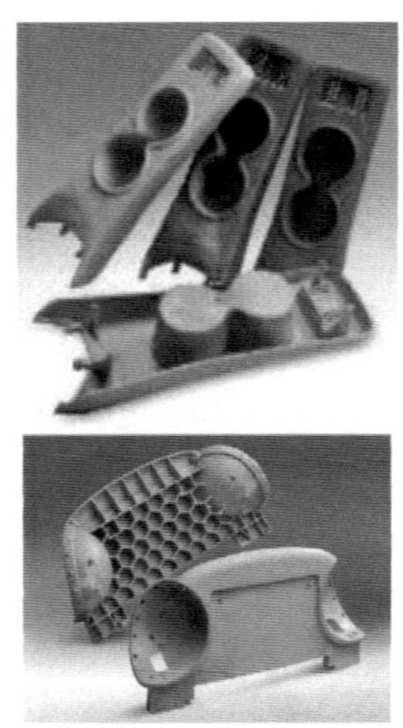

Figure 3. Example of automotive applications for WPC [10]

Although the WPC can be used for injection moulding as a replacement of pure thermoplastics, there is a need to use its property to its full extent. While the market for extruded WPC products is growing with 10% per year [2], the injection moulded WPC products are marginal. There is an unexploited potential for development [4].

Injection moulding is one of the major processing technologies of polymers. Injection moulding is widely used because of its economics to produce high volume of complex plastic articles [18]. It is a process where a plastic or composite is injected into a mould under very high pressure. This is done with an injection moulding machine that consists of two main parts, the injector and the clamping device. The filling behaviour and way the plastic flows into the mould are of paramount importance in determining the quality of the part [5].

The melt flow in the mould can be controlled by the design and manufacturing technology of the mould as well as by the processing conditions in order to obtain the moulded parts with expected morphology, properties, shape, dimensions and surface. The cavity in the injection mould should be filled totally during the injection phase and the way of filling should be laminar and with a wide flow front. The stream flow (jetting phenomenon) should be avoided. Furthermore the rheological phenomena occurring during the melt flow in injection moulds must be known by the mould designers and controlled in the technological process [19].

Today's simulation tools allow the simulation of the filling, packing and cooling process and also a qualitative prediction of the part's shrinkage and warpage for semi-crystalline materials.

It helps to predict the problems that may occur due to wrong tool design or not optimized processing conditions [19]. According to literature [11, 12, 13, 17] the mould filling can be divided into fountain flow and solid flow.

The fountain flow (Figure 4) is typical for thermoplastic polymers and it is a result of wall contact by the melt and occurs due to the parabolic flow front [11]. The skin of the plastic in contact with the cool mould freezes rapidly, while the central core remains molten. When additional material is injected, it flows into this central core, displacing the material already there, which then forms a new flow front. The flow of this displaced material is a combination of forward flow and outward flow. The outward flow contacts the wall, freezes, and forms the next section of skin while the forward flow forms the new molten core. The frozen layer is formed by the flow front inflating, and so is subject to only a low shear stress and, therefore, has a very low level of molecular orientation. Initially, the frozen layer is very thin, so heat is lost very rapidly. This results in more plastic freezing and the frozen layer getting thicker, cutting down the heat flow. After a time, the frozen layer will reach a thickness such that the heat lost by conduction is equal to the heat input from plastic flow and frictional heating, i.e., an equilibrium condition is reached [5].

The solid flow is typical for highly filled polymer compounds and thermosets [7, 8, 11]. By this flow the melt slips on a layer of low viscosity resin. The flow front consists of less compacted melt followed by a compact melt zone. An orientation of reinforcing particles is not possible [11] (see Figure 5).

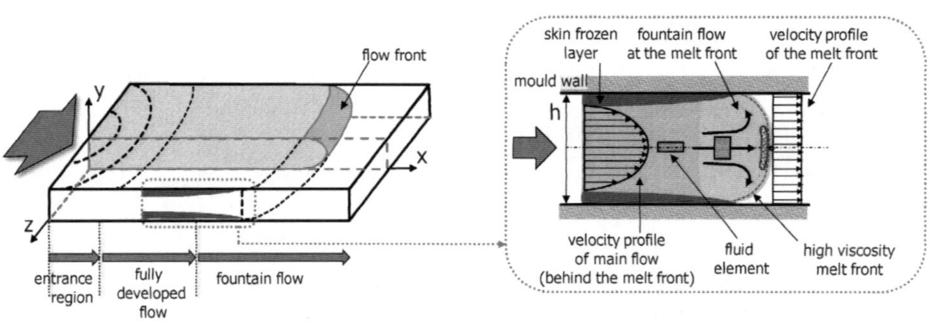

Figure 4. Various flow regimes vs. cross section of the wall thickness; fountain flow (top) [acc. to 5, 14, 15]

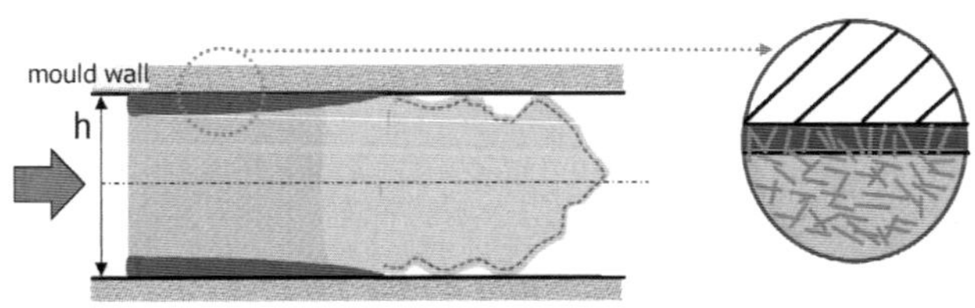

Figure 5. Flow characteristic of highly filled polymers (bottom) [acc. to 11]

In this paper, the focus lies on the numerical simulation of the mould filling process in injection moulding of two WPC types with various compositions using commercial simulation software and the experimental visualisation with the help of a filling study.

2. Materials and methods

2.1. Materials

For the experimental work two different PP-based WPC were used. Figure 6 shows the picture of pellets and light microscopy (LiMi) images of the investigated materials WPC-01 (left) and WPC-02 (right). From the LiMi image the wood-matrix polymer interface can be seen.

WPC-01 has a wood fibre content of 50 wt. %. The viscosity of WPC-02 is approx. 10% higher than the viscosity of WPC-01 and the density is approx. 6% higher. A comparison of viscosity is shown in Figure 7 and of density in Figure 8.

WPC-01 WPC-02

Figure 6. Pellets of the investigated WPC and LiMi images

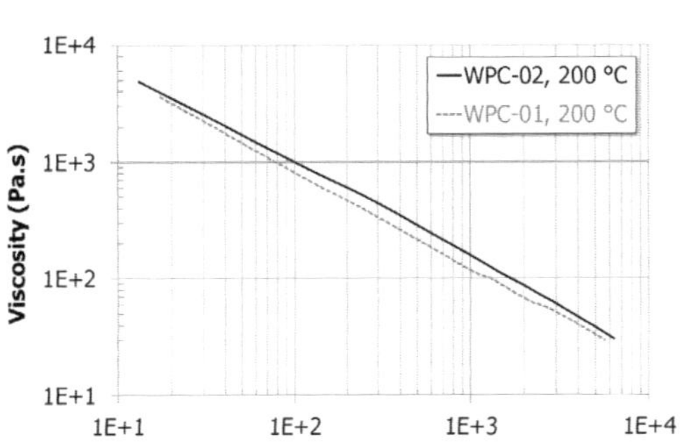

Figure 7. Viscosity

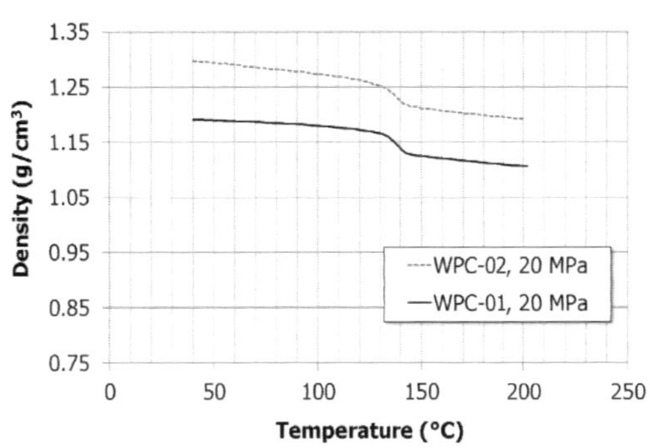

Figure 8. Density

2.2. *Injection moulding experiments*

For comparing the simulation results with real parts several test specimens were produced on an injection moulding machine Arburg ALLROUNDER 470A 1000-400.

The mould used in this study was a box-shaped test part (stacking-box). The cavity is shown in (Figure 9-left). The mould was equipped with a hot runner system and the gate position was at the centre of the box.

The process parameters for the injection moulding tests were selected according to the processing instructions from the material manufacturer. For the filling study the dosing volume was varied between 45 and 150 cm^3 and the holding pressure was not applied. The other machine parameters are summarized in Table 1.

The materials were not pre-dried, they were stored in laboratory under normal storage conditions, at a temperature between 23°C and 25°C and a humidity between 40 to 45%. The residual moisture content of the materials was approx. 3.3%. The moisture content was measured with water content analyser HydroTracer FMX (aboni GmbH für Mess- und Automatisierungstechnik, Germany).

Table 1. The most important parameters of the injection moulding process

Barrel temperatures (°C), [$Z0_{Hopper}$-Z1-Z2-Z3-Z4]	40-180-185-190-195
Nozzle temperature (°C)	200
Mould temperature (°C)	40
Clamping force (kN)	650
Back pressure (bar)	50
Dosing speed (m/min)	20
Injection rate (cm^3/s)	50
Cooling time (s)	25

2.3. *Filling simulation*

The filling of the injection mould was investigated with the commercial software package Autodesk® Moldflow® Insight 2016 (AMI). The process parameters from the experiment (see Table 1) were used for the simulation. Simulations were performed with a 3D-mesh. The gating system for the mould was not modelled, but an injection point was directly set on the part. The model used for computation in AMI is presented in Figure 9-right.

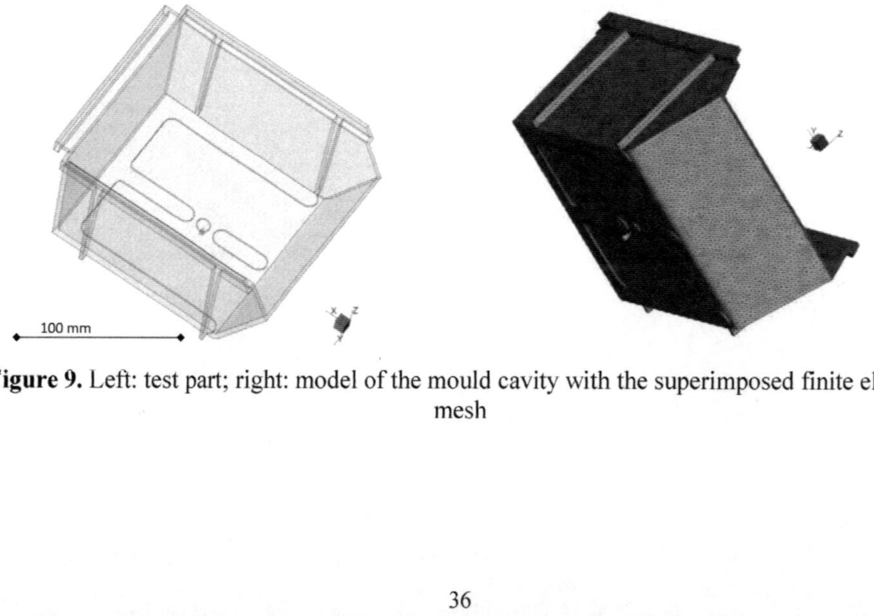

Figure 9. Left: test part; right: model of the mould cavity with the superimposed finite element mesh

The viscosity of the materials was measured with a Göttfert high pressure capillary rheometer using the slit die with flush mounted pressure transducers at two temperatures (190°C, 200°C) in a shear rate range between approx. 10 – 7,000 s^{-1}. The viscosity model (Cross-WLF) was fitted from the corrected data (Weissenberg/Rabinowitsch) of these flow curves.

The viscosity measurements in the lower shear rate range (<10 s^{-1}) using the parallel plate rotational rheometer were unsuitable for generating viscosity data. Due to the wall slip between the smooth plates and the polymer the determination of rheological data was not possible.

3. Results and discussion

The results of the experimental filling studies with WPC-01 and WPC-02 are shown in Figure 10 and Figure 11. It can be clearly seen that for both investigated WPCs no classic fountain flow occurred (see Figure 4-top). The melt front has not the parabolic profile typical for thermoplastics. The flow front increasingly breaks up, is frayed and brittle.

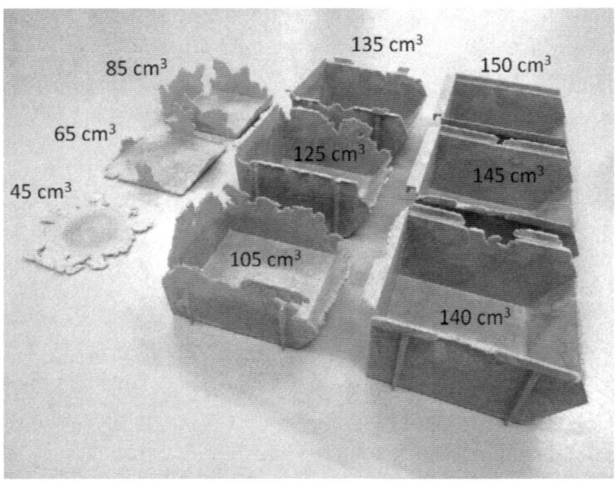

Figure 10. Filling study with WPC-01

Figure 11. Filling study with WPC-02

Instead of the compact melt zone the less compacted melt is folded and only at the end of the flow path compacted.

Furthermore, several repeated injection moulding cycles showed bad reproducibility of the shape of the flow fronts (Figure 12). The flow length was similar, but the shape of the flow fronts was different. In this regard the various compositions of both WPC showed no obvious influence on mould filling.

In the case of pure PP as expected the fountain flow occurred in the cavity and finally the filling characteristics was uniform (Figure 13-left). For the unfilled PP the agreement between simulation and experimental measurement is excellent (Figure 13-right).

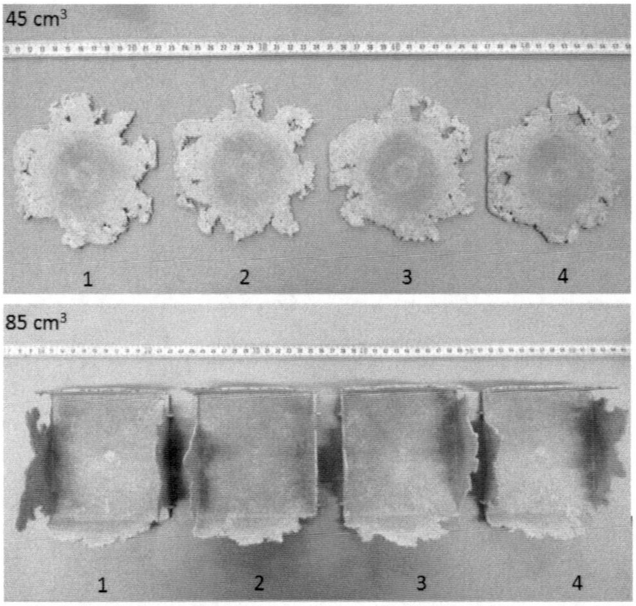

Figure 12. Injection moulded short shots of a WPC-01 in a mould for two different dosing volumes (45 and 85 cm^3); four consecutive cycles

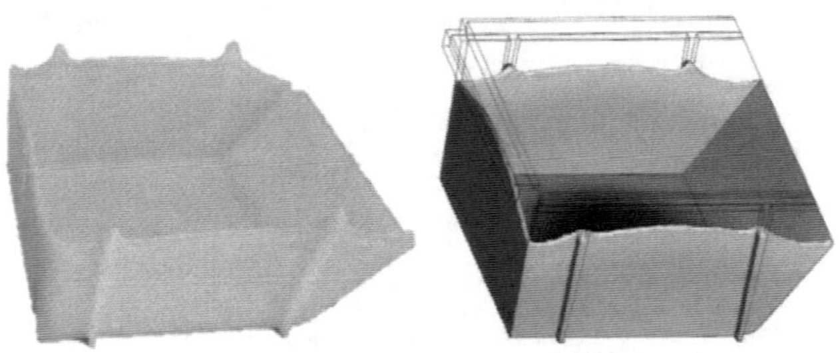

Figure 13. Comparison of filling behaviour for PP

The simulation results of the filling behaviour for WPC-01 are shown in Figure 14-left. The presentation contains the frame-by-frame recordings. The predicted melt front advancement, as well as weld lines are not in good agreement with the advancement pattern observed in the short shots of the moulding. Unfortunately, the shape of the melt front could not be calculated satisfactorily.

The predicted melt-front advancement for the WPC-02 had a similar filling behaviour as the results of WPC-01 (see Figure 14-right).

The reasons for the bad agreement between experiment and simulation and the lack of fountain flow and further the occurrence of the irregular shape of the melt front may be attributed to the melt elasticity and the wall slip, which are not considered in the simulation model.

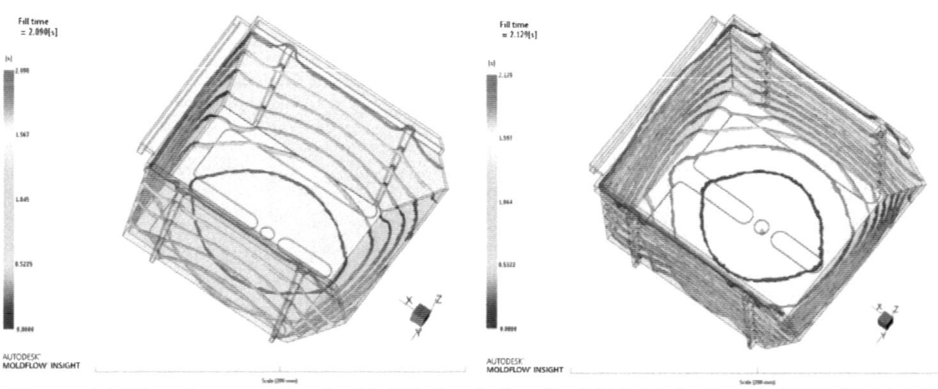

Figure 14. Flow front predicted with 3D-simulation for WPC-01 (left) and WPC-02 (right)

The high viscosity of WPC reduces the melt elasticity so that the fountain flow during the cavity filling often does not develop. The fountain flow effect occurs because the non-slip condition on the mould walls forces the material to flow from the centre to the outward mould walls [6].

Furthermore, Funke [6] shows that the wall slip effect can be intensified by particular process settings. Eg. low melt and mould temperatures increase the material's wall slipping tendency.

As is well known, WPC tend to slip at the wall. In previous work [20, 22] the influence of moisture content on the rheological characteristics of WPC was investigated. With the dried PP-based WPC with lower relative moisture content (0.5%) shear flow occurred, whereas undried WPC with 3.8% relative moisture content showed existence of wall slip (pure plug flow). These results were obtained from by-pass extrusion rheometer and high pressure capillary rheometer at 200°C.

4. Conclusion and outlook

This paper presents the results of numerical simulation and experimental visualisation of the mould filling process in injection moulding of WPC.

Due to low melt elasticity of highly filled WPC the classic fountain flow like for unfilled thermoplastics did not develop. Melt front break and so called finger effects occurred. The shape of the melt front was unpredictable and not reproducible.

The filling behaviour of WPC cannot yet be accurately predicted by using the 3D simulation software. In the case of highly filled thermoplastics like WPC the simulation of mould filling processes requires different adjustments compared to the use of conventional thermoplastics (e.g. an adequate rheological model for wall slipping).

To exactly describe the shape of the flow front, filling stage simulation should use equations for the material properties which take into account the elastic extensional behaviour of the melt.

Further investigations with WPC with different wood content will be done to identify possible effects (eg. wall slip, moisture content) regarding flow and filling behaviour. Furthermore, the influence of the filler size, shape and aspect ratio on the flow behaviour will be investigated.

Acknowledgements

The authors would like to thank battenfeld-cincinnati Austria GmbH for supplying the material used in this work.

References

[1] Renard, C.: Injection molding WPC, bachelor thesis, KTH Industrial technology and management, Stockholm, 2011.

[2] Haider, A, Eder, A: Markets, Applications, and Processes for Wood Polymer Composites (WPC) in Europe, Presented at 1st International Conference on Processing Technologies for the Forest and Bio-based Products Industries (2010), http://conference.fh-

salzburg.ac.at/fileadmin/files/documents/presentations/4B_Haider.pdf [accessed 11. April 2011].

[3] Kahr, S.C.: Wood Plastic Composites - determination and comparison of the specific energy consumption for WPC compounding methods based on two state-of-the-art formulations, master thesis, Montanuniversität Leoben, 2010.

[4] Haider, A.: The market for injection moulded and extruded WPC products, URL: http://www.plastics.gl/consumer/the-right-mix/ [15. August 2015].

[5] Shoemaker, J.: Moldflow design guide, Moldflow Corporation, Framingham, 2006.

[6] Funke, C., Albring, E., Moritzer, E.: WPC has different filling behaviour, Kunststoffe international 8 (2010), 53 - 56.

[7] Thienel, P., Hoster, B., Bayerl, H., Braun, U.: A mould with a window allows development of the filling pattern for thermosetting molding compound, Kunststoffe 80 (1990) 12, 1346-1350.

[8] Pfefferkorn, T.: Analyse der Verarbeitungs- und Materialeigenschaften elektrisch leitfähiger Kunststoffe auf Basis niedrig schmelzender Metalllegierungen, Dissertation, RWTH Aachen, 2009.

[9] Restrepo, J., López, I.D.: Implementation of highly efficient meshless method for the thermal analysis of fountain flow during filling in injection molding, 4SPE ANTEC 2012, Orlando, 2012.

[10] JELU-WERK: WPC - Anwendungen und Produkte, JELU-WERK Josef Ehrler GmbH & Co KG Natural fibres, D-73494 Rosenberg, URL: www.jeluplast.com/wpc-anwendungen/#produktfotos [accessed 10. March 2011].

[11] Englich, S., Scheffler, T., Gehde, M.: Mold filling behavior of highly anorganic filled phenolic injection moulding compounds, 29th International Conference of the Polymer Processing Society (PPS 29), Nuremberg, 2013.

[12] Thienel, P., Hoster, B., Schröder, T., Schröder, K., Kretzschner, J., Ludwig, R.: Duroplast-Spritzgießen mit Gasinnendruck, Kunststoffe 83 (1993), 91 - 95.

[13] Niemann, K., Schröder, K.: Spritzgießen von Duroplasten, Hüthig Verlag, Heidelberg, 1994.

[14] Chung, T.N.: Strömungsanalyse der Bindenahtformation beim Spritzgießen von thermoplastischen Kunststoffen, Dissertation, Technische Universität Chemnitz, 2001.

[15] Dietz, W., White, J.L.: Ein einfaches Modell zur Berechnung des Druckverlustes während des Werkzeugfüllvorganges und der eingefrorenen Orientierung beim Spritzgießen amorpher Kunststoffe, Rheologica Acta 17 (1978) 6, 676 - 692.

[16] Clemons, C.: Raw materials for wood –polymer composites, in Wood-polymer composites, edited by K.O Niska and M.Sain, Woodhead Publishing Limited, Cambridge, 2008.

[17] Li, C.S., Hung, C.F., Shen, Y.K.: Computer simulation and analysis of fountain flow in filling process of injection molding, Journal of Polymer Research 1 (1994) 2, 163 -173.

[18] Sina, L.T., Rahmanb, W.A.W.A., Rahmatb, A.R., Teea, T.T., Beea, S.B., Yua, L.C.: Computer aided injection moulding process analysis of polyvinyl alcohol–starch green biodegradable polymer compound, Journal of Manufacturing Processes 14 (2012), 8–19.

[19] Bociaga, E., Jaruga. T., Experimental investigation of polymer flow in injection mould, Archives of Materials Science and Engineering 28 (2007) 3, 165 - 172.

[20] Duretek, I., Friesenbichler, W., Holzer, C.: New insights into characterization of flow properties of PP based wood plastic composites, 26th International Conference of the Polymer Processing Society (PPS 26), Banff, 2010.

[21] Hristov, V., Takacs, E., Vlachopoulos, J.: Surface tearing and wall slip phenomena in extrusion of highly filled HDPE/Wood flour composites, Polymer Engineering and Science, 46 (2006) 9, 1204 - 1214.

[22] Duretek, I., Selvasankar, R.K., Langecker, G.R., Holzer, C.: Rheology of wood plastic composites using by-pass extrusion rheometer on the twin screw extruder, 9th WPC - Natural Fibre and other innovative Composites Congress and Exhibition, Stuttgart, 19th – 20th June 2012.

Material flow data for numerical simulation of powder injection molding

I Duretek [1, 2], **C Holzer** [1]

[1]Department of Polymer Engineering and Science, Chair of Polymer Processing, Montanuniversitaet Leoben, Leoben, Austria
[2]Polymer Competence Center Leoben GmbH, Leoben, Austria

E-mail: Ivica.Duretek@unileoben.ac.at

Abstract. The powder injection molding (PIM) process is a cost efficient and important net-shape manufacturing process that is not completely understood.
For the application of simulation programs for the powder injection molding process, apart from suitable physical models, exact material data and in particular knowledge of the flow behavior are essential in order to get precise numerical results. The flow processes of highly filled polymers are complex. Occurring effects are very hard to separate, like shear flow with yield stress, wall slip, elastic effects, etc. Furthermore, the occurrence of phase separation due to the multi-phase composition of compounds is quite probable.
In this work, the flow behavior of a 316L stainless steel feedstock for powder injection molding was investigated. Additionally, the influence of pre-shearing on the flow behavior of PIM-feedstocks under practical conditions was examined and evaluated by a special PIM injection molding machine rheometer.
In order to have a better understanding of key factors of PIM during the injection step, 3D non-isothermal numerical simulations were conducted with a commercial injection molding simulation software using experimental feedstock properties. The simulation results were compared with the experimental results.
The mold filling studies amply illustrate the effect of mold temperature on the filling behavior during the mold filling stage. Moreover, the rheological measurements showed that at low shear rates no zero shear viscosity was observed, but instead the viscosity further increased strongly. This flow behavior could be described with the Cross-WLF approach with Herschel-Bulkley extension very well.

Keywords: viscosity, rheology, powder injection molding, simulation, 316L

1. Introduction

Powder injection molding (PIM) is s a multi-step technology for manufacturing complex metal or ceramic parts. It consists of four sequential technological steps – mixing, injection molding, debinding and sintering. All these steps have an effect on the characteristics of the final parts. Optimization of the PIM process for production is a long and expensive procedure. The injection molding process as the first step in part production is very critical. With the help of simulation software the injection molding process of PIM can be successfully optimized. For achieving practically relevant simulation results, accurate thermodynamic material data and especially reliable rheological data are necessary. It is important to know the viscosity of the material in order to predict flow patterns and molding forces.

IOP Conf. Series: Journal of Physics: Conf. Series **790** (2017) 012007 doi:10.1088/1742-6596/790/1/012007

Adding fillers to a neat polymer melt changes its rheology, influencing both the way the melt is processed and the properties of the ultimate product. Key factors are filler size and shape, filler concentration, and the extent of any interactions among the particles.

The consequences of adding fillers are an increase in melt viscosity and a decrease in die swell. Moreover, particle interactions increase the non-Newtonian (i.e. shear thinning) range and cause it to occur at a lower shear rate than for the unfilled polymer melt. Filled polymers (eg. PIM-feedstock) have a higher viscosity at low shear rates, and yielding may occur with increased filler concentration (Figure 1). At higher shear rates the effect of the filler decreases and the matrix contributions dominate. Another effect of the filler is the reduction of the linear viscoelastic range, characterized by the onset of the strain dependent behavior [14].

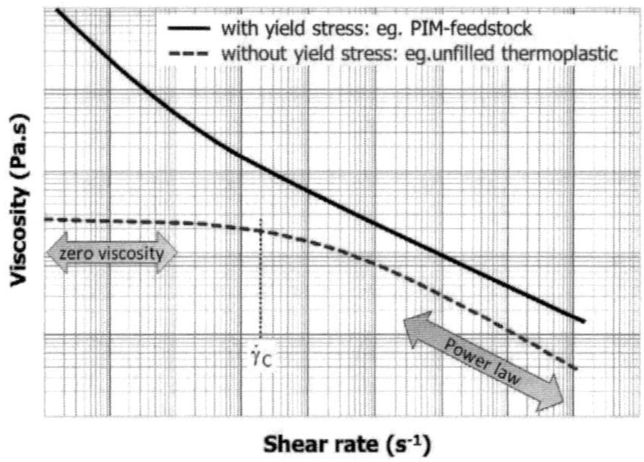

Figure 1. Viscosity curves of a non-Newtonian fluid with (black line) and without yield stress (blue, dashed line)

The simplest description of a non-Newtonian flow behavior is the Power law model, which can capture the constant slope of the decreasing flow curve at higher shear rates. Addition of zero shear viscosity and the onset of shear thinning $\dot{\gamma}_C$ extend the Power law model (Carreau, Ellis, Cross ...).

In contrast to unfilled thermoplastics for PIM-feedstocks no Newtonian plateau is observed at very low shear rates, but instead the viscosity keeps increasing strongly. This indicates the possible presence of the limiting shear stress (flow limit) also known as yield stress or yield point, depending on the material structure and the applied shear stress.

The flow profile differences comparing classical Cross model appropriate for unfilled thermoplastics and the Cross-Herschel-Bulkley model appropriate for PIM-feedstocks are shown in Figure 2. Due to the yield stress the viscosity in the middle of the flow channel (very low shear rates) increases significantly. The result is a flattened velocity profile, well known as plug flow. Here, the whole shear load is concentrated in a very small layer close to the mold wall and the feedstock is "slipping" at this layer of low viscosity, causing typical jetting phenomena. Therefore the standard rheological model for thermoplastics cannot be used to fit the flow behavior of PIM-feedstocks [10].

Shear thinning behavior in combination with the yield stress are generally modeled with the Herschel-Bulkley equation. However for PIM-Feedstocks, the Cross model with Herschel-Bulkley extension and CARPOW equation are more adequate for describing their flow behavior within a given shear rate range (Table 1). CARPOW equation is a combination of the Carreau equation and a power law.

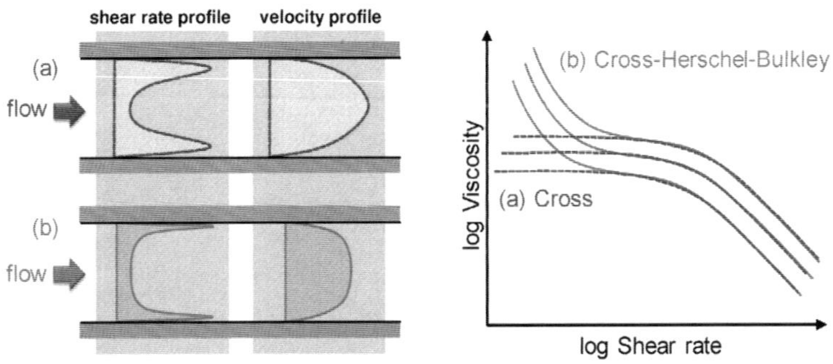

Figure 2. Flow profile differences for (a) the Cross and (b) the Cross-Herschel-Bulkley model [10]

Table 1. Some fitting models for describing the flow behavior within a shear rate range including the yielding

Model	Equation	
Herschel-Bulkley [16]	$\tau = \tau_0 + K \cdot \dot{\gamma}^n$	(1)
Cross-Herschel-Bulkley [12]	$\eta = \dfrac{\eta_0}{1+\left(\dfrac{\eta_0 \cdot \dot{\gamma}}{D_4}\right)^{(1-n)}} + \tau_\gamma \cdot \left(\dfrac{1-e^{(-\alpha\cdot\dot{\gamma})}}{\dot{\gamma}}\right)$	(2)
CARPOW [15]	$\eta = \dfrac{d}{\dot{\gamma}^n} + \dfrac{A}{(1+B\cdot\dot{\gamma})^C}$	(3)

where: τ_0 is the yield stress or yield point, η is the viscosity, $\dot{\gamma}$ is the shear rate, K, n and D_4 are the parameters of flow equations, τ_γ defines the slope point, α is the gradient to the higher viscosities, d and C are flow exponents for low and high shear rate range, A and B are Carreau material constants

High pressure capillary rheometers with round dies are versatile and widely used in the rheological measurement of viscous materials at shear rates higher 100 s^{-1}. However, capillary rheometers with round dies suffer from some limitations when attempting to measure the viscosity of highly filled plastics like PIM-feedstocks.

In general, the pressure curves in Bagley diagrams for unfilled thermoplastic melts are nearly linear, as long as the melt pressure does have a negligible influence on the viscosity.

In highly filled compounds, such as metallic PIM-feedstocks, nonlinear Bagley curves can occur which give physically nonsensical negative pressure entrance lose (see Figure 3).

The reason for nonlinear pressure drop in Bagley diagrams can be the pressure dependence of viscosity [6] and on the other hand error in the pressure measurements using the "pressure hole" technique by round rheological dies. The occurrence of flow phenomena such as wall slip and inhomogeneous particle distribution can also lead to nonlinear Bagley plots. Nyborg [3] and Bilovol [4] also found negative pressure losses by measurements on metal PIM-feedstocks. Nyborg [3] determines this effect for a stainless steel 316L feedstock with LLDPE binder system and Bilovol [4] for Catamold 316L.

To avoid nonsensical negative pressure losses, one should use a quadratic approximation and optionally the zero-length capillary (eg. L/D < 0.5).

Previous works [1, 2, 5, 9] showed that usage of the slit die for the rheological characterization of PIM-feedstocks instead the round die is very beneficial. The major advantage of the silt die is that the pressure transducers can be mounted on the die wall, therefore the pressure drop along the die can be measured directly and the Bagley's correction is not necessary.

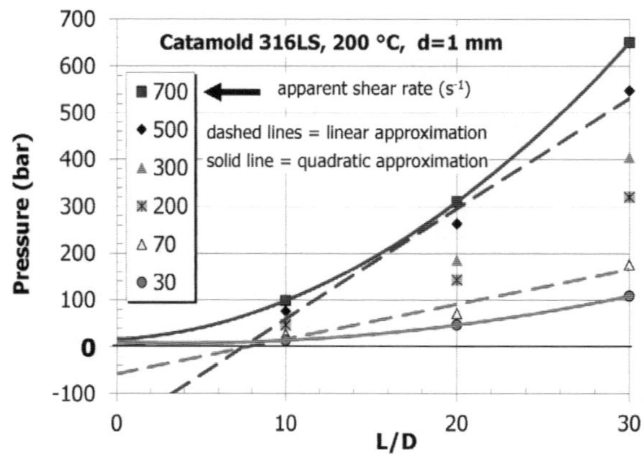

Figure 3. Bagley diagram with linear and polynomial approximation ; stainless steel feedstock 316L with POM - binder system [1]

In this paper we present some experimental techniques to determine the rheological behavior of PIM-feedstock. Furthermore, the filling study and results of 3D non-isothermal numerical simulations during the injection step were compared.

With this input a better understanding of rheological and filling behavior of the PIM-feedstock is given and a successful injection simulation can be achieved.

2. Materials and experimental procedure

2.1. Materials

Two commercial stainless steel feedstocks, Catamold 316LG and Catamold 316LA (BASF) were used for the investigation.

Figure 4 shows morphology of the investigated feedstocks recorded with the scanning electron microscopy (SEM). From the SEM images, it can be seen that both feedstocks contain spherical particles coated by the polymeric binder. Each feedstock has a different particle size as shown in Figure 3.

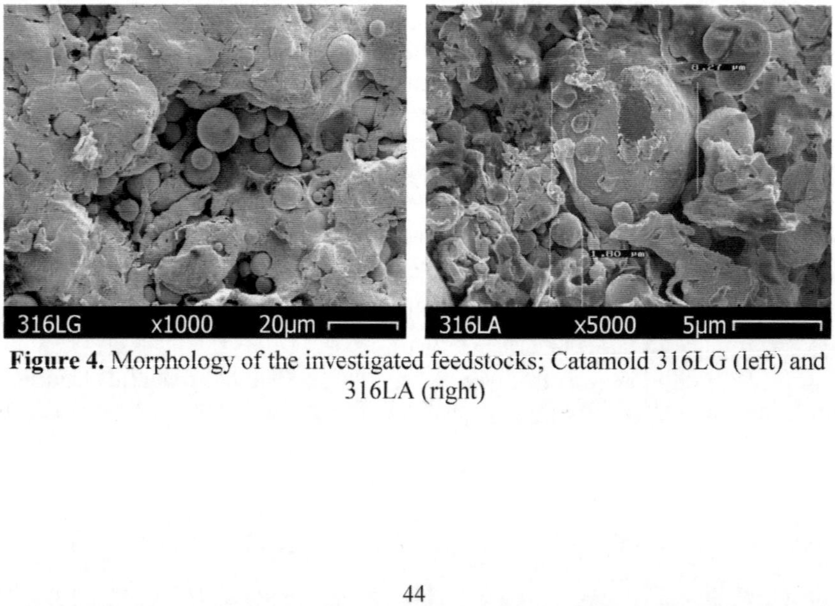

Figure 4. Morphology of the investigated feedstocks; Catamold 316LG (left) and 316LA (right)

2.2. Rheological measurements

Rheological behavior of feedstocks in shear rate range between 10^{-3} and approx. 5 s^{-1} was measured using a controlled shear stress (CSS) rheometer MCR501 (Anton Paar GmbH) in oscillating mode (Figure 4). Please note that the specified shear rates are based on the assumption that the Cox-Merz rule applies. A smooth plate–plate configuration was used with a gap size of 1 mm and a plate diameter of 25 mm. The sample was protected by a dry nitrogen environment to prevent degradation of the binder system.

The measurements at shear rates from approx. 100 s^{-1} to approx. 4,000 s^{-1} were performed with the slit die with conical entry (see Figure 6 a, b) on the high pressure capillary rheometer (HPCR) Rheograph 2002 (Göttfert Werkstoff-Prüfmaschinen GmbH) and on the injection molding machine (IMM) with so-called PIM injection machine rheometer (see Figure 6 c).

Measurements on the PIM injection machine rheometer allowed the rheological characterization under practical processing conditions.

Along the flow path inside the slit die, pressures are measured by five sensors in order to calculate shear stress. The pressure sensors are flush-type sensors, which have direct contact with the melt. The wall temperature is measured at three positions at a distance of 1 mm away from the flow channel wall to identify the wall temperature increases along the flow length due to shear heating.

Width (w) of the slit is 10 mm and height (h) 1 mm, thus they fulfill the requirement of "infinitely wide" (w≥10 h) slit. The true shear rate was obtained by applying the correction for the non-Newtonian behavior of the melt according to Weissenberg-Rabinowitsch [7, 8].

The experiments were performed at 180, 190 and 200°C.

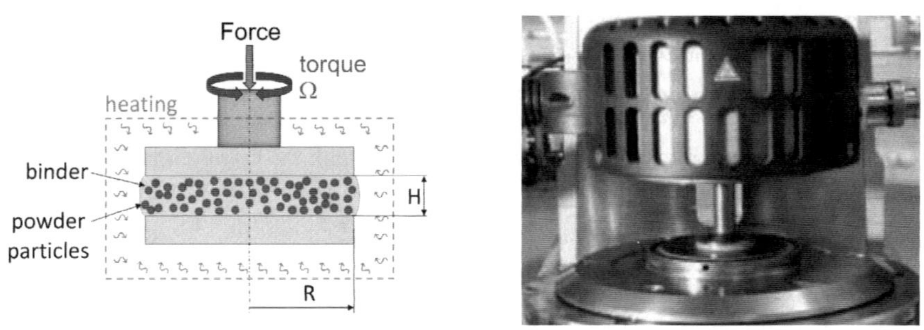

Figure 5. Plate-plate rheometer: principle sketch (left), view into measuring chamber of MCR501 (right)

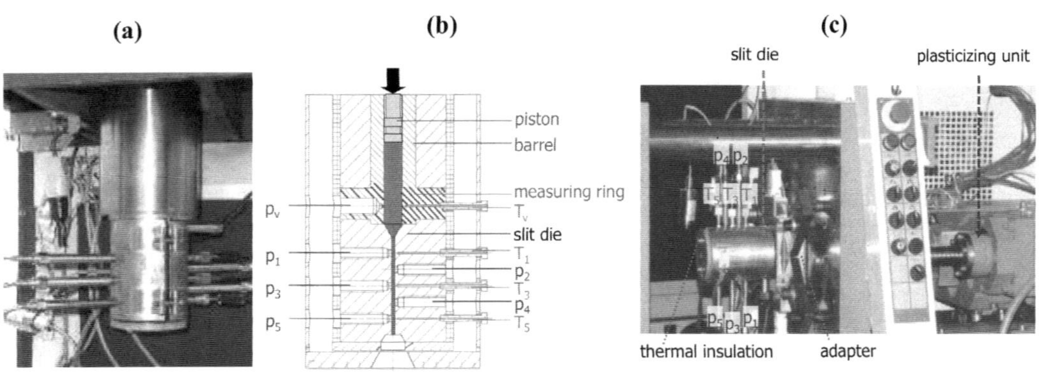

Figure 6. Slit die on high pressure capillary rheometer (a, b) and on injection molding machine (c)

IOP Conf. Series: Journal of Physics: Conf. Series **790** (2017) 012007 doi:10.1088/1742-6596/790/1/012007

2.3. Injection molding experiments and the filling simulation

For comparing the results of the filling simulation a short shot study was performed on an injection machine ARBURG 320C 500-100 with 20 mm diameter screw. The filling patterns provide information about order of fill, potential sites of porosity formation, race tracking and similar defect producing flow structures.

The mold used in this study was a rectangular plate with deflections with cold runner system (Figure 7 left). The plate thickness is 3 mm. The moving half of the mold and the injection molded part with runner and gate are shown in Figure 7 right.

The process parameters for the injection molding tests (short shot study) were selected by processing instructions from the material manufacturer. For the filling study the dosing volume was varied between 3 and 29 cm^3 and the holding pressure was not applied.

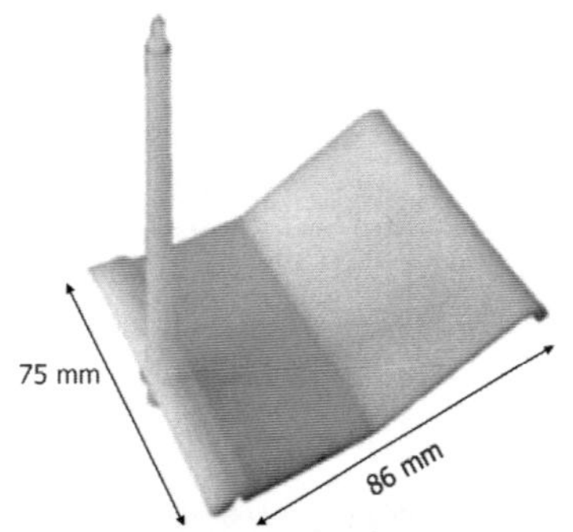

Figure 7. Rectangular plate with deflectors (left) and moving half of the mold (right)

The filling behavior of the injection molding was investigated by commercial software package Autodesk® Simulation Moldflow® Insight (ASMI). Simulations were performed with 3D-model. The model used for computation in ASMI-software is presented in Figure 8.

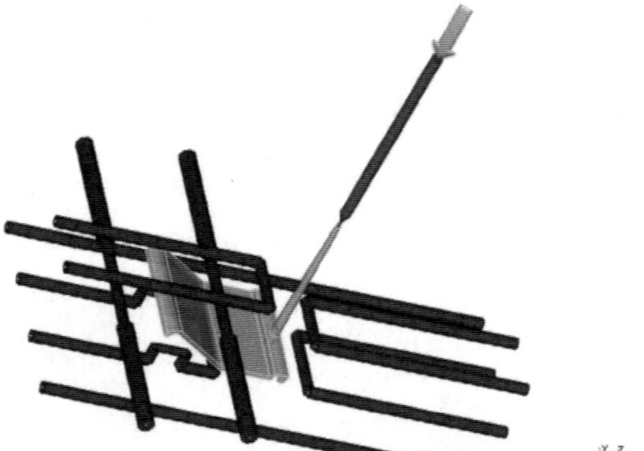

Figure 8. Model of the mold cavity with the cooling circuit (blue) and gating system [9]

3. Results and discussion

The composition of the feedstock and the pre-shearing have great influence on the flow behavior of PIM-feedstocks. For 316LA whose particle size distribution is bimodal, a change in flow characteristic was observed during the viscosity measurement on the high pressure capillary rheometer with the slit die. This was observed as a change in slope of the flow curve after a particular wall shear stress (τ_{crit}) in the flow curve (see Figure 9).

In calculating of the true viscosity using the Weissenberg-Rabinowitsch correction due to the change in slope of the flow curve the viscosity curve shows an implausible discontinuous trend (see Figure 10).

An explanation for this behavior is that the change in the slope of the flow curve is caused by a rearrangement of the powder distribution in the melt when flowing through the flow channel.

Therefore, the flow curve (Figure 9) was divided into two ranges and the true viscosity was calculated separately for each range (see Figure 11).

The slopes of the viscosity curves for both ranges are nearly unchanged. The viscosity above the critical shear rate $\dot{\gamma}_{crit}$ (range B) is slightly higher than the viscosity in the region below (region A).

On the other hand viscosity measurements on the injection molding machine showed no such change in the slope of the flow curve. In the injection molding machine this rearrangement seems to happen in the screw of the plasticizing unit during plasticization of the feedstock. Viscosity measured on the high pressure capillary rheometer (HCPR) is higher than viscosity measured on the injection molding machine (IMM) (see Figure 12). The possible reason is the melt preparation (homogenization) by the screw in the injection molding machines which shows different conditions in comparison to the high pressure capillary rheometer, e.g. pre-shearing and melt temperature control.

The viscosity measurements on Catamold 316LG using a controlled shear stress rheometer (CSS) showed that at low shear rates, as expected, the Newtonian plateau is not present (see Figure 13). Instead the viscosity strongly increases in the direction of the yield point. Out of these investigations the Cross-WLF model with Herschel-Bulkley extension was applied to feedstocks as material model. This model takes into account the shear thinning behavior as well as limiting shear stress. It describes very well the rheological behavior of this investigated feedstock.

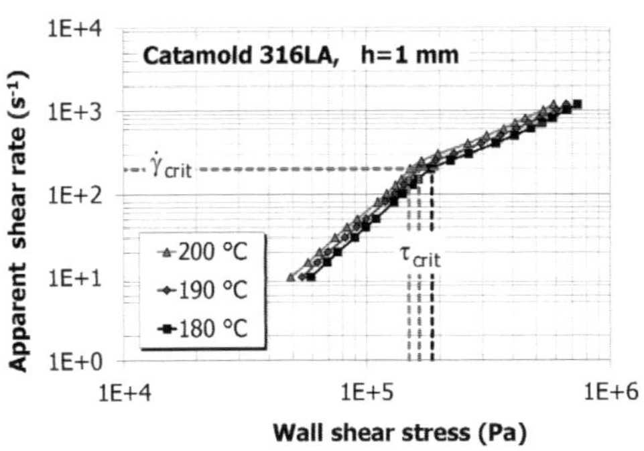

Figure 9. Apparent flow curves of Catamold 316LA, measured with slit die on HCPR at three temperatures

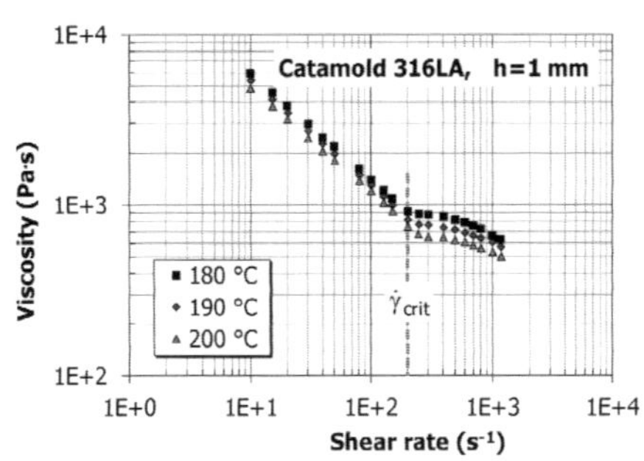

Figure 10. . Viscosity curves of Catamold 316LA, measured with slit die on HCPR at three temperatures

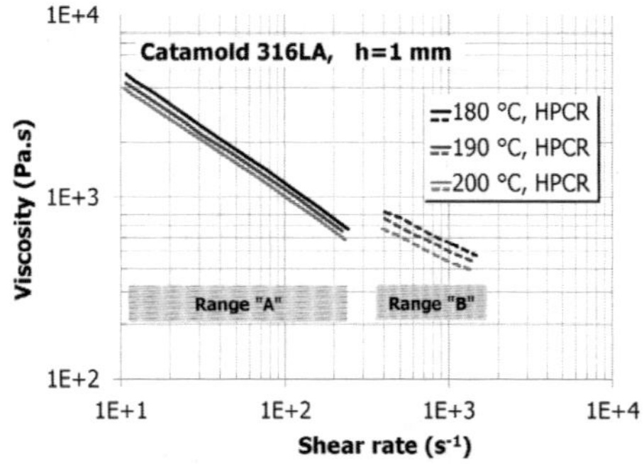

Figure 11. Divided evaluation of the true viscosity of Catamold 316LA

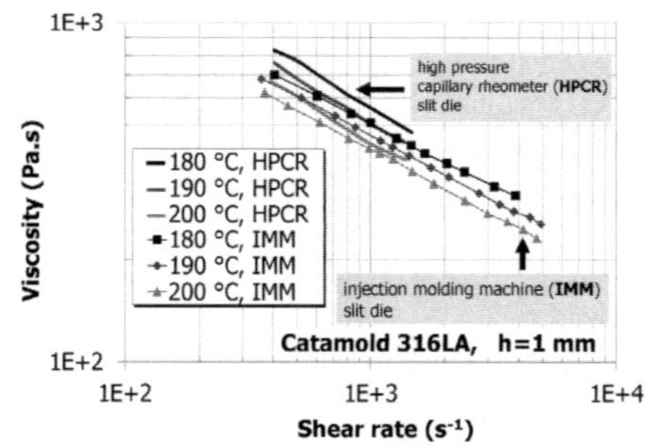

Figure 12. Comparison of the true viscosities of Catamold 316LA, measured with slit die on high pressure capillary rheomerer (HPCR) and injection molding machine (IMM)

Table 2. Cross-WLF model with Herschel-Bulkley extension

Model	Equation	
Cross-WLF model with Herschel-Bulkley extension [12]	$$\eta = \frac{\eta_0}{1+\left(\frac{\eta_0 \cdot \dot{\gamma}}{D_4}\right)^{(1-n)}} + \tau_\gamma \cdot \left(\frac{1-e^{(-\alpha \cdot \dot{\gamma})}}{\dot{\gamma}}\right)$$ $$\eta_0 = D_1 \cdot \exp\left[\frac{-A_1 \cdot (T - D_2 - D_3 \cdot p)}{A_2 + T - D_2}\right]$$	(4)

where: A_1, A_2, D_1, D_2, D_3, and D_4 are material constants, T – temperature, p - pressure.

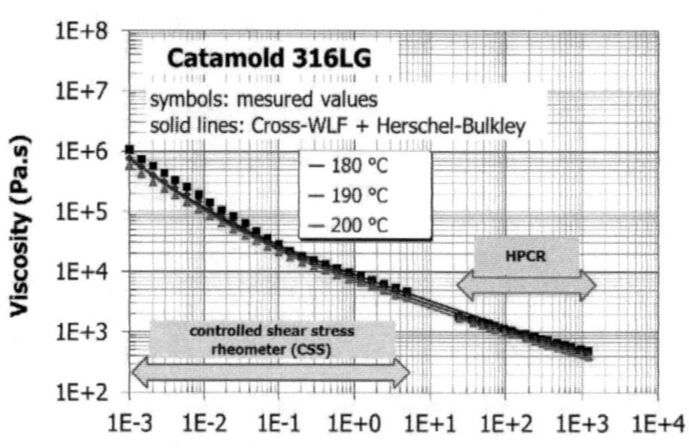

Figure 13. Measured viscosities of Catamold 316LG and approximated with the Cross-WLF model with Herschel-Bulkley extension

Figure 14 shows the comparison of experimentally determined and calculated filling patterns for filling stage between 7% and 89%.

It can clearly be seen, that the fully developed fountain flow which is typical for unfilled thermoplastics does not exist. The shape of the flow front is very coarse and not smooth and over the entire flow front leads to wrinkles formation. The cause for this effect can be the too low mold temperature. The melt front already pushed a cooler region in front of the flow, which will remain visible in the form of flow lines in the filled molding.

Additionally, there is an advance of the melt at the middle of the cavity.

Furthermore, the surface of the molded part in the area of the film gate is marked with "black lines". These "black lines" indicate the powder-binder phase separation of the feedstock; i.e. in this area an elevated binder concentration exists. It is important to point out that phase separation has a major influence on the quality of the green parts.

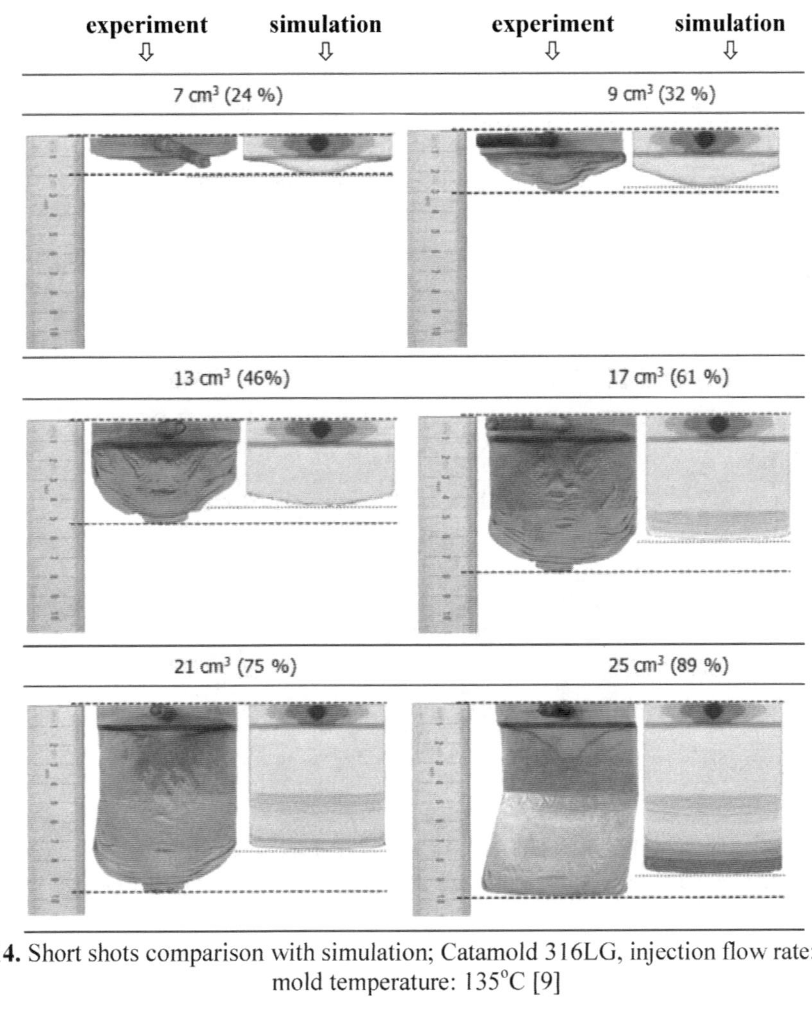

Figure 14. Short shots comparison with simulation; Catamold 316LG, injection flow rate: 14 cm^3/s, mold temperature: 135°C [9]

According to Thornagel [11] the shear gradients are the cause of powder-binder separation and thus two layers of high binder content close to the flow channel wall (Figure 15).

Due to adhesion of the feedstock at the wall of the channel, a significant shear rate peak occurs close to the wall and a plateau at a much lower shear rate level is observed in the middle of the flow domain (Figure 15). Focusing on the powder particles, reaction forces are depending on their current position in the shear rate field and on their size. Particles flowing in the middle of the channel within the plateau of the lower shear rate experience uniform reaction forces and continue their way without changing the direction. Other particles not flowing within the lower shear plateau experience a non-uniform shear rate influence resulting in a rotation of these particles. The rotation speed grows along with the shear stress gradients. Such rotating particles try to leave areas of high shear gradients. Therefore the area of the highest shear rate is characterized by high binder content in the channel while the plateau of the lower shear rate accommodates a powder rich material [11].

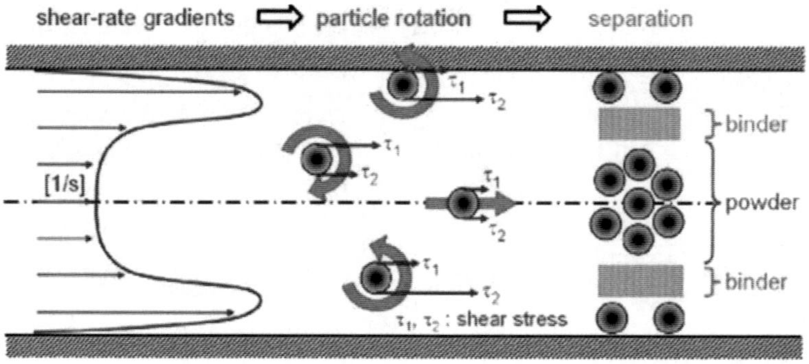

Figure 15. Local shear stress gradients cause powder - binder separation [11]

Usually in the gate area high shear rate gradients occur. The particle concentration in the volume unit increases while the binder concentration decreases. [11]. Figure 16 shows this phenomenon in a simplified way. The feedstock flow transports the separation pattern and changes the pattern continuously.

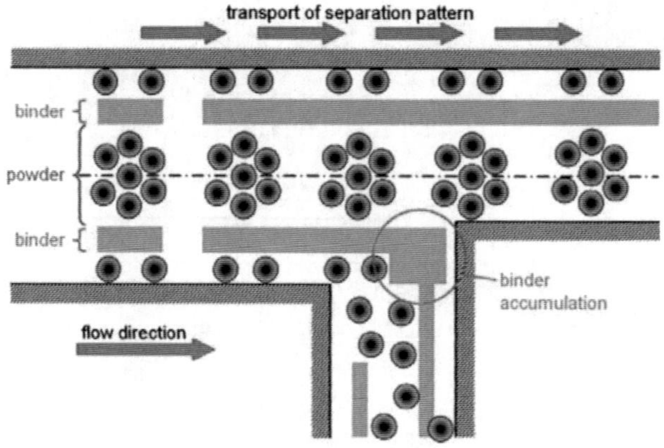

Figure 16. Feedstock flow results in transport of separation pattern [11]

As Figure 14 shows, the simulation cannot predict the phase separation phenomenon, furthermore the predicted shape of the melt flow front and the filing time are not accurate enough.

Many commercial simulations software for injection molding are considering PIM-feedstock as a bulk material by the complete set of 3D Navier Stokes equations. Considering feedstock as a bulk material means that the model cannot differentiate between powder and binder (one-phase simulation). But in this case a set of 3D Navier Stokes had to be formulated for each phase - one for the powder and one for the binder - and both had to be coupled in an appropriate way. This practical approach based on a diffusive flux model is currently being developed by SIGMASOFT [11].

Based on these results, additional filling studies were carried out at the mold temperatures of 180 °C and 200 °C. The molded parts of the filling study are shown in Figure 17. The higher mold temperature 180 °C resulted in an improved fountain flow; nevertheless it was not fully developed and the wrinkling is still present.

A significant improvement with regard to the development of the melt flow front occurs at the mold wall temperature of 200 °C and the wrinkling was completely eliminated (see Figure 17).

Furthermore, at 200 °C a very good agreement was achieved between filling simulation and experiment. The difference between calculated and measured pressures in front of the screw is not more than 7% and in the cavity max. 6% [9].

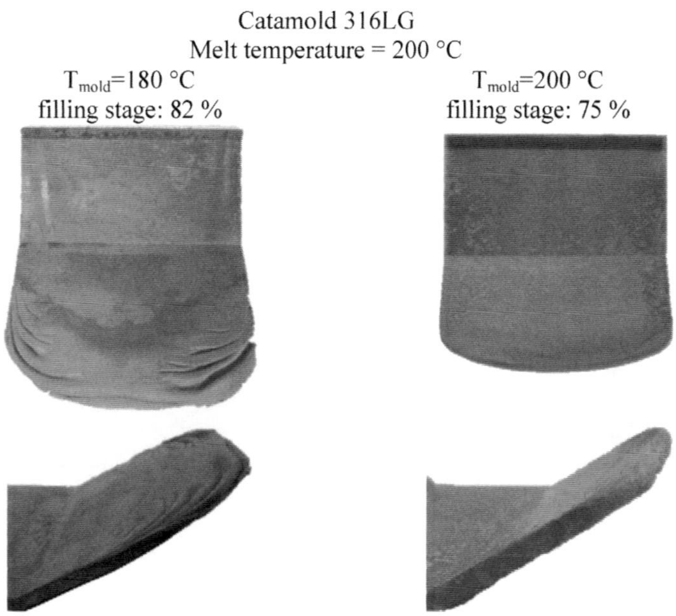

Figure 17. Experimental filling images of the rectangular plate with deflections for Catamold 316LG at mold temperatures of 180°C and 200°C [9]

4. Conclusion and outlook

In the present paper the results of controlled shear stress and slit die rheology of 316L stainless steel feedstock for powder injection molding and the related analysis of the filling stage of injection molding are shown.

Different powder characteristics of Catamold 316LA and 316LG are also reflected in the flow properties of the two feedstocks. For 316LA whose particle size distribution is bimodal, a change in flow characteristic was observed during the viscosity measurement on the high pressure capillary rheometer with the slit die. Possible cause of this flow behavior can be attributed to the rearrangement of the powder distribution in the melt. The measured viscosity curves on the injection molding machine showed no change in the slope of the flow curve.

The control shear stress measurements (CSS) with the plate-plate rheometer showed that for the feedstocks at low shear rates $<1\ s^{-1}$ no Newtonian plateau is observed, but instead the viscosity increases strongly. This flow behavior could be described very well with the Cross-WLF model with Herschel-Bulkley extension.

Using the rheological data, flow simulations were performed and it was observed that the simulations do not accurately mimic the flow response of feedstock. This was attributed to the fact that current software do not take into account that feedstock is a multiphase material.

On the base of the comparison of short shots with injection molding simulation the influence of the observed rheological behavior onto mold filling and the related injection molding simulation was shown. At too low mold temperatures at the melt flow front no fountain flow was developed. The absence of a fountain flow can lead to an inhomogeneous microstructure in the molding or poor dispersion of the filler particles.

The mold temperature is especially important during the filling process. Besides the evolution of the fountain flow at higher mold temperatures; the cycle time, the surface finish of the molded part, the mold filling grade and the mold filling rate are affected [9].

The higher the mold temperature, the better the cavity can be filled properly. Furthermore, the flow resistance is reduced. For the feedstocks which show unsteady flow conditions, the application of the variotherm process in the filling phase of the injection molding process is recommended. This means that during filling the cavity the mold wall temperature is raised quickly to the melt temperature.

The next step in our research will be investigation of the influence of surface texture of parallel plates in rotational rheometry to investigate slip effects during viscosity measurements and subsequently comparing the numerically simulated results with experiments.

Acknowledgements
The authors would like to thank Dr. Joamin Gonzalez-Gutierrez and Prof. Thomas Lucyshyn for their useful technical discussion on this topic. Furthermore, we thank the Polymer Competence Center Leoben, PCCL for providing the rheometer for the rheological measurements.

References

[1] Friesenbichler, W., Duretek, I.: Rheologische Charakterisierung hochgefüllter PIM-Feedstocks und Holzcompounds, Tagungshandbuch zum 17. Leobener Kunststoffkolloquium, Leoben, 20. und 21. November 2003.

[2] Duretek, I., Lucyshyn, T., Langecker, G.R.: Powder-Injection-Moulding (PIM) / Stoffdatenbestimmung und Simulation, 52. Jahrestagung der Österreichischen Physikalischen Gesellschaft, Leoben, 23. - 26. September 2002.

[3] Nyborg, L., Johansson, E., Niederhauser, S.: Assessment of rheology and related properties of PIM feedstock material, proceedings of 2nd european symposium on powder injection moulding, EPMA, October 2000, Shrewsbury, 2000, 67 - 74.

[4] Bilovol, V. V.: Mould filling simulations during powder injection moulding, doctoral thesis, TU Delft, 2003.

[5] Kukla, C., Friesenbichler, W, Duretek, I., Thornagel, M.: New insights into feedstock behaviour and injection moulding simulation for PIM, proceedings of the international powder metallurgy congress and exhibition – EURO PM 2008, Mannheim, Germany, 29th September – 1st October 2008.

[6] Laun, H.M.: Polymer rheology with a slit die, *Rheologica Acta* 22 (1983) **2** 181 – 185.

[7] Eisenschitz, R., Rabinowitsch, B., Weissenberg, K.: Mitteil Deutsch, *Materialspruefungsamt*, Sonderheft 9 (1929) **91** 91 – 94.

[8] Rabinowitsch, B.: Über die Viskosität und Elastizität von Solen, Zeitschrift für physikalische *Chemie* (1929) **145**, 1 - 26.

[9] Duretek, I.: Untersuchungen zum Fließverhalten von Metall-, Keramik- und Holz-Kunststoffverbunden, Dissertation, Montauniversitaet Leoben, 2013.

[10] Thornagel, M.: Injection moulding simulation: New developments offer rewards for the PIM industry, *Powder Injection Moulding International* (2012) **6** 65 - 68.

[11] Thornagel, M.: Simulating flow can help avoid mould mistakes, Materials Today (5th October 2010), Elsevier Ltd., [accessed 2015 August 29] http://www.materialstoday.com/molding-and-pressing/features/simulating-flow-can-help-avoid-mould-mistakes/#

[12] Sigmasoft V4.8 - new functionality, SIGMA Engineering GmbH, Aachen, 2012.

[13] Hartwig, T., Kukla, C. Duretek, I., Thornagel, M., Caballero, M., Breitenmoser, G.N., d'Jeff Van der Ven, Bihr, J.C., Vicenci, B., Witzleben, M., Gornik, C., Jegust, S., Birth, U., etal.: New material laws for powder filled injection moulding feedstocks - final report, European Union CRAFT Programme Framework 6, project acronym: MATLAW, project no. 033006, 2009.
[14] Franck, A.: Understanding rheology of thermoplastic polymer, AAN013, TA Instruments, 2013.
[15] Geiger, K.: Ein neues heuristisches Fließgesetz, 21. Stuttgarter Kunststoffkolloquium, Stuttgart, 2009.
[16] Dapčević, T., Dokić, P., Hadnađe,v, M., Pojić, M.: Determining the yield stress of food products – importance and shortcomings, Food Processing - Quality and Safety (2008) **35** 143 - 149.

Rheological study of copper and copper grapheme feedstock for powder injection molding

N Emira Binti Azaman[1], M Rafi Raza[1,2], N Muhamad[1,*], M Niaz Akhtar[1,3] and A Bakar Sulong[1]

[1]Department of Mechanical and Materials Engineering, Universiti Kebangsaan Malaysia, Bangi, Selangor Malaysia
[2]Department of Mechanical Engineering, COMSATS Institute of Information Technology,Sahiwal Pakistan
[3]Department of Physics, COMSATS Institute of Information Technology,Lahore Pakistan

E-mail: m.norhamidi@gmail.com

Abstract. Heatsink is one of the solution to optimize the performance of smart electronic devices. Copper and its composites are helping the electronic industry to solve the heating problem. Copper-graphene heat sink material with enhanced thermal conductivity is the ultimate goal.Powder injection molding (PIM) has advantages of high precision and production rate, complex shape, low cost and suitabality for metal and cremics.PIM consists of four sub sequential steps; feedstock preparation, molding, debinding and sintering. Feedstock preparation is a critical step in PIM process. Any deficiency at this stage cannot be recovered at latter stages. Therefore, this research was carried out to investigate the injectability of copper and copper graphene composite using PIM. PEG based multicomponent binder system was used and the powder loading was upto 7vol.% less than the critical powder loading was used to provide the wettability of the copper powder and graphene nanoplatelets (GNps). Corpper-graphene feedstock contained 0.5vol.% of GNps . To ensure the homogeneity of GNps within feedstock a unique technique was addopted. The microscopic results showed that the feedstock is homogeneous and ready for injection. The viscosity-shear rate relationship was determined and results showed that the addition of 0.5vol.% of GNps in copper has increased the viscosity upto 64.9% at 140°C than that of pure copper feedstock. This attribute may be due to the large surface area of GNps. On the other hand, by increasing the temperature, viscosity of the feedstock was decreased, which was recommended for PIM. The overall viscosity and share rate lies within the range recommended for PIM process. It is clear that both feedstocks showed pseudo plastic behaviour which is suitable for PIM process. In the pseudo plastic behaviour, the viscosity decreases with the shear rate. It may be due to change in the structure of the solid particles or the binder. The molding results showed that both copper feedstocks were successfully molded and free from the physical defects.

1. Introduction

Nowadays, the electronic industry has a serious concern to achieve optimal performance of electronic chips which attain high temperature during operation resulting reduced performance. To overcome this issue, heat sinks are developed with metal composites. These composites are reinforced with highly conductive reinforced materials results in enhanced the thermal conductivity and these composites are consider more attractive for heat sink applications [1]. These heat sinks are attached to the back of the chips to pull thermal energy away from the chips and transfer it into the surrounding air. PIM is a new technology which modifies the basic molding process of plastics and ceramics and now applicable for metals and ceramic materials. PIM process has emerged as a key technology due to its ability to produce small and complex components at low cost with high production volume [2, 3]. Various materials are used for thermal management and copper is one of the most recommended materials because of its high thermal conductivity. In order to produce the heat sink with complex shape using PIM, a flow able feedstock in molding machine is required[4]. Besides, the fabrication of defects free metal matrix composites using PIM has been reported [5, 6]. In order to achieve a high electrical conductivity in polymeric composites with low cost and easy processing, graphite based materials are getting more research attraction [7]. Apart from that, attempt to incorporate graphene nanoplatelets (GNps) into the feedstock is a new approach to enhance the performance of electronic devices. Graphene is one of the most fascinating materials being studied today, and received world-wide attention due to its potential applications [8, 9]. Graphene is a fast emerging material due to its unique structure and excellent mechanical, thermal and electrical properties [9, 10]. Recent research has shown that the graphene-based materials can have a great impact on electronic and optoelectronic devices, chemical sensors, nanocomposites and energy storage [9]. Graphene has demonstrated excellent mechanical properties with Young's modulus of 1 TPa and high thermal conductivity, about 5000 W/mK for freely suspended samples, which is among the highest of any known material[11-13].

However, in Cu-graphene composite the uniform dispersion of GNps into the feedstock is a challenge for researchers. The developed high performance smart materials for thermal management will provide extensive application of PIM to enhance the performance of electronic chips[14].

This research studied investigates the uniform dispersion of GNps with in copper matrix and their effects on flow behaviour of Cu feedstock by using a lower viscosity binder system. Based on viscosity results suitability of Cu-GNps composite for PIM and appropriate moulding temperature were identified.

2. Materials and Procedures

Gas atomized copper powder with an average particle size of 22μm with purity 99.9% was supplied by Sandvik Materials Technology, and graphene nanoplatelets (XGnP-M grade) with average thickness of 6 to 8 nm, surface area of 120 to 150 m^2/g and average particle diameters of 5, 15 or 25 μm was provided by XG Sciences, Inc.. Binder system contains polyethylene glycol (PEG) as the major component, polymethyl methacrylate (PMMA) as the backbone polymer and stearic acid (SA) as the surface active agent. The physical properties of the binder components are listed in Table 1.

Table 1. Physical properties of binder components used in this study

Binder ingredient	Density (g/cm^3)	Melting Temperature (°C)
PEG	1.21	61-66
PMMA	1.16	160
Stearic acid	0.96	67-69

Fig. 1 shows the FESEM micrograph of Cu copper powder which shows that the particles are spherical in shape.

On the other hand Fig.2 shows the flake nature of the GNps.

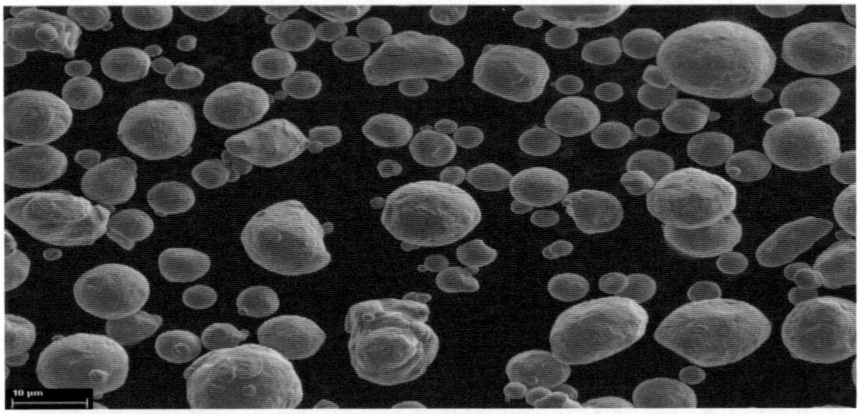

Figure 1. FESEM image of Cu shows spherical nature of the particles

Figure 2. FESEM micrograph showing the flakes of graphene GNps

2.1. Preparation of Copper and Copper-Graphene Feedstock

The binder system used in this study consists of 73vol.% PEG, 25vol.% PMMA and 2vol.% SA. Critical powder loading was determined 67vol.% by using Brabender shown in Fig.3. In order to determine the characteristics and the flow behaviour, two formulation of feedstock have been prepared and shown in Table 2. Formulation C-1 have solid loading 60vol.% while in formulation C-2, powder loading was reduced to 58vol.% due to the presence of 0.5vol.% GNps, having large surface area and caused to raise the viscosity.

Table 2. Feedstock formulations developed for viscosity measurement in this study

Copper mixture	Powder loading (vol. %)	Graphene (vol.%)
C-1	60	0
C-2	58	0.5

Two formulations of copper powder were prepared using a Brabender mixer machine shown in Fig. 3. In order to get homogenous dispersion results of GNps in the Cu matrix, sonication has been applied as the dispersion technique of GNps. Initially, GNps were immersed in distilled water and sonicated for 1 hour at temperature of 50°C follows by the dissolution of PEG in GNps and water solution with the help of magnetic stirrer. Finally the mix was dried in the oven for 24 hours. The dried GNps-PEG mixture was then mixed with the other components of the binder system (PMMA, SA) and copper powder in Brabender at temperature of 150°C shown in Figure 3. This step allowed GNps to be mixed uniformly within feedstock.

Figure 3. Compounding of Cu-Graphene using Brabender

2.2. Viscosity Measurement
The viscosity of both feedstocks were measured using capillary rheometer Shimadzu CFT-500D at various temperatures from 130 to 150°C and data was acquired to study the effects of temperature and GNps on viscosity of copper.

2.3. Microstructural Analysis
The shape of copper powder, GNps and to ensure the homogeneity of GNps within feedstocks, filed emission electron microscope (FESEM- ZEISS SUPRA 55VP) was used.

3. Results and Discussion

3.1. Flow behaviour of Pure Copper and GNps Reinforced Copper composite Feedstocks
Two different volume contents of copper 58 vol. % and 60 vol. % were tested for viscosity over a wide of temperature. The obtained results showed that both formulations exhibit pseudoplastic behaviour which is recommended for PIM [15]. Fig. 4 and Fig. 5 show the relation between viscosity and shear rate for pure copper (C-1) and graphene reinforced copper composite (C-2) feedstock respectively. Formulation C-2 showed higher viscosity as compared to C-1 even at low solid loading. It may be due to the presence of GNps which create hurdles and resist against the movement of copper powder particles. At lower temperature 130°C the addition of GNps, the viscosity of C-1 is 90.39 Pa.s – 231.6 Pa.s while the viscosity of C-2 is 81.4 Pa.s – 308 Pa.s. At this temperature, the viscosity of composite mixture Cu/Graphene is higher up to 32.76% than that without graphene. The increasing factor in viscosity was assumed due to large surface area of graphene nanoplatelets. On the other hand, the interstitial type of packing within the powder particles attributed to the increase of viscosity of the mixture. At the temperature of 140°C, the viscosity value of C-1 was measured to be 77.5 Pa.s – 141.3 Pa.s while the value of C-2 is 56.5 Pa.s – 233 Pa.s. For this temperature, the difference in viscosity is up to 64.9%. At higher temperature (150°C), the viscosity values for both feedstocks are apparently low. For C-1 feedstock, the viscosity value is 69 Pa.s – 150 Pa.s while for C-2, the viscosity value is 33.7 Pa.s – 152 Pa.s. This is occurred because at higher temperature, the binder particles begin to soften and entering the spaces between the metal powder thus facilitating the outflow when pressure is applied. From this rheology test, it was found that the viscosity of the feedstock was decreased as the temperature was increased, which is a good criteria required in PIM.

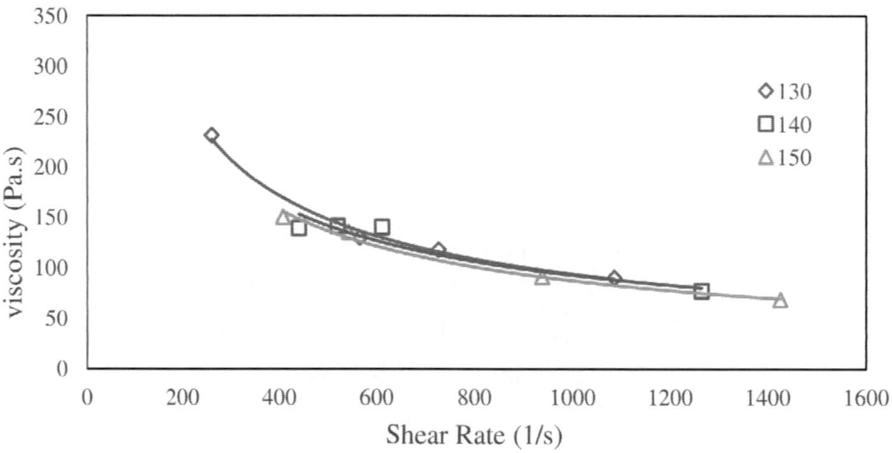

Figure 4. Viscosity vs. shear rate of copper powder (C-1) feedstock showing pseudo plastic behaviour

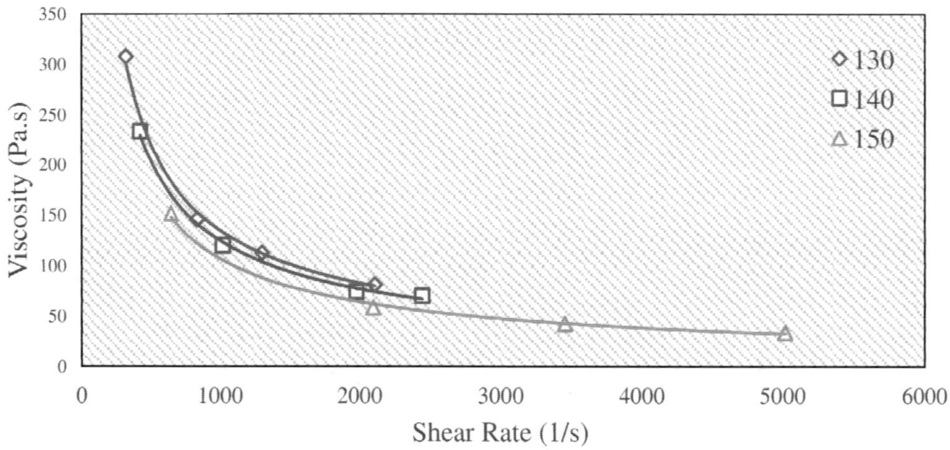

Figure 5. Viscosity vs. shear rate of graphene reinforced copper (C-2) feedstock shows pseudo plastic behaviour

Formulation C-1 has powder loading 60vol.% that is 2 vol.% higher than C-2 (58vol.%). In formulation C-2, lower powder loading was used due to the presence of GNps. The lower powder loading caused to reduce the friction among the powder particles and GNps due to the excessive binder [16]. However, in this case, the addition of graphene nanoplatelets to the C-2 feedstock contribute to the increase of viscosity even though it has lower powder loading because of large surface area of the graphene nanoplatelets.

3.2. Effects of GNps on flow index (n) and activation energy (E)

The value of flow index ‘n’ at various temperatures is given in Table 3. The results revealed that the value of ‘n’ is less than 1 which means feedstock has pseudo plastic behaviour as clear from Fig.4 and Fig.5. On bases of these results both feedstocks are considered suitable for injection molding. Table 3 shows the values of flow behaviour index ‘n’ at a range of temperatures. The values of flow index ‘n’ is not uniform it may be due to the orientation of GNpsand low viscosity of binder at higher temperature.

Table 3. n-values of feedstock at different temperatures

Feedstock	Temperature (°C)	N
C-1	130	0.344
	140	0.391
	150	0.357
C-2	130	0.3192
	140	0.3233
	150	0.2922

Temperature dependency of viscosity is very important to judge the flow of feedstock. Fig. 6. showed that formulation C-2 has a higher value of activation energy which is 17.71 kJ/mol, compared to feedstock C-2 which is 4.96 kJ/mol. Higher value of E means, feedstock has a strong temperature dependency to viscosity [17]. Therefore, feedstock C-2 required a more accurate temperature control during injection molding compared to feedstock C-2. But for this case, even the value of C-2 is higher as compared to C-1, but it is still in acceptable range.

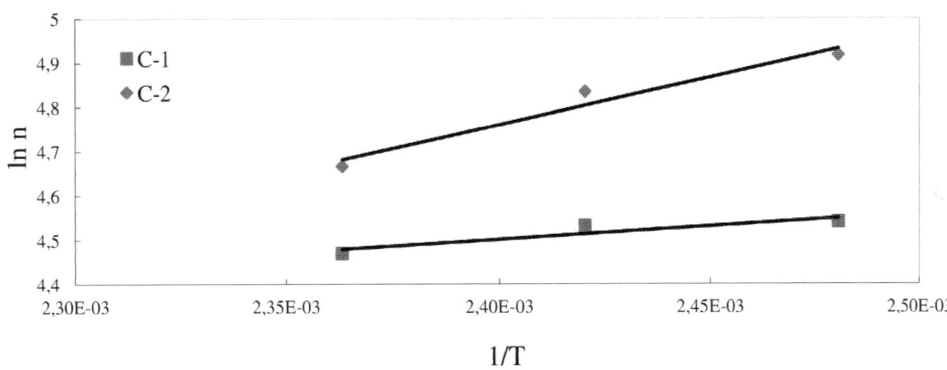

Figure 6. Relationship between viscosity and temperature at shear rate of 1000 s^{-1}

3.3. Molding analysis of feedstock

The feedstocks were successfully molded andmolded parts are free from the physical defectsas shown in and Fig. 7.

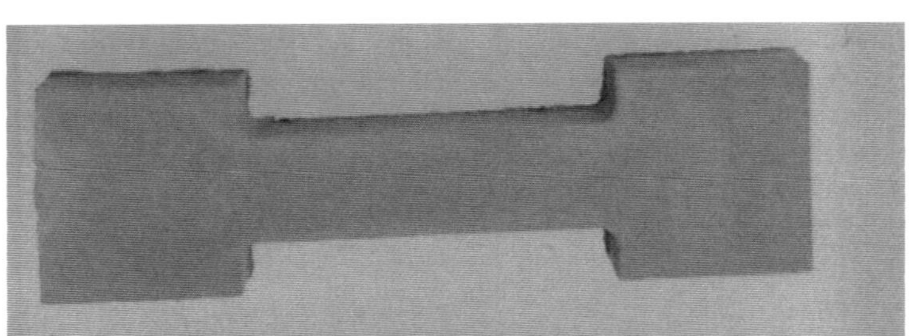

Figure 7. Physically defects free Cu-GNpscomposite parts produced via PIM

3.4. *Microstructural analysis of Feedstock*

FESEM analysis was performed to ensure the coating of PEG on GNps. Fig. 8 showed that the GNps were evenly coated with PEG after stirring. The SEM results in Fig.9 revealed that the binder was uniformly distributed within the matrix and it covered the metal and GNps. With the addition of GNps no evidence of clustering was observed.

Figure 8. FESEM of shows the coating of PEG on GNps

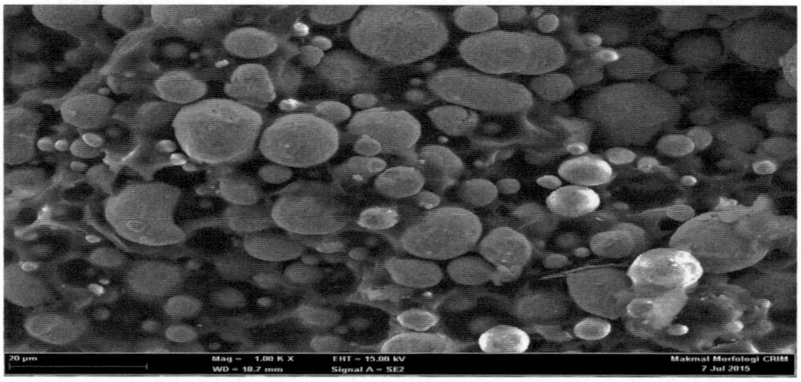

Figure 9. Scanning electronic microscopy image of Cu-Graphene feedstock showing uniform dispersion and spherical particles of Cu

4. Conclusions

The adopted procedure to disperse GNps within the Cu matrix is suitable to achieve uniform dispersion and recommended for PIM. It is found that both formulations exhibit pseudo plastic behaviour and can be injected successfully through injection molding process. The results of rheological studies showed that small amount of GNps increased the viscosity of the composite it is recommended that for Cu-GNps composites the solid loading should be more than 5% less that of critical powder loading.

Acknowledgment

The authors would like to thank the UniversitiKebangsaan Malaysia for the financial support (Grant Nos. GUP-2015-018 and DIP-2012-29).

References

[1] Zweben C 2006 Thermal Materials Solve Power Electronics Challenges. In: *Power Electronics Technology* www.powerelectronics.com) p 8

[2] Kate K H, Enneti R K, Onbattuvelli V P and Atre S V 2013 Feedstock properties and injection molding simulations of bimodal mixtures of nanoscale and microscale aluminum nitride *Ceramics International* **39** 6887-97

[3] Heaney D F 2012 *Handbook of metal injection molding*: Elsevier)

[4] Muhsan A S, Ahmad F, Mohamed N M and Raza M R 2011 Flow behavior of Cu/CNTs feedstocks for powder injection molding *International Journal of Applied Physics and Mathematics* **1** 199

[5] Loh N, Tor S and Khor K 2001 Production of metal matrix composite part by powder injection molding *Journal of Materials Processing Technology* **108** 398-407

[6] Raza M R, Ahmad F, Omar M, German R and Muhsan A S 2012 Defect Analysis of 316LSS during the Powder Injection Moulding Process. In: *Defect and Diffusion Forum*: Trans Tech Publ) pp 35-43

[7] Jiang X and Drzal L T 2012 Reduction in percolation threshold of injection molded high-density polyethylene/exfoliated graphene nanoplatelets composites by solid state ball milling and solid state shear pulverization *Journal of Applied Polymer Science* **124** 525-35

[8] Marques P, Sousa A, Gonçalves G, Grácio J, Singh M, Almeida N and Cruz S 2011 *Functionalized graphene nanocomposites*: INTECH Open Access Publisher)

[9] Singh V, Joung D, Zhai L, Das S, Khondaker S I and Seal S 2011 Graphene based materials: past, present and future *Progress in Materials Science* **56** 1178-271

[10] Potts J R, Dreyer D R, Bielawski C W and Ruoff R S 2011 Graphene-based polymer nanocomposites *Polymer* **52** 5-25

[11] Eric Pop V V, and Ajit K. Roy 2012 Thermal properties of graphene: Fundamentals and applications *MRS Bull.***37** 28

[12] Lee C, Wei X, Kysar J W and Hone J 2008 Measurement of the elastic properties and intrinsic strength of monolayer graphene *science* **321** 385-8

[13] Balandin A A, Ghosh S, Bao W, Calizo I, Teweldebrhan D, Miao F and Lau C N 2008 Superior thermal conductivity of single-layer graphene *Nano letters* **8** 902-7

[14] Muhsan A S, Ahmad F, Mohamed N M and Raza M R 2013 Nanoscale Dispersion of Carbon Nanotubes in Copper Matrix Nanocomposites for Thermal Management Applications *Journal of Nanoengineering and Nanomanufacturing* **3** 248-52

[15] Amir Khalid B M, Erween Abdul Rahim, Waluyo Adi Siswanto and Kamil Abdullah Oct. 2014 Rheological Properties of Copper-Graphite MIM Feedstocks Prepared with Waste PET Binder *Applied Mechanics and Materials* **660** 209-13

[16] Ahmad F 2005 Rheology of metal composite mixes for powder injection molding *International journal of powder metallurgy* **41** 43-8

[17] Amin M, Muhamad N, Jamaludin K R, Abdullah S and Rahman M N A 2012 Rheological investigation of mim feedstocks prepared with different particle sizes

IOP Conf. Series: Journal of Physics: Conf. Series **790** (2017) 012009 doi:10.1088/1742-6596/790/1/012009

Microstructural and rheological analysis of fillers and asphalt mastics

R Geber[1], A Simon[1], I Kocserha[1], and A Buzimov[2]

[1]University of Miskolc, Institute of Ceramics and Polymer Engineering, Hungary
[2]Tomsk State University, Department of Mechanics of Deformed Solids, Russia

E-mail: femgeber@uni-miskolc.hu

Abstract. Pavements are made of different grades of mineral aggregates and organic binder. The aggregates are sorted in different sizes and different amount which are mixed together with bitumen. The finest mineral fraction (d<0.063 mm) is called filler. This component has an important role in asphalt mixture - it fills the gaps between the aggregates and if mixed with bitumen (which is called asphalt mastics) it sticks the larger particles together. Particle size, microstructure and surface properties of fillers highly affect the cohesion with bitumen, therefore the aim of our research was to investigate the microstructure of mineral fillers (limestone, dolomite) which are used in Hungarian road constructions with the use of different techniques (particle size distribution, scanning electronmicroscopy tests, mercury intrusion porosimetry, BET specific surface tests, determination of hydrophobicity). After the tests of fillers, asphalt mastics were prepared and rheological examinations were obtained. These examinations served to observe the interaction and the effect of fillers. The stiffening effect of fillers and the causes of rutting were also investigated. Based on our results, it can be stated that particle size, hydrophobic properties and the amount of fillers highly affect the rheological properties of mastics.

1. Introduction

Asphalt mixtures are world-wide used pavement materials. They consist of mineral aggregates in different particle sizes and bitumen, as binding material [1]. One of the most important components of asphalt are mineral fillers (d<0.063mm). The mixture of bitumen and fine mineral particles generates asphalt mastic which is the main bonding phase of asphalt mixture. It affects both the cohesion between coarser particles, which carry capacity to a great extent, and the strength and stiffness of pavement [2, 3, 4, 5, 6].

Nowadays, not only international researches [7, 8, 9] but also Hungarian projects deal with the improvement of pavement performance. Significant results were achieved with the use of different alternative aggregates and other fibre-based asphalt modifiers [10, 11, 12, 13]. Since the performance of pavements highly depend on environmental circumstances (mainly temperature) and on traffic load, it is also important to examine the rheological properties [14, 15].

A fundamental point of the relation formed with bitumen is the mineral origin of the filler. The most widespread used filler is limestone and Hungary has rich and good quality mineral resources of it. Numerous researches have dealt with the investigations of mineral fillers and asphalt mastics to reveal the contact mechanisms between mineral aggregates and bitumen [16, 17, 18]. Our current research tried to widen the knowledge in this topic. As the strength of interaction between bitumen and filler depends on the properties of components, the Authors have divided the examinations into two parts; analysis of raw materials (filler, bitumen) and analysis of asphalt mastics.

Considering the above mentioned, the aims of this research were the followings:

- complex testing procedures (X-ray powder diffraction, scanning electronmicroscopy, BET specific surface tests, mercury intrusion porosimetry, determination of hydrophilic coefficient) were done on mineral fillers.
- with the use of fillers, asphalt mastic mixtures were created in which effect mechanism of type, particle size and quantity of fillers (volume fraction) can be observed at the same time. Rheological tests were made on the mastics in high temperature ranges in order to describe the behaviour of asphalt pavements in summer.
- summarizing the above mentioned: the aim of this research was to observe rutting, the typical defect of pavements with the use of different mineral fillers.

2. Materials and methodology

2.1. Sample preparation

Two types of mineral materials were tested during the research, which originate from Hungary (limestone /Mexikóvölgy/ and dolomite /Pilisvörösvár/). The following sample preparation process were done on mineral fillers: units of mineral materials that are smaller than a given particle size (d<0.063 mm and d<0.045 mm) were created using standard sieves. The fractions were then dried to weight-constancy in a drying chamber. After the samples were cooled down, they were hermetically sealed in containers to avoid moisture uptake until further investigation. Dolomite was supplied in two different particle size ranges (d<0.3 mm, and d<0.045 mm, respectively) which were previously fractioned by the manufacturer.

2.2. Testing procedures

Determination of particle size distribution of fractioned fillers were made with Horiba LA-950V2 instrument. The tests were done by using wet method, therefore to dispergate mineral fillers sodium-pyrophosphate and ultrasonic treatment was also used.

For morphological tests Carl Zeiss EVO MA10 scanning electron microscope (SEM) was used. Micrographs were taken in different magnifications for further observation. Morphology and geometrical features of fillers were observed with this technique.

Specific surface area of the fillers (by BET-method) were determined by nitrogen adsorption method, using TriStar 3000 instrument (manufactured by Micromeritics). In order to determine the pore structure and pore size distribution within the mineral material mercury intrusion porosimetry (MIP) was used by PASCAL 140 and PASCAL 440 instruments (manufactured by Thermo Scientific).

Hydrophilic coefficients of fillers were determined by sedimentation method. The test procedure was described elsewhere [18].

During the research asphalt mastic samples were prepared with the use of mineral fillers and standard bitumen (Százhalombatta) with penetration grade 50/70. The ratios of filler to mastics on a volumetric base (i.e. filler volume/mastics volume; F/M ratio) were selected at 5%, 13% and 20%. Mineral fillers were mixed with bitumen in a mixer in which temperature was kept at 160 °C and mixing speed was 500 rpm.

Haake RheoStress RS80 dynamic shear rheometer with a standard 25 mm diameter parallel plate testing geometry was used to reveal the rheological properties of mastics in the linear viscoelastic region. Considering the particle size of mineral fillers, gap size of rheometer was set to 1.5 mm. Rheological tests were performed between 60 °C-135 °C considering the pavement temperature in summer and paving temperatures. During the research program flow and viscosity curves, master curves, as well as creep-recovery curves were determined.

3. Results and discussion

3.1. Test results of mineral fillers

Table 1. contains all the test results performed on mineral fillers. It was proved by the results of XRD tests that as regards to their mineral composition, fillers are pure materials, free of contamination. Limestone filler contains only CaCO3, while dolomite contains MgCa(CO3)2.

Table 1. Results of different tests of fillers

Material property	unit	Limestone d<0.045 mm	d<0.063 mm	Dolomite d<0.045 mm	d<0.063 mm
Mineral composition	wt%	100 % CaCO3		100 % MgCa(CO3)2	
Bulk density	.3	2.717		2.842	
BET specific surface area	2 -1 m^{2}g	1.11	1.02	1.10	0.38
Hydrophilic coefficient	-	0.74	0.76	0.73	0.89
Average pore diameter	^m	2.04	2.47	1.25	2.57
Median pore diameter	^m	3.78	3.80	2.49	10.49

Bulk density of fillers was also determined by pycnometer method. The determination of density was important to create asphalt mastic mixtures. It can be concluded that dolomite has a higher bulk density value.

From the minerals available for the Authors, fractions of different particle sizes were formed. Having defined their particle size distribution it was found that these mineral materials can be characterized as polydisperse systems and they contain fine and coarse particles in different quantity by fraction.

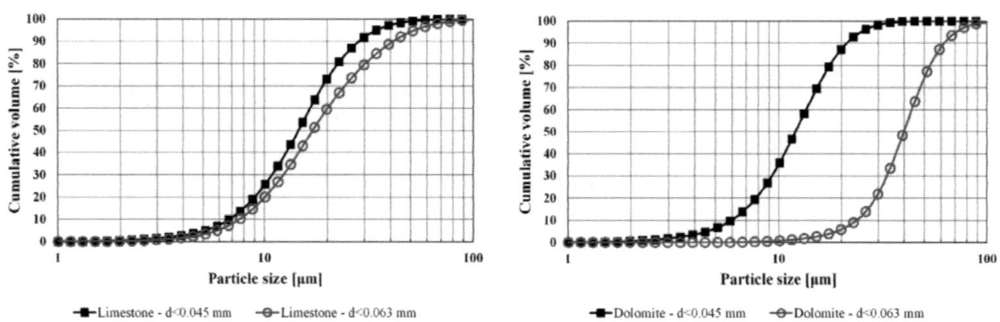

Figure 1. Particle size distribution of fillers

Fig. 1. well illustrates the differences of PSD between the mineral materials. Both fractions of limestone contain much more fine particles (typically particles below 10 microns) than dolomite, which contains quantitatively more of bigger particles. In dolomite (d<0.063mm) there are less fine (d<10 microns) particles. However, in the d<0.045 mm fraction of dolomite the presence of fine particles was highly significant.

During the research BET specific surface area were also determined. It was found that the fractions of fillers, except the coarser dolomite fraction, have a relatively low BET value (~ 1 m^2g^{-1}), which is favourable in asphalt technology in the viewpoint of bitumen consumption.

Hydrophilic coefficient of fillers were also determined. It can be concluded that all of the fractions are hydrophobic because their hydrophilic coefficients are under 1. As carbonated minerals can form a better contact with bitumen, the use of these fillers are highly favourable.

Pore size distribution, pore diameters were examined by mercury intrusion porosimetry (MIP). The results shows that the average pore diameter of fillers are about 2-3 ^m. The results indicate that in coarser particles there are bigger pores. If median pore diameters are inspected, pore diameter of dolomite (d<0.063 mm) has the highest value. This phenomena was also observed later on during scanning electronmicroscopy studies.

Fig. 2. shows the microstructures of fillers (magnification=200X). The SEM micrographs confirm the results of particle size distribution tests. Both of the fillers can be characterized as a polydisperse system. As it can be seen, the larger fraction of dolomite does not contain much fine particles.

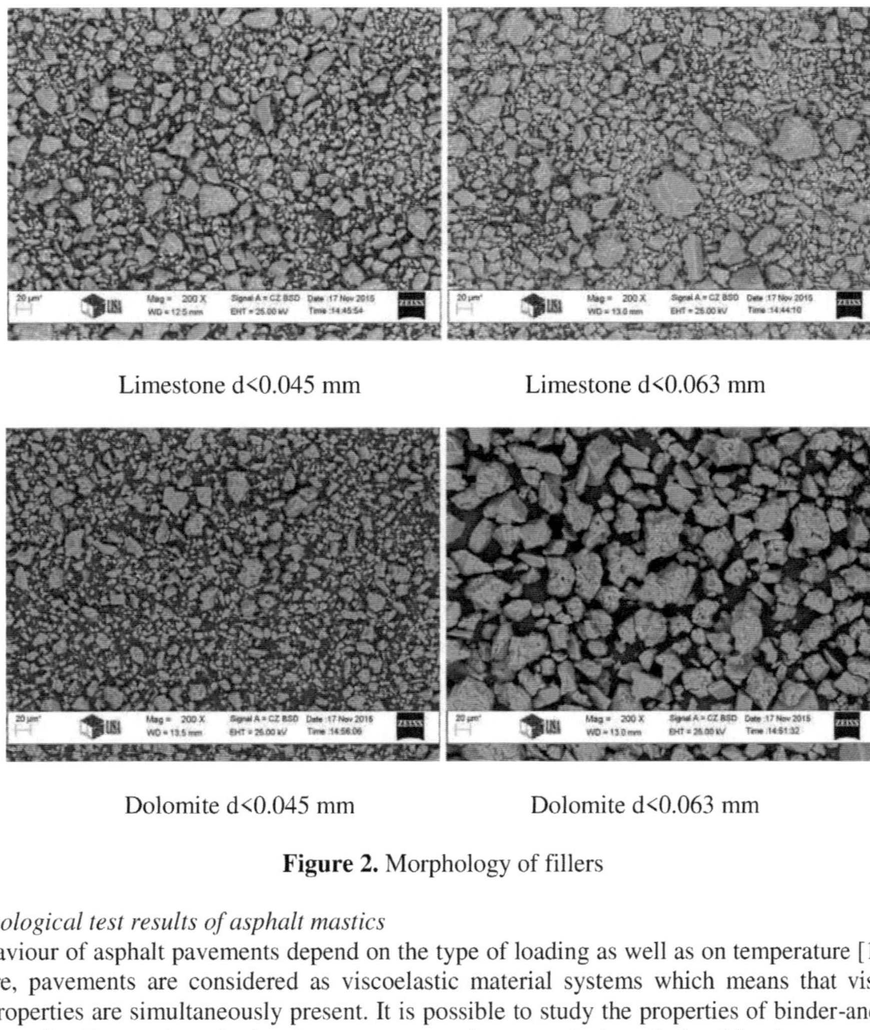

Limestone d<0.045 mm Limestone d<0.063 mm

Dolomite d<0.045 mm Dolomite d<0.063 mm

Figure 2. Morphology of fillers

3.2. Rheological test results of asphalt mastics

The behaviour of asphalt pavements depend on the type of loading as well as on temperature [1, 14, 20]. Therefore, pavements are considered as viscoelastic material systems which means that viscous and elastic properties are simultaneously present. It is possible to study the properties of binder-and-asphalt-mixtures under linear viscoelastic circumstances and to reveal the relationships between them [1]. Because of the page limitation of this article, the following diagrams are just examples of the whole research results. On the forthcoming figures asphalt mastics (made with limestone and dolomite fillers in different particle size ranges /d<0.045mm and 0.063 mm/ with F/M ratio of 20%) are shown. Former research results were published elsewhere [22, 23].

During rheological investigations creep-recovery tests were done on fillers. Testing conditions were:

T=60°C and **t**=1 Pa. Using these parameters all of the tests were made in the linear viscoelastic regime. (Abbreviations on the diagrams are: LS=limestone; D=dolomite; 45 μm/63 μm=particle size ranges of fillers; 025=F/M ratio).

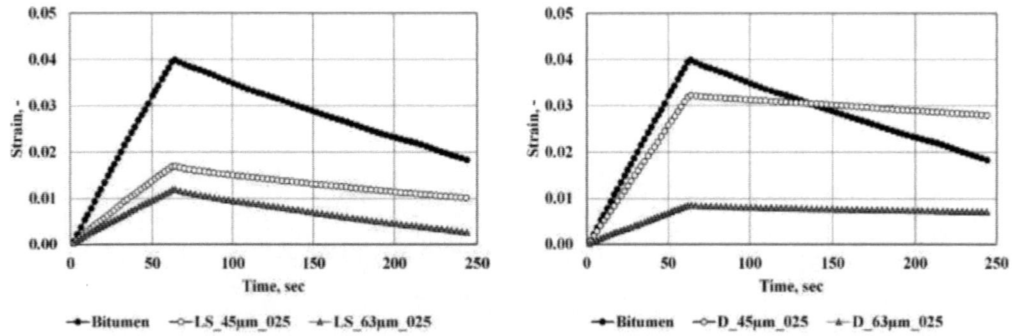

Figure 3. Flow curves of asphalt mastics at 60 °C and 135 °C

Figure 3. shows the creep-recovery curves of asphalt mastics. It can be concluded that with the use of fillers the strain of binder material can be reduced. The pure bitumen has the highest deformation during loading, and at the same time the coarser fraction of fillers will produce the lowest strain in material structure. After unloading in the recovery phase the lowest permanent deformation was produced by limestone.

To examine the flow phenomena of asphalt mastics, flow curves and viscosity curves were determined. At the lowest testing temperature (summer conditions), because of the high viscosity of mastics, shear rate was set up to 20 s-1. At elevated testing temperatures (compacting temperature) shear rate region was extended up to 200 s-1. The determined flow curves of mastics at 60 °C and 135 °C are shown in Figure 4.

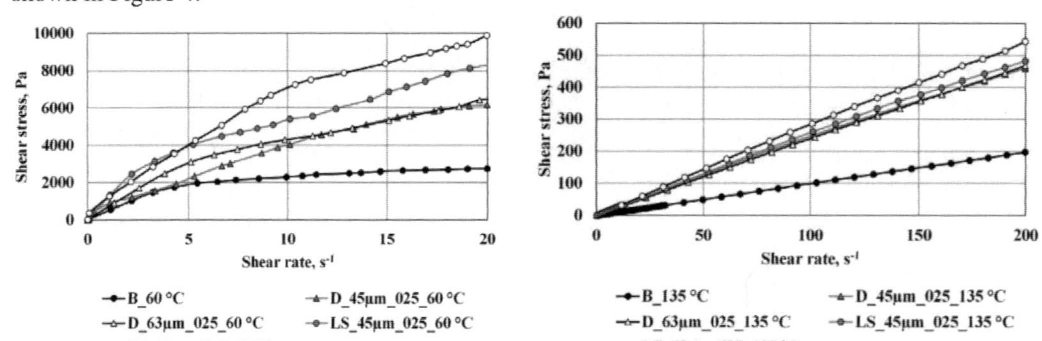

Figure 4. Flow curves of asphalt mastics at 60°C and 135°C

As bitumen and asphalt mastics have yield stress, e.g. the mixtures are pseudoplastic and can be described at lower testing temperature with Herschel-Bulkley model. At higher testing temperatures flow phenomena were changed, Newtonian flows were observed above the yield stress, therefore the mixtures behave as Bingham materials.

Dynamic oscillatory tests were also obtained on mastics. During these tests frequency ranges were chosen 0.1 Hz-10 Hz, while temperature was changed by 15°C. Dynamic parameters (storage and loss modulus, complex modulus, phase angle and loss tangent) were determined and using the WLF-equation master curves of mastics were constructed. Reference temperature were chosen at 60°C. Figure 5. shows the complex modulus and phase angle master curves of mastics.

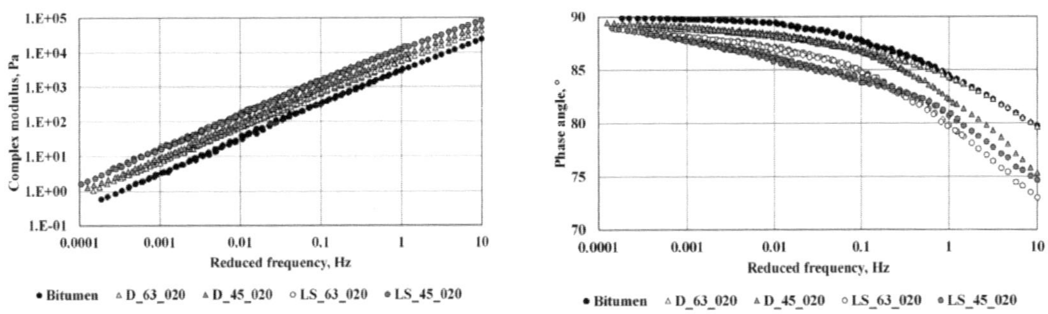

Figure 5. Complex modulus and phase angle master curves of asphalt mastics (T_{ref}=60°C)

The master curves show well the effect of particle size and quality of fillers on the complex modulus (properly speaking the stiffness). It can be concluded that limestone can create a stronger contact with bitumen than dolomite. Coarser particles increased the modulus, therefore using these particle sized fillers elastic properties can also increase. On closer examination of phase angle master curves the results prove the above mentioned. Limestone, especially the coarser fraction, can improve the elastic behaviour of mastics much better than dolomite. However, using both fillers phase angle will decrease, therefore the elastic properties of mastics will increase. The usage of these mineral materials as fillers can decrease rutting of pavements.

4. Conclusions

During this research work two types of fine particle mineral materials were tested as asphalt fillers. Examinations of raw materials as well as asphalt mastics were made by different techniques.

Having defined their particle size distribution it was found that these mineral materials can be characterized as polydisperse systems and they contain fine and coarse particles in different quantity by fraction.

By material structure test the Authors have observed the morphological features of fillers. The form of particles is diversified, their surface is broken up by sharp borders due to comminution processes, which increase the internal friction of the mineral skeleton and pavement. Polydispersity of mineral materials ensures good space filling and through this greater compactness and strength for asphalt mixtures.

During the examination of the material structure signs of minimal porosity and small specific surface area was observed. Porosimetry tests showed that coarser particles contain bigger pores within the material structure.

It was proved that the particle size distribution of the fractions as well as the quantity of fine and coarse particles affect the material features in each case. Presence of fines result in the increase of specific surface area but at the same time decrease the value of the hydrophilic coefficient, improving the bitumen stabilizing ability of mineral materials.

By the use of fillers the Authors have created asphalt mastic mixtures in which effect mechanism of type, particle size and quantity of fillers (volume fraction) can be observed at the same time. Rheological tests on the mastics were made with which the behaviour of asphalt pavements in summer can be described. With the measuring methods used, the purpose was to observe rutting, the typical defect of pavements.

During the analysis of creep-recovery features the behaviour of mastics were investigated. In view of the results it was also stated that the coarse particles in mastics increase the elasticity of mixtures in each case and thus decrease strain that is developed to the effect of load and also the amount of deformation remaining from recovery. It was also proved that during the use of limestone minor deformations developed which relates to the fact that limestone is able to create stronger relation with the binder.

To examine the flow phenomena occurring during the operational use of pavements in summer and the compacting of asphalt mixtures the real flow curves and viscosity curves of mastics were determined. As mixtures have yield stress, besides the viscoelastic nature of mixtures, there is also plasticity to certain extent in the material system. Having this in mind the mixtures were characterized with two rheological models in the testing temperature ranges. It was proved that at lower testing temperatures mastics behave as pseudoplastic materials and they could be described with Herschel-Bulkley model. At higher testing temperature the materials behave as Bingham materials. With measuring results it was reinforced that with the increase of the amount and particle size of fillers pseudoplastic nature is strengthened, while at the highest testing temperature - which is in fact the compacting temperature of asphalt pavements - mastics can be characterized as Bingham materials one and all.

With dynamic tests the dependence of asphalt mastics particle size and quantity of fillers the elastic properties will dominate and as a consequence stiffening effect will increase. It was stated that due to the strong interaction between limestone and bitumen stiffness will be greater than during the use of dolomite. All in all, we can say that the use of the observed mineral materials - especially limestone - as asphalt fillers is advantageous in any case as due to their favourable features the stiffness of pavements at high traffic temperature can be improved to a great extent, their resistance to deformation can be increased and finally the possibility of rutting formation can be decreased.

Acknowledgement
This work has been carried out as part of the TÁMOP-4.2.1.B-10/2/KONV-2010-0001 and TÁMOP-4.2.2/B-10/1-2010-0008 project within the framework of the New Hungarian Development Plan. The realization of this project is supported by the European Union, co-financed by the European Social Fund.

References

[1] J. Chen: *Journal of Materials, Concrete Structures and Pavements* **36** 269-277
[2] W Grabowski, J Wilanowicz: *Materials and Structures* **41** 793-804 http://dx.doi.org/10.1617/s11527-007-9283-4
[3] P S Kandhal, C Y Lynn, F Parker: Characterization tests for mineral fillers related to performance of asphalt paving mixtures *NCATReport* No. 98-2
[4] J Chen, P Kuo, P Lin, C Huang, K Lin: *Materials and Structures* **41** 10151024
[5] Y R Kim, D N Little: *Journal of Materials in Civil Engineering* **16** 122-132
[6] A Zulkati, W Y Diew, D S Delai: *Journal of Transportation Engineering* **138** 902-910
[7] M Chen, J Lin, S Wu, C Liu: *Construction and Building Materials* **25** 1532-1536
[8] K Yan, H Xu, H Zhang: *Construction and Building Materials* **48** 622-627
[9] A Khodaii, A Mehrara: *Construction and Building Materials* **23** 2586-2592 doi :10.1016/j .conbuildmat.2009.02.015
[10] Cs Tóth and Z Soós: *Építőanyag-Journal of Silicate Based and Composite Materials* **67** 126-131 http://dx.doi.org/10.14382/epitoanyag-jsbcm.2015.20
[11] Z Soós: *Az aszfalt* **20** 5559
[12] K Almássy and L A Joó: *Journal of Silicate Based and Composite Materials* **61** 5559 http://dx.doi.org/10.14382/epitoanyag-jsbcm.2009.10
[13] A Kavussi, R G Hicks: *Journal of Association of Asphalt Paving Technologists* **66** 153-186
[14] L A Gömze, Á Kovács *Építőanyag- Journal of Silicate Based and Composite Materials* **57** 34-38 http://dx.doi.org/10.14382/epitoanyag-jsbcm.2005.?
[15] L A Gömze, L N Gömze, S N Kulkov, L I Shabalin, I Gotman, F Pedraza, G L Lecomte, T Mayorova, E Kurovics, A Hamza: *Építőanyag-Journal of Silicate Based and Composite Materials* **67** 143-149 http://dx.doi.org/10.14382/epitoanyag-jsbcm.2015.24
[16] N V Maidanova, D A Rozental: *Russian Journal of Applied Chemistry* **79** 1386-1387
[17] J Craus, I Ishai, A Sides: *Journal of Association of Asphalt Paving Technologists* **47** 558588.
[18] R Géber: *Journal of Silicate Based and Composite Materials* **59** 73-76 http://dx.doi.org/10.14382/epitoanyag-jsbcm.2007.11
[19] D Lesueur, D Little: *Transportation Research Record* **1661** 93-100

[20] A Pérez-Lepe, F J Martínez-Boza, C Gallegos, O González, M E Munoz, A Santamaria: *Fuel* **82** 1339-1348

[21] B M Harris, K D Stuart: *Journal of Association of Asphalt Paving Technologists* ***64*** *54-95*

[22] R Géber, A Apkaryan, S Kulkov, L A Gömze: *Materials Science and Engineering-A publication of the University of Miskolc,* **39** 5-11

[23] R Géber, L A Gömze: *Materials Science Forum* **752** 209-216

Calibration problems with the viscosity measurement of liquid metallurgical slags

H P Heller[1], M Schürmann[2], K Scholl[3], N Haustein[1], B Lychatz[1] and J Falkus[4]

[1] TU Bergakademie Freiberg, Institute of Iron and Steel Technology, Freiberg, Germany
[2] Hüttenwerke Krupp Mannesmann GmbH, Duisburg, Germany
[3] ArcelorMittal Eisenhüttenstadt GmbH, Eisenhüttenstadt, Germany
[4] AGH Krakow, Department of Ferrous Metallurgy, Krakow, Poland

E-mail: heller@iest.tu-freiberg.de

Abstract. The viscosity of slag is an important characteristic of liquid slags regarding its lubricating effect and mass transfer. For measurement, however, they exhibit considerable differences in the values reported. Therefore, the rotation method, mostly used for high temperatures areas, is investigated regarding the impacts of any geometric inaccuracies. Furthermore, problems in the centering and use of calibration slags are discussed. It appears that, with the use of a more precise rheometer with air bearing, an error of less than +/- 3 % is possible in compliance with geometric critical values and online monitoring of the central operations. The verification was carried out with a blast furnace slag, which is also proposed as a calibration slag.

Keywords: rheometer, viscosity, slags, calibration, geometry errors

1. Introduction

In metallurgy, the viscosity of slags is an important parameter. It plays a major role for transport and reaction velocities, for emulsification of slag in steel or inversely and for emptying or filling of metallurgical vessels. In continuous casting process, viscosity of the casting powder slag is responsible for lubrication and detachment of hot strand from the cold mold [1-3].

For measurements in high-temperature areas, which can be between 900 and 1800°C, values of determined viscosities of metallurgical slags show considerable deviations [4-6]. On one side, this can be explained by different measuring methods and the various crucible materials and other components of the slag contact (graphite, molybdenum, platinum). On the other side, geometrical problems can also cause higher errors in measurement. Due to the large temperature range, the thermal expansion cannot be neglected. Thus, all works have to be performed with a wide gap (more than 2 mm), leading to higher revolution speeds (rpm) and respectively low torque.

Moreover, a disadvantage is that the measuring arrangement is not visible during experiments. It is not sure, if the furnace meets the requirements. Another source of error can be changes within the slag, due to evaporation of fragile components or due to reactions of the slag components. Such reactions can also cause changes in geometry. Therefore, an appropriate inert material, an inert gas and a suitable method have to be chosen depending on the slag investigated.

2. Device

The most commonly used method of measuring viscosity is Couette's rotation method. At the Institute of Iron and Steel Technology (IEST), TU Bergakademie Freiberg, this rotation method is mostly used for slags in addition to the oscillating method [7].

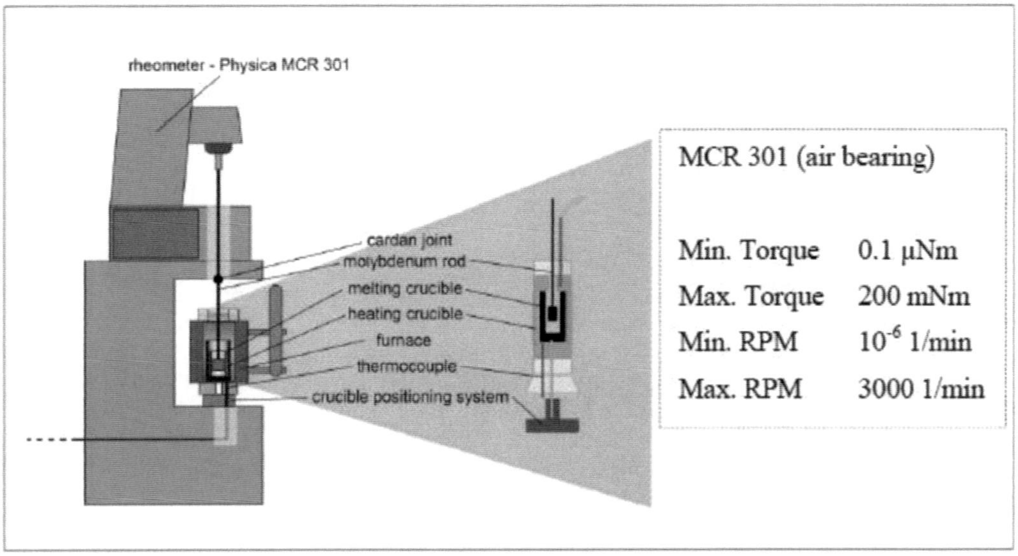

Figure 1. Scheme of the high-temperature-viscometer at IEST

The viscometer consists of a commercial rheometer MCR 301 from Anton Paar and a selfengineered high temperature device. As shown in figure 1, the furnace is approximately 0.7 m away from the rheometer head, which is conditioned by heat reasons. For the measurements of homogeneous slags, the middle shaft is equipped with a cardan joint to reduce the influence of imprecise centering. The MCR 301 is a precision-rheometer with air bearing and can measure small torques down to below 1 µNm over a wide range of revolution speeds. The advantages of this high temperature viscometer are:

- There is no calibration necessary, since the torque and rpm are measured in absolute values.
- Due to the measureable low torque, low rpm are applicable.
- Rotation and oscillation are possible.
- By employing a high frequency induction furnace, temperatures up to 1750 °C are possible.

3. Geometry

In order to estimate the influence of geometrical errors on measurements with a cardan joint, experiments with silicone oils from wacker were carried out at room temperature. These oils were found to be the most stable in regards of time dependencies. Oils between 50 and 20000 mPas were available. The manufacturer's data for these oils was confirmed by a cone-plate- measuring system with Peltier element. Since the viscosity is strongly depending on the temperature, temperature-viscosity-curves were obtained for each oils used. For all of them, a uniform dependency of dq/dT = 2 %/K at 25°C could be found and the influence of temperature on cold experiments respected.

Series of measurements were carried out for the influence of a curved shaft, an inclined rotating bob, an inclined crucible, a wrong immersion depth and for a bad alignment/centering. The most important results are displayed in figure 2. On the left side, the impact of an inclined crucible by itself and in combination with a tilted rotating bob is displayed. On the right side, the effect of an imprecise centering is shown.

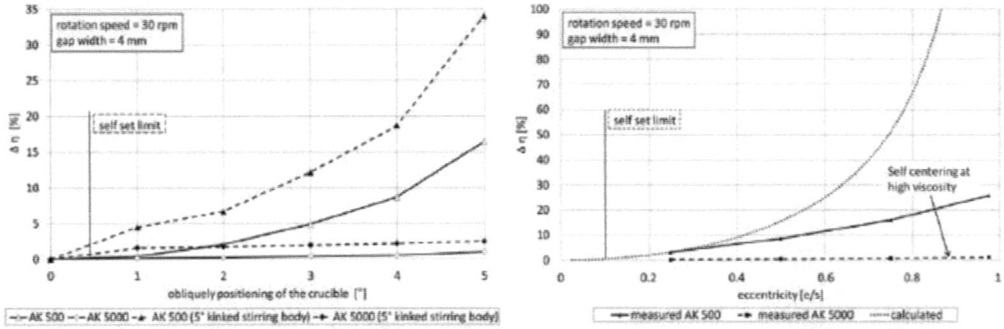

Figure 2. Influence of geometric malposition to measured viscosity

As visible, small malpositions cause acceptable measurement errors. For high viscosities, the error is not extreme since the cardan joint allows a self-centering (Figure 2, curves for AK 5000 with a viscosity of approximately 5 Pas). At low viscosities (AK 500 with approximately 500 mPas), the error quickly exceeds the acceptable limits. Therefore, it is necessary to limit malpositioning below a self-set threshold value, especially for precise measurements of thin slags. Thus, measurements at the IEST are carried out with followinglimits for geometry, since they will lead to an expected geometrical error of malposition of maximum 5%.

- Obliquity of crucible < 0.5° (inclined position)
- Obliquity of rotating bob < 1°
- Eccentricity < 0.5 mm

4. High temperature trials

In order to meet the requirements during hot temperature trials, the following procedure is maintained: prior to the heating of the furnace, the position of the crucible and the centricity of the rotating bob are controlled and adjusted if necessary. For that, a calibration tool is exchanged for the rotating bob which allows an adjustment of the crucible and a centered run. Afterwards, the rotation bob is brought into position by video recording and image processing to ensure no eccentricity or tilting. Figure 3 shows the tool for crucible adjustment on the left side and the image processing applied on the right side. Afterwards, the furnace is put over the crucible without changing its position and another visual control of the centric run of the rotation bob is done. Subsequently, the furnace is closed and the experiment continues with heating and melting the slag.

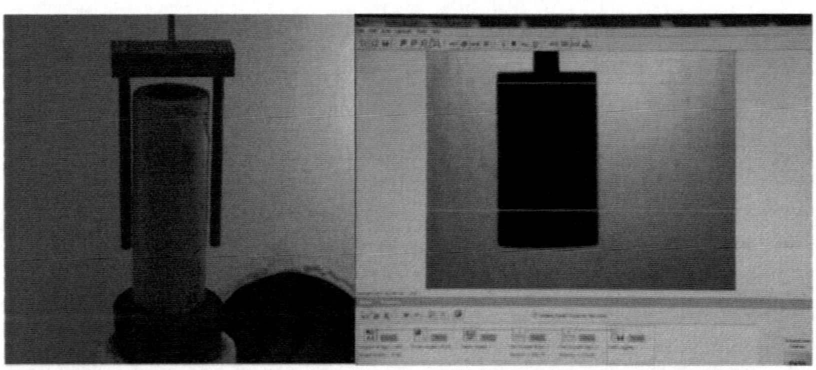

Figure 3. Tool for crucible adjustment (left), image processing (right)

To obtain the correct immersion depth, the rotating bob is slowly immersed in the crucible, while the torque is detected. An increase in torque defines the first moment of touching the slag clearly visible. Based on this height, the measuring depth is calculated (slag 7 mm above bob) and the bob is driven into position. Usually, three cooling curved are measured at one sample, which reach from a definded maximum temperature until solidification, respectively exceeding a maximum viscosity. The cooling rates are commonly 10 or 20 K/min and the rotation speed 15 or 30 rpm.

5. Reproducibility

For demonstrating the reproducibility, measurements were performed on a blast furnace slag using molybdenum as crucible and rotating bob material. Blast furnace slags crystallized stable and does not react with the crucible material. The selected composition (39.98% CaO, 35.93% **SiO2**, 12.32% **M2O3**, 8.76% MgO, 0.78% **Fe2O3**, 0.16% MnO, 0.06% **K2O**, 0.57% TiO_2, < 0.01% Na_2O) shows a low viscosity and a crystallization that starts below 1250 °C. Thirteen measurements were performed on this slag, as shown in figure 4. Large deviations appear only on two cooling curves. These deviations were due to the not centric rotating bobs. The rotation of the gauge bar was monitored by a video camera and an online image processing in order to record the data of the experiment.

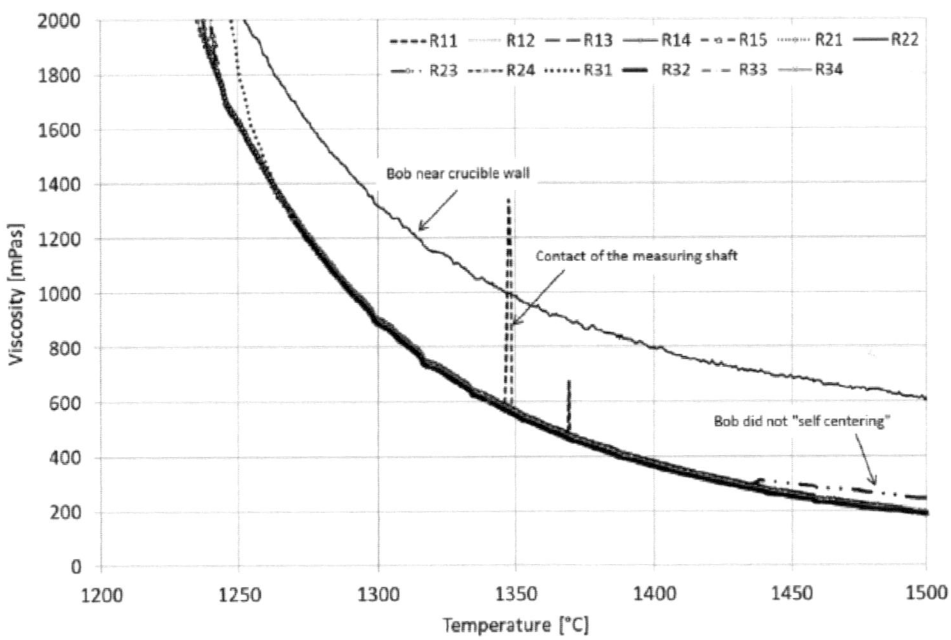

Figure 4. Viscosity temperature curves of a blast furnace slag with some centering problems

When the bob rotated at the crucible wall, an extreme error of 200% is observed. Also, small inaccuracies in centering can produce larger errors. This means that the centric rotation should always be supervised when using a cardan joint. If the incorrect curves are eliminated, the image in figure 5 is generated.

The deviations between the biggest and the smallest data lie below 5%, which appears to be sufficient for high temperature measurements.

Simultaneously, the blast furnace slag proves to be a suitable reference material. Measurements of a reference material should be done at regular intervals. A common granulated blast furnace slag (CaO-SiO_2-Al_2O_3-MgO) with low viscosity (0.2 Pas at 1500°C) should be employed. This way, no changes in composition are expected and some existing models for calculation of viscosity can also be used.

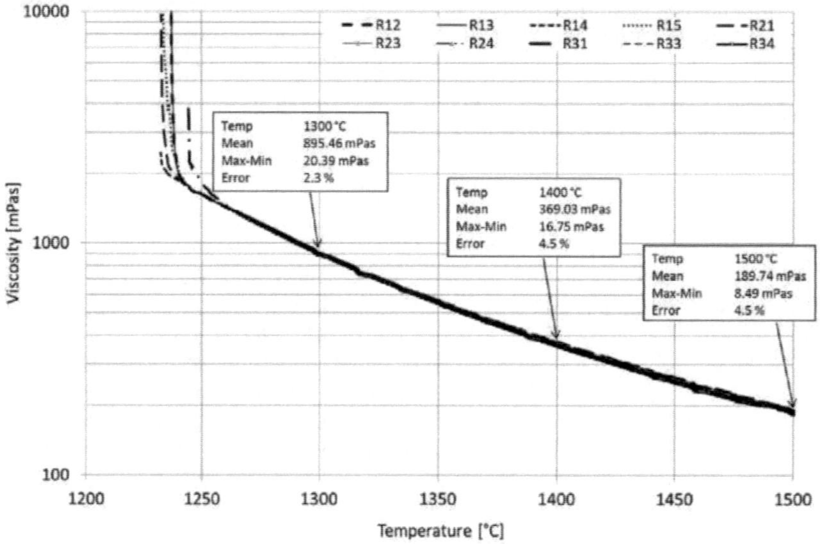

Figure 5. Measured viscosity of a blast furnace slag

Other proposals for reference materials are:

- Calibration glass by Physikalisch-Technische Bundesanstalt.
- There are three kinds available. However, the lowest viscosity is about ~16 Pas at 1400°C and it is containing PbO. Therefore, an oxidizing atmosphere and Pt as material should be used.
- Calibration slag by Mills based on SiO_2-Al_2O_3-Li_2O.
- The viscosity is about 2 Pas at 1400°C, which is the maximum temperature recommended (lowest viscosity 0.5 Pas at 1600°C).
- Continuous casting slags with CaO-SiO_2-Al_2O_3-MgO-CaF_2-Na_2O having a lowest viscosity of < 0.2 Pas at 1500°C. Yet, the high CaF_2 and Na_2O content make changes at high temperatures possible.

All the above being considered, a blast furnace slag seems to be the best choice.

6. Conclusions

Accurate viscosity measurements using the rotation method (Couette) on liquid slags to 1700°C are possible if there is a sufficient adjustment of the crucible and rotating body and a supervision of rotating shaft by online image processing (by using of two cameras for monitoring rotation outside the furnace). Furthermore, inert materials and protection gases have to be used. However measurements of a reference material should be done at regular intervals.

Reference

[1] Kromhout J A Melzer S Zinngrebe E W Kamperman A A and Boom R 2008 *Steel Research Int.* **79** 143
[2] Lamut J Falkus J Jurjevec B and Knap M 2012 *Archives of metallurgy and materials* **57** 319
[3] Schulz T Lychatz B Haustein N and Janke D 2013 *Metallurgical and Materials Transactions B* **44** 317
[4] Broadbent C B Franken M Gould D and Mills K C 1992 *4th Int. Conf. on Molten Slags and Fluxes, Sendai* 427
[5] Seetharaman S Mukai K and Sichen Du 2005 *Steel Research Int.* **76** 267
[6] Mills K C Chapman L Fox A B and Sridhar S 2001 *Scand. J. Metall.* **30** 396
[7] Heller H P Hötzel M Lychatz B and Haustein N 2013 *Steel Research Int.* **84** 982

Modification of the rheological properties of screen printing ceramic paints containing gold

P Izak[1], J Mastalska-Poplawska[1], J Lis[1], A Stempkowska[2]

[1]AGH University of Science and Technology, Faculty of Materials Science and Ceramics, Krakow, Poland
[2]AGH University of Science and Technology, Faculty of Mining and Geoengineering, Krakow, Poland

E-mail: izak@agh.edu.pl

Abstract. This work presents the results of modification of rheological properties of screen printing paints containing gold. 15 wt% glossy gold paste and 15 wt% glossy liquid gold were used as modifiers containing gold. The study showed that the gold paint for screen printing can be obtained by evaporation of the 15 wt% liquid gold and the golden luster. The compaction process of liquid gold by evaporation is slow and easy to perform in industrial conditions.

The second way to adapt the 15 wt% gold ceramic paint for screen printing application depended on adding the aniseed oil and the pine oil. The course of the flow curve of the gold paste without modification indicates that it is shear thinning and shows the desired effect of thixotropy, and even anti-thixotropy, at low shear rates ($<50^{-1}$ s^{-1}). The introduction of the essential oils eliminates this phenomenon and the paste converts itself from the non- rheostable to the rheostable liquid.

1. Introduction

From the rheological point of view, typical screen printing paints based on nonaqueous solvents show Newtonian or pseudoplastic behaviour, i.e. they are shear-thinning and have yield point [1-8]. Flow curves courses and equation models of typical rheostable fluids are shown in Fig.1 [4].

Flow curves are usually presented as a change in shear stress or a change of apparent viscosity versus shear rate gradient. These courses are equivalent.

As the practice shows, screen-printing pastes should also have little thixotropy, which protects them of spreading on the printed surface [6-9]. However, the courses of flow curves of rheo-unstable fluids (time dependent rheologies) are more complicated and it is not possible to describe them using dimple mathematic formulas (Fig.2) [4].

Generalization of these curves by plotting of the so-called equilibrium curves or by the averaging of their courses at increasing and decreasing shear rates allows these typical rheostable fluid models to characterize the non-rheostable fluids (see also Fig. 1).

The second important aspect consists in adjusting the measurements conditions (shear rate) of these fluids to the conditions occurring during printing.

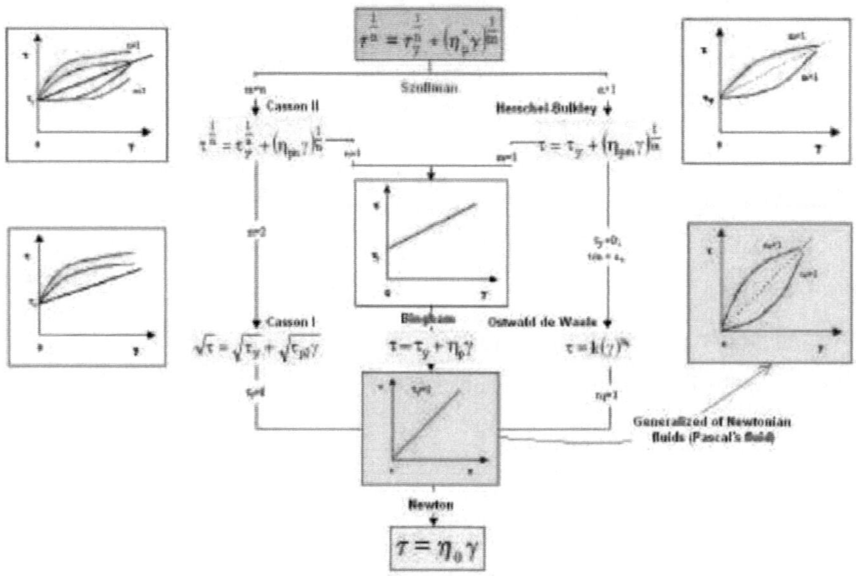

Figure 1. Flow curves courses and equation models of typical rheostable fluids [4]

By addition of the so called additives and other simple technological procedures (thickening, dilution) the rheological properties of ceramic paints (yield point, flow curves courses, thixotropic properties etc.) can be changed significantly, and thus the print quality [9- 13]. Non-polar surface-active substances, various drying retardants, antistatic agents, hardeners etc. are used as auxiliary agents. Such paints cannot be usually used after drying and/ or curing (cross-linking). The problem of drying during printing becomes more important when it concerns expensive gold-based paints. This applies both to the changes in rheological properties of paint and, above all, when the reuse of ore is taken into consideration.

2. Materials and method

The aim of this study is to determine the adjustability of the rheological parameters of paints containing gold in terms of using them for screen printing. The following materials were used for tests:

- 15 wt% glossy gold paste - used for painting strips (PMP);
- 15 wt% liquid glossy gold - used for hand stamping (PMP);
- gold luster (Degussa- Gold Bronze 36);
- anise oil- colourless essential oil with a characteristic odour; melting temperature 1519 °C; containing 80-95 wt% of anethole; received from water distillation of Pimpinello anisum; soluble in alcohol and ether but practically insoluble in water;
- pine oil- colourless essential oil; containing pinene, kanfen, felandrem, kuminal, anisaldehyde and barneol; received from the water distillation of pine needles; soluble in alcohol.

The industrial practice has shown that screen printing paints in operating conditions should have apparent viscosity in the range 600 - 1000 Pas, and shear rate in the range 5001000 s^{-1}. Because liquid gold has its viscosity of 94.8 Pas and gold luster has its viscosity of 7.9 Pas, they were thickening by evaporation at 100 °C. In contrast, gold paste with viscosity of 2606 Pas was diluted by the addition of essential oils.

Apparent viscosity measurements were performed on Rheotest II viscometer, equipped with a cone-plate system, in the wide range of shear rates.

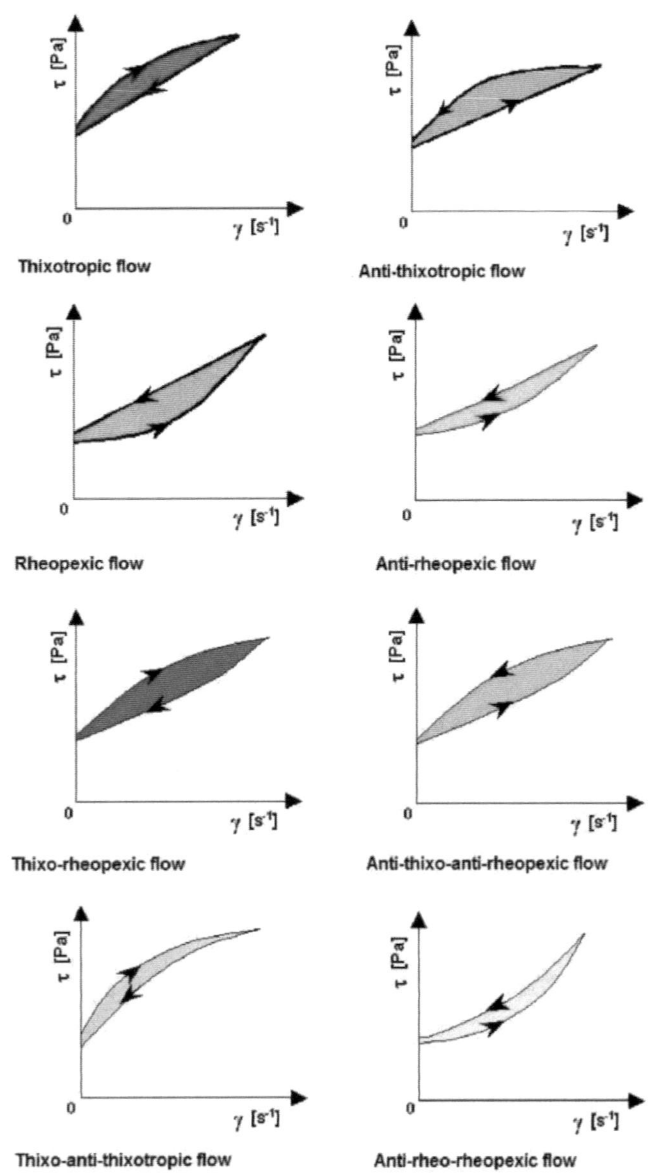

Figure 2. The flows of rheo-unstable fluids and their generalizations [4]

3. Results and discussion

Flow curves of the screen printing gold paints before and after drying are shown in Fig. 3, while the changes of apparent viscosity during drying are shown in Fig. 4. Ceramic liquid gold with the concentration of 15 wt% was labelled as Au15%fluid and Au15%fluid.eva after evaporation. Consequently, the ceramic golden luster was labeled as Au6% luster and Au6% luster.eva after drying.

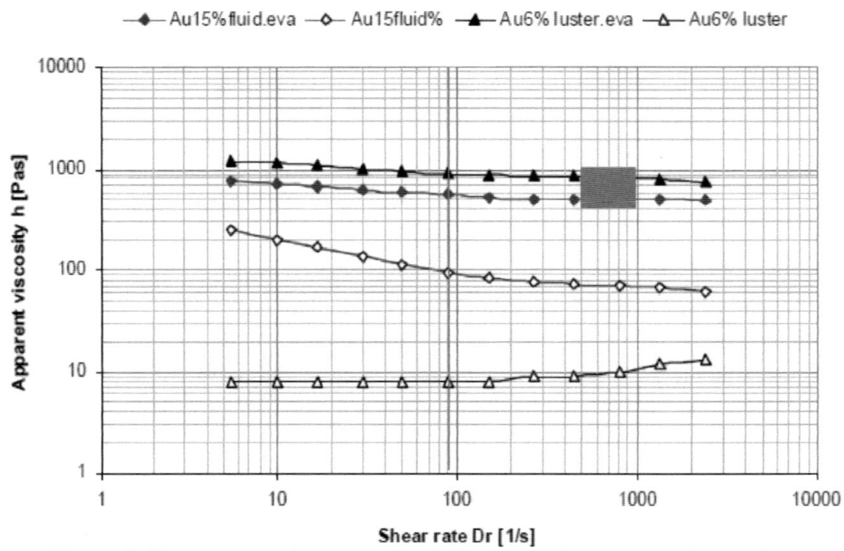

Figure 3. Flow curves of the screen printing paints before and after drying

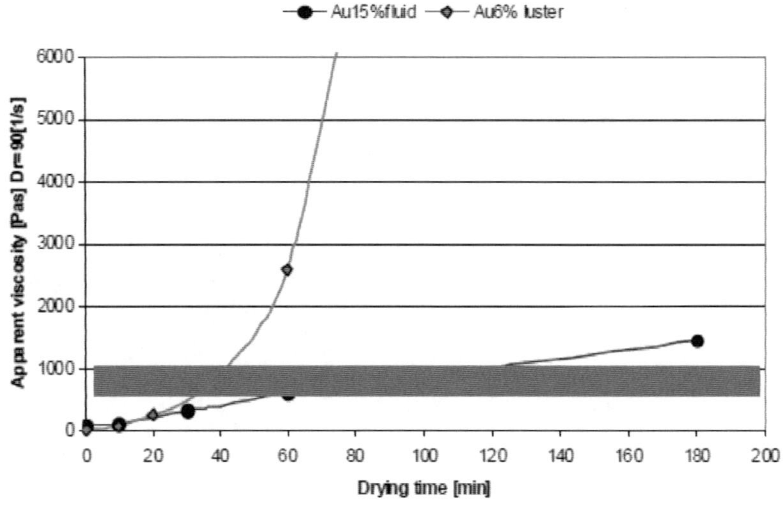

Figure 4. Changes of apparent viscosity of the screen printing paints during drying

The tests have shown that gold paint for screen printing obtained by evaporation can be made of 15 wt% liquid gold and gold luster. The densification process of liquid gold by evaporation is slow and relatively easy to carry out in industrial conditions. Such effects were not obtained during densification of the luster paints because in this case the evaporation process is fast and it is difficult to obtain the screen printing paint of the required viscosity. This also means that under manufacturing conditions, i.e. at 18-25°C, gold paint is constantly compacted by evaporation. From the technological point of view, if this process is slow, it is a positive phenomenon. During the fast evaporation, in addition to the changes in the rheological properties, also the content of ore changes, which can affect the properties of decorated products.

The rheological measurements have shown that flow curves of both slurries, i.e. 15 wt% liquid gold as well as 6 wt% gold luster, after densification by evaporation are similar to the Newtonian flow with a slight yield point. Generally speaking, before and after densification by drying they are still rheostable fluids with constant apparent viscosity as a function of shear rate. The drying process of gold paints causes an increase in flow resistance without any significant changes in the courses of flow curves.

Table 1 shows the fitting of the curves to the Newtonian/Bingham and Ostwald de Waele flow models before and after drying.

Table 1. Parameters of rheological equations of paints containing gold

Sample	Rheological equation	
	Newton/Bingham	**Ostwald de Waele**
Au15%fluid	$\tau = 0.62\gamma + 31.71$ $R^2 = 0.9969$	$\tau = 3.11\ \gamma^{077}$ $R^2 = 0.9934$
Au6%luster	$\tau = 0.10\ \gamma$ $R^2 = 0.9928$	$\tau = 0.16\ \gamma^{092}$ $R^2 = 0.9931$
Au15%fluid.eva	$\tau = 7.47\ \gamma + 218.64$ $R^2 = 0.9969$	$\tau = 13.23\ \gamma^{093}$ $R^2 = 0.9997$
Au6%luster.eva	$\tau = 4.82\ \gamma + 61.64$ $R^2 = 0.9997$	$\tau = 8.36\ \gamma^{092}$ $R^2 = 0.9992$

It should be noticed that both rheological equations, Newton/Bingham and Ostwald de Waele describe very well the flow curves of the tested paints. Fitting power of the measurement results to the equations is above 99%. The practical choice of a model depends on the expected parameters of the rheological equations (consistency, flow index, yield point, viscosity).

Before evaporation the ceramic paint with liquid gold exhibits shear-thinning behaviour. It originates from the fact that during shearing of the liquid gold, polymer chains of the so-called mordant change their conformation. After densification by evaporation this phenomenon practically disappears and the flow coefficient increases from 0.77 to 0.92. During drying in 100°C probably a partial degradation of the organic mordant occurs, which changes the courses of the flow curves. These changes may also result from increasing of the concentration of suspensions containing gold. The luster ceramic paint, as opposed to the liquid gold, practically does not change the properties of thinning. In this case flow coefficients are about 0.92.

It should be noticed that consistency coefficients of the tested paints increased a couple of times (abort 10). Consistency is the measure of the internal structure of the paint slurry, i.e. polymer conformation of the mordant, grains aggregation, concentration etc., which affects the course of the flow.

It can be also noticed that the gold luster paint dries faster and because of that it is harder to obtain proper apparent viscosity for the screen printing purposes.

The second way to adapt 15 wt% gold ceramic paste for the screen printing application was based on the addition of aniseed and pine oils. Depending on the amount and nature of additives, the authors obtained liquids with different apparent viscosities (Fig. 5).

The initial glossy gold paste containing 15 wt% of ore has high viscosity, about 2600 Pas, which hinders the screen printing application on the decorated surface.

The investigations have shown that suspensions of lower viscosity can be obtained from the ceramic gold paints of high viscosities by the addition of oils and they can be adapted to the screen printing. The flow curve of the gold paste without any modification shows that it is shear-thinning and has a desirable thixotropy and even anti-thixotropy at low shear rates (<50 $^{-1}$) (Fig. 6). From the point of view of non-rheostable fluids such a flow curve course is classified as thixotropic-anti-thixotropic.

The introduction of essential oils eliminates this phenomenon and the non-rheostable fluid is converted into a rheostable liquid. It may be also noticed that in the shear rate range of -100 s^{-1} the paste modified with aniseed oil gets slightly concentrated and is shear-thinned above 1000 s^{-1}. In the case of pine oil the flow curve does not change in all the range of shear rates. Because of their structure, essential oils probably do not allow the change of the conformation of the mordant polymer chains in the paste to occur, and thus alter its rheological properties. Perhaps the reason is the mutual insolubility of polymers, i.e. the so- called mordant and oil. There is, however, no detailed information about this.

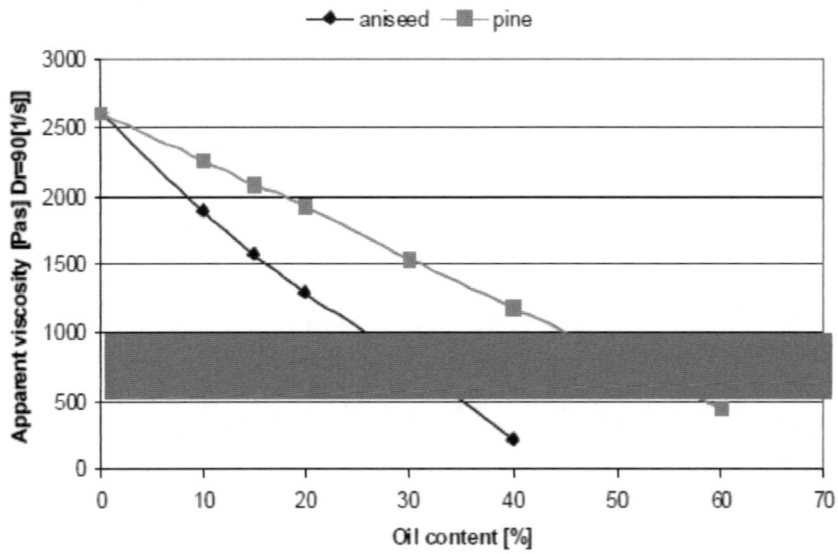

Figure 5. Changes of apparent viscosities of the gold paint modified with aniseed and pine oils

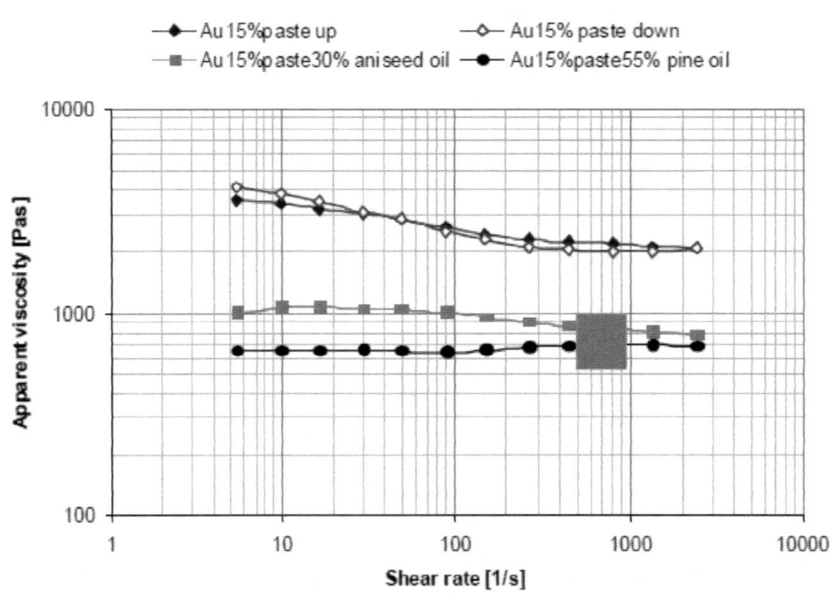

Figure 6. Flow curves of the screen printing paints after modification with essential oils

IOP Conf. Series: Journal of Physics: Conf. Series **790** (2017) 012011 doi:10.1088/1742-6596/790/1/012011

The tests have shown that the addition of the essential oils reduces the effect of shear- thinning. Flow coefficient increases its value from 0.88 to 0.94 in the case of aniseed oil and to 1.00 in the case of pine oil. Diluting the gold paste with the essential oils reduces the consistency coefficient several times. This also applies to the yield points (Table 2).

Table 2. Parameters of the rheological equations of the gold paint modified with essential oils

Sample	Equations of the rheological models	
	Newton/Bingham	**Ostwald de Waele**
Au15%paste (equilibrium)	$T = 20.11y + 374.54$ $R^2 = 0.9995$	$T = 45.21\ Y^{088}$ $R^2 = 0.9999$
Au15%paste 35% aniseed	$T = 7.59\ Y + 218.97$ $R^2 = 0.9980$	$T = 12.14\ Y^{094}$ $R^2 = 0.9995$
Au 15%paste 55%pine	$T = 6.30\ Y + 39.38$ $R^2 = 0.9992$	», 1.00 $T = 6.37\ Y$ $R^2 = 0.9999$

Generally, screen printing paints with pine or aniseed oil do not have thixotropic properties. It is known that the most favorable technological effects are obtained when the screen printing inks are shear-thinning with slight thixotropy [10]. In practical terms aniseed oil is a friendlier solvent because it dries more slowly. Hence, it should be recommended for screen printing techniques, although at higher contents in the paste it can cause characteristic microcracks of the gold layer applied to the ceramic product (Fig. 7) and may change the colour of gold (Fig.8).

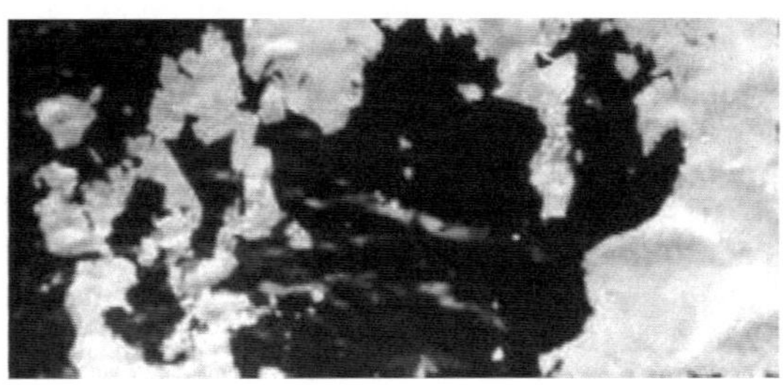

Figure 7. Microcracks of the diluted gold paint after burning (magnified 100-times)

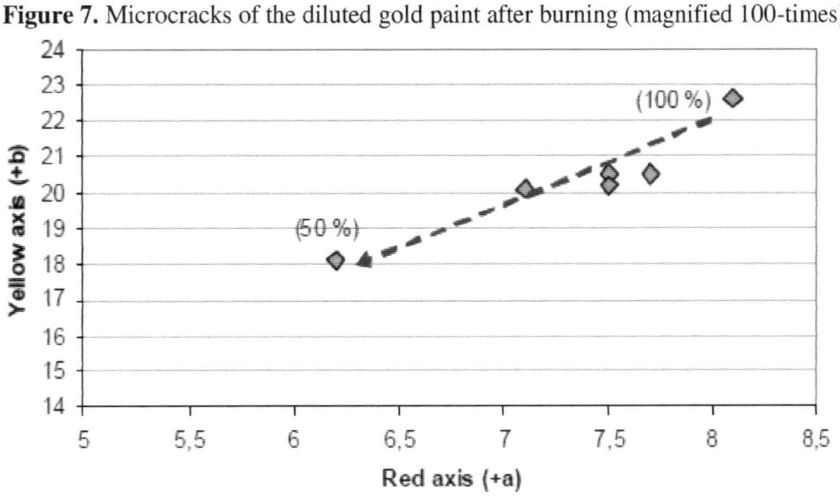

Figure 8. The change of gold paste colour after dilution

It was mentioned that at low shear rates a small effect of dilatancy occurs for the suspension of the gold luster. This phenomenon is unfavourable because it makes difficult correct clarity of an outline drawing to be obtained. When printing with such inks the speed of the squeegee should be reduced, because undesirable characteristic textures of the screen can be obtained. This demonstrates the inability to melt-in-time during printing (Fig. 9).

Figure 9. Screen texture on the flat surface caused by dilatancy (magnified 100 times)

It is worth noting that apparent viscosities of the tested paints at shear rates of 500 - 1000 s^{-1} are about 600-1000 Pas. If the speed of the squeegee corresponds with the shear rates of the ceramic paint during the measurement, the obtained results will be useful for screen printing purposes. Otherwise, the results will not be sufficient to assess the suitability of screen printing inks.

The presented methods of adjusting the viscosity of the inks for screen printing, however, change the gold content in the paint. These results are shown in Table 3.

Table 3. The content of gold in a defined area of printing

Sample number	Content of gold [%]
Au15%fluid	15
Au15fluid eva	16
Au 6% luster	6
Au6%luster eva	8
Au15%paste	15
Au15%paste +30% aniseed	9
Au15%paste + 55% pine	7

To determine this effect, thickness of the gold layer on the fired product was measured. The test was performed by the colorimetric method, using the formation of a complex of trivalent ions of gold with Rhodamine B. Gold was taken into the solution from a well-defined surface of the product (1cm^2) with the use of aqua regia.

The use of essential oils to reduce viscosity of screen printing gold inks is associated with a reduction in gold content. In particular, the biggest reduction can be seen with the use of pine oil (over 45%).

Abrasion resistance of gold coatings is the most representative parameter to assess their properties [14]. Increased susceptibility to abrasion was shown by the paint with pine oil, which directly corresponds to the content of gold (Fig. 10).

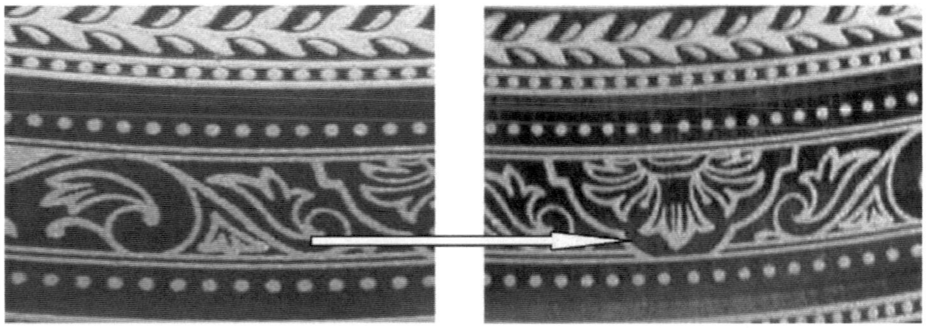

Figure 10. Effect of abrasion of the gold coating after 40 cycles according to the BN-69/7822-01 [14]

Similar effects were obtained for modified gold luster.

4. Conclusions

1. Ceramic glossy gold paints of different viscosities can be adjusted to the screen printing purposes both by thermal thickening and by dilution with the use of essential oils.
2. In the case of dilution the best results are achieved by the use of aniseed oil. Paint dilution leads to the lower content of gold and adversely affects gold coating abrasion.
3. Due to the high drying rate the thermal thickening technique is not recommended for gold lusters.
4. The addition of essential oils changes the rheological properties of screen printing paints. In particular, it eliminates the beneficial effects of thixotropy and pseudoplasticity.
5. Apparent viscosity of the screen printing paints should be adjusted to the speed of the squeegee shift.

References

[1] Kemblowski Z and Petera J 1980 *Rheol. Acta* **19** 529-538
[2] Gliniak A and Izak P 2001 *Glass and Ceramics* **52** 6-9 (in Polish)
[3] Kemblowski Z 1973 *Rheometry of non-Newtonian fluids* (WNT, Warsaw) (in Polish)
[4] Izak P 2013 *Rheology of ceramic slurries* (AGH Publishing House, Krakow) (in Polish)
[5] Izak P 2015 *Rheology in Ceramics* (AGH Publishing House, Krakow) (in Polish)
[6] Izak P 1991 *Glass and Ceramics* **42** 3-11 (in Polish)
[7] Izak P Stobierska E 2004 *Glass and Ceramics* **5** 12-15 (in Polish)
[8] Izak P Depa P 2004 *Ceramics* **84** 271-276 (in Polish)
[9] Guerrieri G and Panini P 2000 *Glazing and decorating of ceramic products* (ACIMAC ICE, Warsaw) (in Polish)
[10] Ziolkowski A 1994 *Poligrafika* **9** 34-35 (in Polish)
[11] Kampert M 1992 *Poligrafika* **12** 24-28 (in Polish)
[12] Kampert M 1993 *Poligrafika* **12** 12-17 (in Polish)
[13] Kampert M 1994 *Poligrafika* **5** 35-38 (in Polish)
[14] Polish norm BN-69/7822-01 *Abrasiveness of ceramic gold* (in Polish)

Modelling of powder consolidation using electro heating assisted by mechanical loading

A Knyazeva[1,2] and S Sorokova[1]

[1] Tomsk Polytechnik University, Tomsk, Russia
2 Institute if strength physics and material science SB RAS, Tomsk, Russia

E-mail: anna-knyazeva@mail.ru

Abstract. The model of the process of reactive sintering assisted by mechanical loading is suggested. The conjugate heat exchange of powder mixture is taken into account. The powder mixture motion is described as viscous liquid with effective viscosity. Mechanical sub problem is one dimensional because friction near the wall is assumed negligible small. Conjugate thermal conductivity problem includes thermal conduction equations for various materials (reactive mixture and walls of the camber. Heat release is possible due to external electrical heating, viscous dissipation and chemical reactions. Kinetical equations correspond to detailed reaction scheme. The problem is solved numerically with special algorithm. As a result the composition of the mixture is obtained for different time moments. The final composition is not uniform.

1. Introduction

There are a lot of technologies of new materials synthesis assisted by mechanical loading together with the heating by different methods. Process of powder consolidation using electricity as the main energy component is widely known and is used in various technologies. Depending on material types, experimental conditions, geometry, preliminary preparation, the way of loading, ones distinguish hot press sintering (HP); hot isostatic pressing (HIP); spark plasma sintering (SPS), and etc [1]. As a result, we come to composite materials with different structure, composition and properties. The most accepted mechanism of the pulsed electric current sintering includes the heating in the gap between neighbouring due to the micro-spark discharge, the heating by joule heat due to electrical current flowing from particle to particle and mass exchange between particles initiated by temperature gradient, electro migration and inelastic strains. The possibility of liquid phase formation in the contact between particles was experimentally confirmed for example in [2]. Numerous physical and mathematical models exist in this field allowing describing the various details of the sintering. Mathematical modelling allows to understand physical phenomena accompanying the new materials production and to give the preliminary prognosis for the choosing of technological parameters [3].

The first major publication on computer simulation of sintering appears in 1965 [4], where finite-difference method has been applied for the sintering of spheres. Nowadays, various approaches are used for the modelling of sintering assisted by pressure and electric field: Monte Carlo, finite difference, discrete element, finite element, neural network etc. Even so, when commercial software's are available, there are a lot of problems in the modelling [3].

Based on physical lows and continuum mechanics, many authors describe the temperature change in complex fields, residual stresses, density, porosity viscosity and mechanical properties evolution, the junction formation between individual particles etc. [5-14].

Theory and technologies of sintering are describes in [9, 12, 15-17] Special attention was paid to the generation and evolution of new ideas, elements of the physical theory of sintering during different time periods. The key role in these theories belongs to theoretical physicists, Frenkel, Herring, and Lifshits.

One of the most popular sintering theory [9, 12, 14] includes the relations between stress tensor's components σ_{ij} and strain rate tensor's component ε_{ij}

$$\sigma_{ij} = 2\eta_0\left[\varphi\varepsilon_{ij} + \left(\psi - \frac{1}{3}\varphi\right)e\delta_{ij}\right] + p_L\delta_{ij} \tag{1}$$

where η_0 is the shear viscosity of a fully dense material, φ and ψ are the normalized shear and bulk viscosities, δ_{ij} is the Kronecker symbol ($\delta_{ij} = 1$ if i = j and $\delta_{ij} = 0$ if $i \neq j$), $e = \varepsilon_{kk} = \varepsilon_{11} + \varepsilon_{22} + \varepsilon_{33}$. Effective sintering stress p_L is the product of the local sintering stress $p_{L0} = 3\alpha/G$ (α is the surface tension, G is the average particle radius) and of the normalized effective sintering stress $p_L = (1-\theta)^2/2$.

The normalized shear and bulk viscosities φ and ψ are defined as:

$$\varphi = (1-\theta)^2;\ \psi = \frac{2}{3}\frac{(1-\theta)^3}{\theta}.$$

Porosity changes in accordance of kinetic equations depending on problem under study.

The finite element simulation [3 and 18] predicts several phenomena, including grain growth, densification, and distortion based on constitutive equations applicable to a linear viscous compressive material. Temperature is given as uniform. In these papers, the relations between strains and stresses are taken in form

$$\varepsilon_{ij} = \frac{1}{2\eta}\sigma'_{ij} + \frac{1}{3K}\left(\frac{\sigma_{kk}}{3} - \sigma_s\right)\delta_{ij} \tag{2}$$

where σ_{ij}' - deviatoric part of stress tensor, σ_s - sintering stress, η and K are effective shear and bulk viscosity. The values σ_s, η, K depend on relative density, surface energy corresponding to some laws depending on sintering stage.

Note, that equations (1) and (2) are practically equivalent. Immediately from (1), we can obtain

$$\varepsilon_{ij} = \frac{1}{2\eta_0\varphi}\left[\sigma_{ij} - \frac{\sigma_{kk}}{3}\right] + \frac{1}{3\psi}\frac{1}{2\eta_0}\left(\frac{\sigma_{kk}}{3} - p_L\right)\delta_{ij}$$

Hence, $\eta = \eta_0\varphi$; $K = 2\eta_0\psi$ and $\sigma_s \equiv p_L$.

One can suggest the generalization for (1) and (2), based on known analogy between viscous and elastic stresses [ ] and taking into account

A majority of the existing papers connecting with the sintering is devoted to the compaction stage. Only few attempts have been made to develop the approach combining the compaction and sintering stages. For example, in the papers [19-21], sintering process is modeled within an idealized Representative Volume Element. The intrinsic deformation of both the solid phase and the melt phase are taken into consideration, whereby elastic deformation is ignored. The macroscopic properties of sintered material are found via computational homogenization of the representative element.

However, not uniform temperature field connecting with the conjugate heat exchange in real technological conditions, and chemical conversions which lead to composition and properties change together with porosity evolution (for example, for intermetallic systems [22,23]) are not taken into account in macroscopic continual models.

In this paper we study the stress-assisted sintering process for the case of intermetallic composite formation in the conditions of conjugate heat exchange.

2. Problem formulation

Problem formulation corresponds to Fig.1.

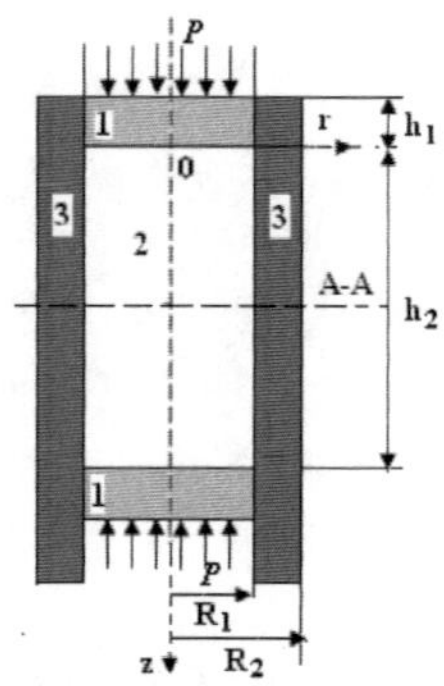

Figure 1. Illustration to the problem formulation

and is based on following assumptions: the wall friction absents; the flow is one-dimensional; the powder mixture flow is similar to viscous fluid; the thermal problem is two dimensional due to conjugate heat exchange; the coordinate system is cylindrical; the heating is carried out by Joule heat; the porosity changes in accordance with kinetical equation; the chemical heat release is possible and corresponds to the simplified reaction scheme:

$$Ti + Ni \rightarrow TiNi$$

$$2Ti + Ni \rightarrow Ti_2Ni$$

$$Ti + 3Ni \rightarrow TiNi_3$$

$$TiNi + 2Ni \rightarrow TiNi_3$$

This gives the features of the model in comparison with [24, 25].

The sub problem

$$\sigma = \left(\zeta + \frac{4}{3}\mu_V \right) \frac{dV}{dz} \tag{3}$$

$$\rho\rho_s \left(\frac{\partial V}{\partial t} + V\frac{\partial V}{\partial z} \right) = \frac{\partial \sigma}{\partial z} \tag{4}$$

$$z = \xi(t):\ -p(t) = \sigma = \left(\zeta + \frac{4}{3}\mu_V \right) \frac{dV}{dz} \tag{5}$$

$$z = \frac{h_2}{2}:\ V = 0;\ \sigma = 0, V = 0, t = 0 \tag{6}$$

describes the flow in the region with moving boundaries. Here $p(t)$ is the given law of external load change; ρ - is relative density, connecting with the porosity, $\rho = 1 - \theta$; $\sigma = \sigma_{zz}$ is stress tensor component in the loading direction; V is the component of velocity vector in this direction.

Porosity changes in accordance with the equation [17]

$$\frac{d\theta}{dt} = k(1-\theta)\frac{\partial V}{\partial z}. \tag{7}$$

In general case, coefficient k can depend on pressure, temperature and other parameters

$$k = k(P, T, Y_k),$$

where $P = \frac{1}{3}\sigma_{kk}$, and Y_k are the species concentrations ($k = 1,2,3,4,5$ for $Ti, Ni, TiNi, Ti_2Ni, TiNi_3$).

In the powder medium, the processes of heat conductivity, structure formation and chemical transformations take a place. Due to heat losses to the walls from reacting mixture, thermal processes are two-dimensional. Thermal conductivity equation for the reacting mixture contains the various volume heat sources:

$$c_2\rho_2\rho_s\left(\frac{\partial T_2}{\partial t} + V\frac{\partial T_2}{\partial z}\right) = \frac{\partial}{\partial z}\left(\lambda_2\frac{\partial T_2}{\partial z}\right) + \frac{1}{r}\frac{\partial}{\partial r}\left(\lambda_2 r\frac{\partial T_2}{\partial r}\right) + W_{ch} + W_H + \sigma\frac{\partial V}{\partial z}. \tag{8}$$

Here term W_H describe the electrical heating

$$W_H = R_e I^2(t),$$

where $I(t)$ is the current, R_e is effective electrical resistance depending on temperature and medium structure. This claims a special investigation. In this paper, we believe that dependence of resistance on time is known from experiment [26]. The last term in (8) corresponds to viscous dissipation. Stress and velocity fields follow from mechanical sub problem. Chemical heat release includes the heat of all reactions.

Corresponding to the reaction scheme, we have five kinetical equations:

$$\frac{dY_k}{dt} = \sum_{i=1}^{4}\omega_i\nu_{ki}, \quad k = 1,2,3,4,5 \tag{9}$$

where

$$\omega_i = z_i\varphi_i(Y_j)exp\left(-\frac{E_k}{RT_4}\right) \equiv \omega_i(z, r, t).$$

It is enough for numerical solution only four equations, because $Y_1 + Y_2 + Y_3 + Y_4 + Y_5 = 1$.

For four reactions we have

$$\omega_1 = Y_1Y_2, \ \omega_2 = Y_1^2Y_2, \ \omega_3 = Y_1Y_2^3; \ \omega_4 = Y_3Y_2^2.$$

The values z_i can take into account the reaction retardation by summary product, similarly to [27, 28].

Hence, for chemical heat release we shall find

$$W_{ch} = \sum_{i=1}^{4} Q_i\omega_i, \quad Q_i = -\sum_{k=1}^{5} h_k\nu_{ki}.$$

Parameters for kinetical equations, heats for the reactions were found based on thermodynamical data similarly to [29].

Usual thermal conductivity equations take a place in the areas 1 and 3 (Fig.1):

$$c_k\rho_k\frac{\partial T_k}{\partial t} = \frac{\partial}{\partial z}\left(\lambda_k\frac{\partial T_k}{\partial z}\right) + \frac{1}{r}\frac{\partial}{\partial r}\left(\lambda_k r\frac{\partial T_k}{\partial r}\right), \quad k = 1,3. \tag{10}$$

Boundary conditions between materials also correspond to Fig.1.

We believe that ideal thermal contact exists in interfaces «1-2»; «1-3» and «2-3». The symmetry conditions are assumed in the axis $r = 0$ and in the plane $z = \frac{h_2}{2}$ (cross-section A-A).

Heat exchange with environment is carried out by radiation, for example

$$z = -h_1 + \xi(t): \ \lambda_1 \frac{\partial T_1}{\partial z} = -\sigma_0 \varepsilon_1 \left(T_1^4 - T_0^4\right);$$

$$r = R_3: \ \lambda_3 \frac{\partial T_3}{\partial z} = -\sigma \varepsilon_3 \left(T_3^4 - T_0^4\right).$$

At the initial time moment we have:

$$T = T_0, \sigma = 0, V = 0, \eta = \eta_0(z), \ \xi = 0;$$

$$Y_1 = Y_{10}; \ Y_2 = Y_{20}; \ Y_3 = 0; \ Y_4 = 0.$$

The position of mobile coordinate $\xi(t)$, connecting with the plunger movement, are calculated during numerical solution of the problem. Numerical algorithm of mechanical part of the problem and thermal conductivity problem in the area with variable size are described in [24-26]. Kinetical equations were solved with Euler method.

3. Problem formulation

This model allows investigating the evolution of the temperature, velocity, stresses, density, porosity and composition during sintering assisted by external mechanical loading. We can vary the size of the reagent, the initial density distribution, initial composition of the powder mixture, limiting stages for the reacting scheme, mechanical load and times, when the heating rate changes and external load acts. As a result we will obtain various composition of the final material. These calculations can be classified as numerical experiment.

Because the comparison of porosity evolution with experiment for chemically inactive system were been demonstrated in [25, 26], here we demonstrate only the possibilities of the model that appear when chemical reactions are taken into account.

The figures 2 illustrate the composition change for powder mixture with the initial state $Y_1 = 0.5;$ $Y_2 = 0.5$. The concentration surfaces correspond to times $t = 1. - 3{,}5$ c; 2 - 4 c; 3 - 4,5 c; 4 - 5 sec.

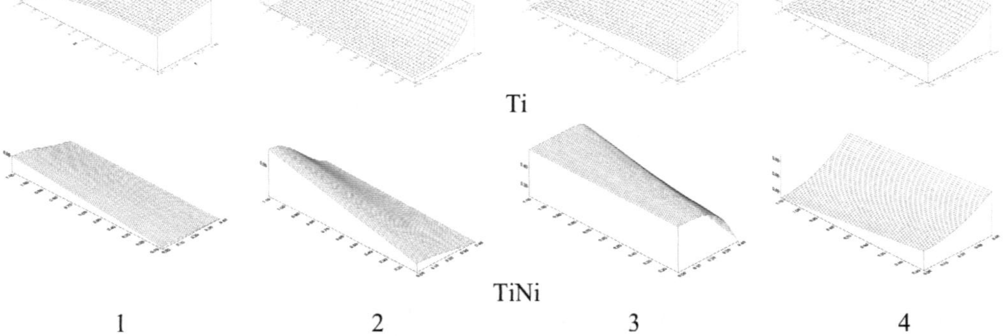

Figure 2. Evolution of the Ti and TiNi in time and in space.

For used here parameters, the chemical reactions proceed quite not uniformly, and concentrations values Due to the conjugate heat exchange, heat losses to the walls retard the reaction and temperature growth. Chemical reaction rates change with time not monotonically. In this case it is not possible present the concentrations in the same scale, and the pictures have only qualitative character. After the sintering, the material contains basically $TiNi_3$ and Ti_2Ni. Other phases stay because the conditions were irreversible and temperature field was not uniform. .

The average composition in time is shown in the Fig.3. We see really that TiNi concentration growths firstly, and then decreases. Additionally note that stresses in the material rises when porosity and composition change. Then, during cooling, stresses relax because external load does not apply.

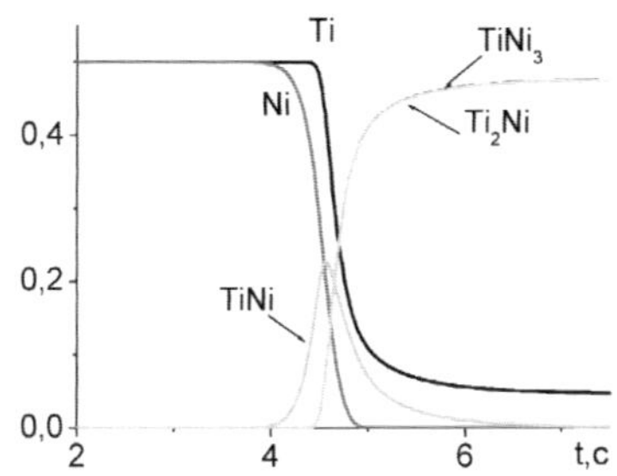

Figure 3. Average composition of the material in time during reactive sintering

Conclusion

Thus, the model of sintering process assisted by mechanical loading in closed volume was suggested. The stages of chemical conversion are taken into account. It was demonstrated that the composition of the final product can be non uniform due to irreversible conditions with not uniform temperature field. The model can predict the composition change with technology parameters variation. However the stresses in the reaction zone can appear due to temperature gradient and composition change. In this case the mechanical sub problem will two-dimensional, and one can expect new features appearance in the process course.

Acknowledgments

This work was supported by the Russian Science Foundation, grant No.14-08-90037.

References

[1] Sintering applications, Edited by Burcu Ertuğ / Published by In Tech (Janeza Trdine 9, 51000 Rijeka, Croatia ISBN 978-953-51-0974-7). 2013. - P.319-324
[2] Andruscshik L.O., Spak A.P., Oshkaderov S.P.: *Nanosystems, Nanomaterials, Nanotechnologies* 2008, V. 6, № **1**, P. 153—215
[3] Suk Hwan Chung, Young-Sam Kwon, Seong Jin Park, Randall M. German: *ASM Handbook*, Volume 22B, Metals Process Simulation D.U. Furrer and S.L. Semiatin, editors. 2010. – P.323-334
[4] Nichols F.A. and Mullins W.W.: *J. Appl. Phys.*, V. **36**, 1965, p 1826–1835
[5] Wilkinson D. S., Ashby M. F.: *Sintering and Catalysis*. Volume 10 of the series Materials Science Research. 1975. pp 473-492
[6] Gutmanas E., Rabinkin E., Roitberg M.: *Scripta Metallurgica*, 1979, V. **13**, Issue 1. – P. 11-15
[7] Dorri B., Kadambi V., Staub F. W., Lynch D. R.: *International Journal of Numerical Methods in Engineering*, 1987, V. **24**, P. 47-57
[8] Wilkinson D.S.: *Journal of Glaciology*, 1988, V. **34**, No. 116. – P.40-45

[9] Olevsky E. and Skorohod V.: *IPMS UNAS*, 1988 . – P. 97–103
[10] Hwang K. S., German R. M., Lenel F. V.: *Powder Metallurgy International*, 1991, V. **23**, No. – 2, P. 86-91
[11] Zhou H., Derby J. J.: *Journal of the American Ceramic Society*, 1998, V. **81**, – P. 533-540
[12] Olevsky E. A.: *Materials Science and Engineering,* 1998, Vol. **R23**, –.P.41-100
[13] Wakaia F., Aldinger F.: *Science and Technology of Advanced Materials*, 2004, **V.5**. –P. 521–525
[14] Tesfaye Tadesse Molla, Henrik Lund Frandsen, Rasmus Bjørk, De Wei Ni, Eugene Olevsky: *Journal of the European Ceramic Society*, 2013, **V.33** – P.1297–1305
[15] Skorokhod V. V.: *Powder Metallurgy and Metal Ceramics*, 2015, **V. 53**, No 9-10, P.529
[16] Skorokhod V.V.: *Powder Metallurgy and Metal Ceramics*, 1999, V. **38**, No. 7-8. – P. 350-357
[17] Skorokhod V.V. Rheology basis of sintering theory (in Russian) / Kiev: Naukova dumka. 1976. – 152 pp.
[18] Seong Jin Park, Suk Hwan Chung, John L. Johnson and Randall M. German: *Materials Transactions (Special Issue on Advances in Computational Materials Science and Engineering* IV), Vol. **47**, No. 11 (2006) pp. 2745 -2752
[19] Ohman M., Runesson K., Larsson F.: *Technische mechanik*, 2012, V.**32**, – P. 463 – 483
[20] Öhman M., Larsson F., Runesson K.: *Computer Methods in Applied Mechanics and Engineering*, 2013, V. **266**. – P. 219–228
[21] Öhman M., Runesson K., and Larsson F.: *Advanced Modelling and Simulation in Engineering Sciences*, 2015. **2:1** - 29 pp (Springer Open Journal)
[22] Ergin N.and Ozdemir O.: *Acta Physica Polonica* A , 2013, V. **123**, No. 2. – P 248-249
[23] Yoruk G., Ozdemir O.: *Intermetallics*, 2012, V.**25**. – P. 60-65
[24] Sorokova S. N., Knyazeva A. G., Pobol A. I., and Goranskyi G.G.: *Advanced Materials Research*, 2014, V. **1040** – P. 495-499
[25] Sorokova S. N., Knyazeva A. G., Pobol A. I., and Goranskyi G.G.: *Applied Mechanics and Materials*, 2015, V. **756**. – P. 534-539
[26] Sorokova S. N., Knyazeva A. G., Pobol A. I., and Goranskyi G.G.: *Chimicheskaya fisica and mesoscopyia chemical physics and mesoscopy*, 2015. V. **17**. № 2. - P. 239-252
[27] Knyazeva A.G., Sorokova S.N.: *Bulletin of the Tomsk Polytechnic University*, 2009, V. **314**, № 2. - P. 96-101
[28] Sorokova S.N., Knyazeva A.G: *Theoretical Foundations of Chemical Engineering*. – 2010. - V.**44**. - № 2. - P. 172–185
[29] Kryukova O.N., Knyazeva A.G:. *Russian Physics Journal*. – V. **57** – №9/3. – 2014. P. 113–117

Characterisation of the wall-slip during extrusion of heavy-clay products

I Kocserha[1], A L Gömze[1], S Kulkov[2], E Kalatur[2], S P Buyakova[2], R Géber[1], A Y Buzimov[2]

[1] Institute of Ceramic and Polymer Engineering, Faculty of Materials Science, University of Miskolc
[2] Tomsk State University, Institute of Strength Physics and Materials Science SB RAS and Tomsk Polytechnik University

E-mail: istvan.kocserha@uni-miskolc.hu

Abstract. During extrusion through the extrusion die, heavy-clay compounds are usually show plug flow with extensive slip at the wall of the die. In this study, the viscosity and the thickness of the slip layer were investigated. For the examination a brick-clay from Malyi (Hungary) deposit was applied as a raw material. The clay was characterised by XRPD, BET, SEM and granulometry. As the slip layer consists of suspension of the fine clay fraction so the clay minerals content of the clay (d<2μm) was separated by the help of sedimentation. The viscosity of suspension with different water content was measured by means of rotational viscosimeter. The thickness of the slip layer was calculated from the measured viscosity and other data obtained from an earlier study with capillary rheometer. The calculated thickness value showed a tendency to reach a limit value by increasing the extrusion speed.

1. Introduction

The production of the traditional fired brick product demands the preparation and shaping of the largest volume of plastic ceramic pastes. Additive materials are often mixed into the clay based compounds in order to change an unbeneficial behaviour of the clay (i.g. too high plasticity) or to improve the properties of the fired products [**1-5**]. Results from rheological and tribological studies established that not only the clay but the clay-compounds can also be described as Bingham-fluid at low shear rate [**6,7,8**]. However it was also observed that a thin layer built up from clay-suspension between the forming die and the clay body [**9**]. Accordingly the extrusion behaviour of the these materials is influenced by the slip layer beside the shear stress. The aim of this study is the examination of the layer and the calculation of the layer thickness based on former studies.

2. Background

During extrusion the clay-water suspension fills up the void space between clay particles and owing to the rapidly increasing pressure it flows in the direction of the die wall [**6, 10**]. The rigid particles of the clay behaves as a filter and it retain the rougher fraction of the suspension so a thin layer builds up on the surface of the extrudate (or on the die wall if we approach from the other side) which has its own viscosity and thickness. This two behaviours are the function of more parameters such as the composition of the clay; the extrusion pressure and surface roughness [**11**]. In the extrusion process the layer affects the volume of the slip and share fraction of the extruded body. A typical flow of a clay or a clay based compounds can be seen on the figure 1. The liquid layer at the wall developed not only by clay-water systems but the at the production of technical ceramic pastes [**12-14**], drugs [**15**] products of the food industry [**16**], polymers [**17**] or at the flow of foams, cements through a pipe [**18**]. For the examination of the layer beside the rheological methods technique based on magnetic resonance are also applied [**19**]. The determination of the rheological parameters and wall slip based on shaping technology demands a sophisticated method like capillary rheometry [**11**]. From such a measurement the true flow curve can be obtained with the description of the flow behaviour.

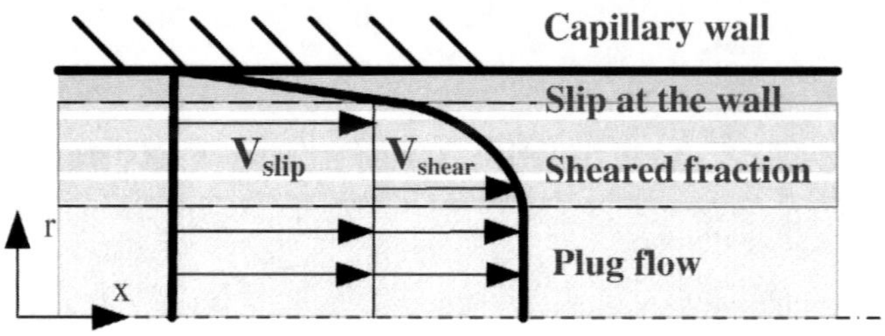

Figure 1. Generalized capillary flow profiles of the clay or clay-based compounds

The development of the slip layer is the function of the reological behaviour of the liquid phase, particle morphology and size distribution of the clay or ceramic pastes [**15**]. For their alumina paste, Gracyk and Gleissle [**20, 21**] found that although the flow through the capillary was entirely plug flow, there was a dependency of the wall shear stress upon the diameter of the capillary. As thickness of the slip layer compared to the diameter of the capillary is negligible, it was supposed that the slipping speed (v_{slip}) and the thickness of the layer (s) is linearly proportional as follows:

$$\frac{v_{slip}}{s} \approx \cdot\dot{\gamma}_s \quad (1) \qquad \text{and} \qquad \frac{\tau}{\dot{\gamma}_s} = \cdot\eta_s \quad (2)$$

where γ_s is the shear rate in the layer, η_s is the viscosity of the layer and τ is the wall shear stress.

Equating the two expression Eq. (1) and Eq. (2), a quantity was introduced: reference slip layer thickness: s_r (Eq.3) Higher s_r value means an increase in s and decreases in η_s.

$$s_r = \frac{s}{\eta_s} = \frac{v_{slip}}{\tau} \quad (3)$$

s_r is easily calculable as v_{slip} and τ are obtainable from a measurement with a capillary rheometer. The method was applied by Khan et all [**22**] for the measurement of alumina paste.

According to Eq. 2 the real thickness of the slip layer (s) remained unknown. Supposing that the value η_s from the Eq.3 is measurable then the real layer thickness becomes calculable because the shear stress and the slip velocity is available [**8**]. Based on former experiments [**8**, **10**] the following conditions were assumed:

- the slip layer build up from a suspension of water and the d<2 μm component of the clay;
- the moisture content of the layer changes according to the clay mixture;
- the thickness of the layer increases as the pressure increases;
- slipping speed (v_{slip}) and the thickness of the layer (s) is linearly proportional.

Some related studies supposed some μm for thickness of the layer. Applying Eq. (3) and supposing 1 μm and 5 μm for the thickness of the slip layer, the calculation with v_{slip} = 100mm/s data from measurements with capillary rheometer [**8**] resulted the shear rate in the layer:

$s_1 = 1$ μm ⇨ $\dot{\gamma}_{s1}$ = 100/0,001= 100000 1/s

$s_2 = 5$ μm $\dot{\gamma}_{s2}$ = 100/0,005= 20000 1/s.

Consequently the shear rate can be as high as 100000 1/s in the layer assuming Newtonian fluid. Higher shear rates usually measure with capillary rheometer but it is non usable for clay-suspension. Applying rotational viscometer the measuring limit can be elevated over 1000 1/s and the behaviour of the suspension became observable.

3. Materials and methods

An ordinary brick clay with high plasticity (Atterberg number: 21.5) from the Mályi clay deposit (Hungary) was chosen as the raw material for the examinations. Before preparation of the material, it was characterised by the XPRD, BET, granulometry and morphology (Table 1.). The mineral composition of the yellow clay was analysed by XRPD applying a Bruker D8 Advance X-ray diffractometer. The quantitative result was calculated by Rietveld profile refinement. The yellow clay consisted of 35.2% clay minerals. Roentgen amorphous materials could be considered to be plastic components in the clay because these materials are very fine graded aluminium- and iron-oxide-hydroxides. Oligoclase and microcline represented the feldspar content in an amount of 9.8%. In addition to the 31.1% free quartz, there were a small amount of dolomite and calcite. Because of the clay minerals and relatively high amount of roentgen amorphous materials, the clay body showed a highly plastic and sticking behaviour. The mica content was 15.3%. The particle size distribution was determined by Horiba LA950 laser granulometer and the specific surface was measured on a Micrometrics Tristar 3000 instrument. In order to get the finest particle of the clay, the particles was separated by sedimentation into seven fractions including: $5\mu m<d<15\mu m$; $2\mu m<d<5\mu m$; $d<2\mu m$. Analysing morphology by means of a Hitachi TM-1000 tabletop SEM found the particles of the clay have mainly anisotropic shape. The morphology of two fraction can be seen on the Fig 2.

Table 1. Mineralogical composition (wt%), grain size distribution and specific surface area BET surface area of the clay

Oligoclase	Microcline	Calcite	Dolomite	Quartz	Mica	Illite	Mont-morillonite	Kaolinite	XRD amorphous
5,4%	4,4%	1,2%	0,9%	31,1%	15,3 %	23,1 %	7,9%	4,2%	6,4%
Granulometry (μm)	$d_{0,1}$	$d_{0,5}$	$d_{0,9}$	BET in fraction (μm)		d<2	2<d<5	5<d<15	15<d<45
	1,9	11,9	50,8	(m^2/g)		~37	~48	~34	~15

Figure 2. Morphology of the clay fractions

According to the above declared conditions, two clay suspension with different moisture content was prepared. For the examination, the fraction of $d<2\mu m$ was applied and the relative moisture content was adjusted to w=28% and w=50% by adding water to the suspension. The mixture with lower moisture content was highly plastic while the other was relatively fluent. The rheological tests were carried out on a rotational viscometer (Fig.3. Haake RS80; head: CP 2Ti with 25mm diameter). The gap was set to 0,1mm and the ambient temperature was 23°C. During measurement slow ramp time was applied up to the maximal shear rate of 1600 1/s.

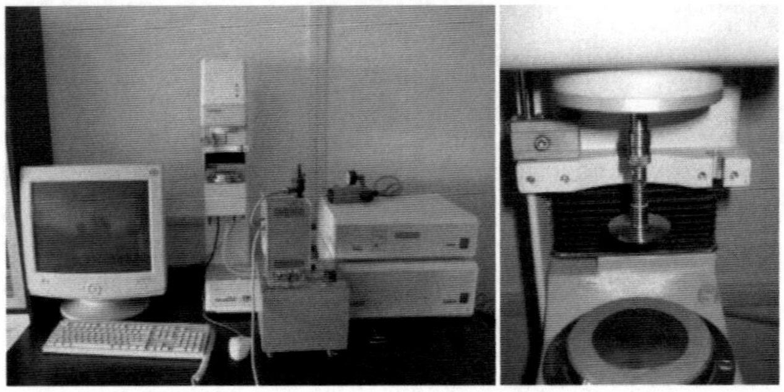

Figure 3. Viscometer RS80 and the CP 2Ti 25mm measure head

4. Results and Discussion

The results of the measurements were the flow curves of the suspensions. Each test was carried out three times and an average of viscosity value was calculated. The shear stress-shear rate and viscosity curves are presented on the Fig. 4. The Fig 5 represents the average stress and viscosity values.

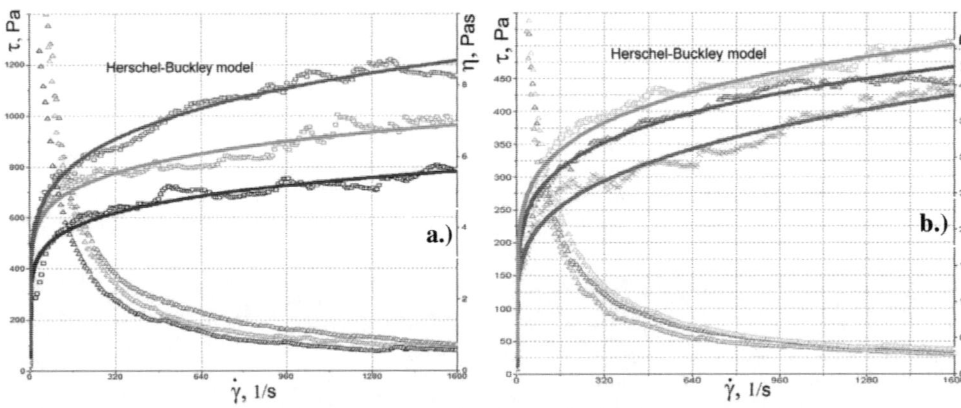

Figure 4. Flow curves for fraction below 2 µm, moisture content w=28% (a.) and w=50%-os (b.)

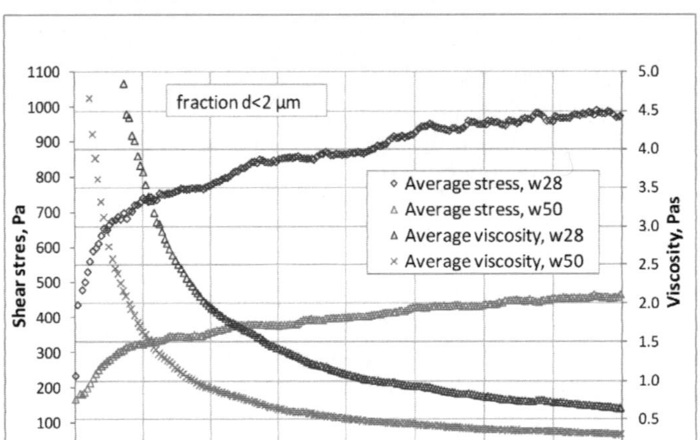

Figure 5. Averaged flow curves for fraction below 2 µm, moisture content w=28% (a.) and w=50%-os (b.)

The viscosity results are summarized in the Table 2. It is visible in Fig 5 that as the shear rate increases the viscosity of the suspension tends to reach a limit value. The viscosity of the suspension was below 1 Pas for both moisture content. It follows from this that the viscosity of the slip layer can be changed between the viscosity of the pure liquid phase (viscosity of water is 0.001 Pas) and around 0,64 Pas. Supposing that the non-compressibility of the suspension and take the viscosity value as constant at high shear rates, it can be extended over 10000 1/s. Results also show that the clay fraction (d<2μm) followed the Herschel-Buckley model (Eq.4) for both water content. The model parameter and coefficient of determination are summarized in table 3.

$$\tau = \tau_0 + K\dot{\gamma}^n \qquad (4)$$

Table 2. Viscosity of clay-water suspension

Fraction	Water content	Viscosity over 1000 1/s,
	w, %	ή, Pas
d < 2 μm	28	0,64
d < 2 μm	50	0,27

Table 3..Model parameters

	τ_0	K	n	r^2
	Pa	*Pas*		
w=28%	76,20	259,40	0,166	0,94
w=50%	37,74	96,22	0,179	0,96

Applying the viscosity value in Eq.3, the “s” layer thickness was calculated. The table 4 shows the values of the shear stress and the slip velocity which were the basis of the calculation. They were obtained from the analysis of the same yellow clay with capillary rheometer and they were presented in other study [**8**]. The ram speed affected the slip velocity and the shear stress for all applied capillary diameter (d=3; 4 and 5mm). The Fig.6 illustrates the calculated slip layer thickness as a function of extrusion (ram) speed for three different capillary diameter. During shaping, the calculated thickness of the slip layer resulted in some micrometer.

Table 4. The shear stress and the slip velocity data for different capillary

τ	$v_{cs\ d3}$	$v_{cs\ d4}$	$v_{cs\ d5}$
Pa	mm/s	mm/s	mm/s
0,27	25,32	4,69	1,79
0,29	35,57	9,23	4,28
0,31	46,42	13,55	6,74
0,33	57,40	17,82	9,18
0,35	68,41	22,06	11,61
0,37	79,45	26,28	14,04
0,39	90,50	30,50	16,47
0,41	101,56	34,71	18,90

During extrusion of clay and clay-based compounds the layer behaves similar to this. Fig. 6 shows that the layer was also thicker as the extrusion speed increased. As extrusion speed and the extrusion pressure increase parallel it facilitates the flow of the suspension toward the die wall. The thickness is tends to reach a limit and this limit value is the function of the mineral and water content and also the extrusion pressure and surface roughness. Former studies proved **[6, 7, 8]** that clay and the clay with additives (clay compounds with pore former and opening agents) behaves similar i.e. Bingham model. However the thickness of the layer was calculated point to point from the measurement data and it showed non-linear tendency.

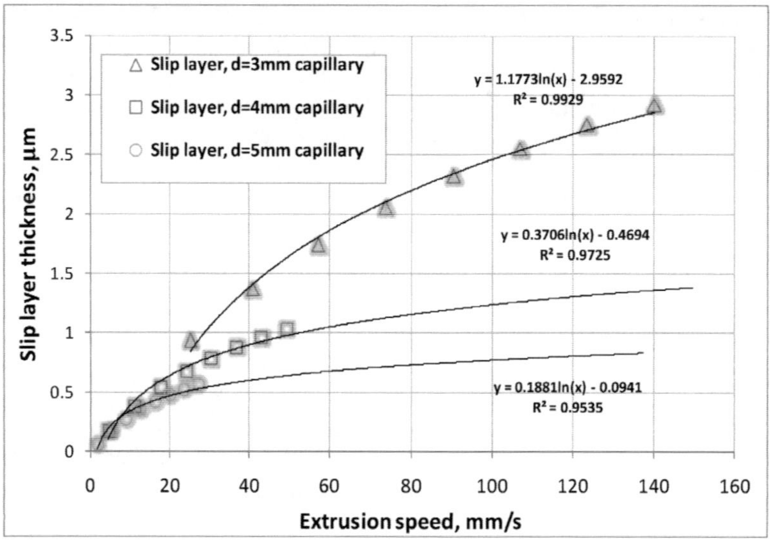

Figure 6. The change of thickness of the slip layer as the function of extrusion speed

Summary
In this study the slip layer was examined which is usually built on the surface of a heavy-clay products during shaping. The layer is composed of suspension of the finest clay particles and water. The basic condition for formation of the layer was defined and it was characterised by means of capillary and rotational rheometer. The viscosity of the clay fraction (d<2μm) was determined and based on the results of earlier tests of the clay the layer thickness was calculated..

Acknowledgment
The authors would like to acknowledge the support by Malyi Brickworks Ltd., and the TÁMOP-4.2.1.B-10/2/KONV-2010-0001 project.

References
[1] V. Ducman, T. Kopar (2007), *Materials and technology* **41/6** pp. 289–293
[2] K. Junge (2000) *Ziegelindustrie International* **12**, pp. 25-39
[3] Kutassy L. (1982).*Építőanyag-JSBCM* **34**. 9, pp. 332-335
[4] Z. Xuanye (2003) *Ziegelindustrie International* **4** pp. 22-27
[5] M. Sveda (2000) *Ziegelindustrie International* **11** pp.29-35
[6] W. Hoffmann. H. Berger (2001) *Ziegelindustrie International* **4** pp. 24-34
[7] F. Laenger: Designing an extruder with allowance for the properties of ceramic extrusion compounds. Part 1-5. *German Ceramic Society Report*, vol 69/3-8, 1992
[8] Kocserha István (2012) *Anyagmérnöki Tudományok, A Miskolci Egyetem közleménye* ISSN 2063-6784, **37/**1 pp. 199-210
[9] M. Mooney (1931) *Journal of Rheology* **30** pp.210–222.
[10] Kocserha, L.A. Gömze (2010) *Applied Clay Science*, Elsevier **48/3** pp. 425-430
[11] J.J. Benbow, J. Bridgwater: Paste Flow and Extrusion. *Clarendon Press*, 1993
[12] P.J. Martin, D.I. Wilson, P.E. Bonnett (2004) *Journal of the European Ceramic Society* **24** pp.3155–3168
[13] R.N. Das, C.D. Madhusoodana, K. Okada (2002) *Journal of the European Ceramic Society* **22** pp.2893–2900
[14] J.J. Benbow, S. Blackburn, H. Mills (1998) *Journal of Materials Science* **33** pp.5827-5833
[15] S.L. Rough, J. Bridgwater, D.I. Wilson (2000) *International Journal of Pharmaceutics* **204** pp.117-126
[16] C.I. Hicks, H. See (2010) *Rheologica Acta* **49** pp.719–732
[17] Y.C. Lam, Z.Y. Wang, X. Chen, S.C. Joshi (2007) *Powder Technology* **177** pp.162–169
[18] R. Alfani, N. Grizzuti, G.L. Guerrini, G. Lezzi (2007) *Rheologica Acta* **46** pp.703–709

IOP Conf. Series: Journal of Physics: Conf. Series **790** (2017) 012013 doi:10.1088/1742-6596/790/1/012013

[19] J. Götz, W. Kreibich, M. Peciar (2002) *Rheologica Acta* **41**(1-2) pp.134–143
[20] J. Graczyk, W. Gleissle: Rheology of extrudable alumina paste. In: C. Galassi, Editor, Euro–Ceramics IV, Vol. 2, Gruppo Editoriale Faenza Editrice S.p.A,, Italy (1995) pp.63-70
[21] J. Graczyk, W. Gleissle: Rheology of extrudable alumina paste. In: C. Galassi, Editor, Euro–Ceramics IV, Vol. 2, Gruppo Editoriale Faenza Editrice S.p.A,, Italy 1995 pp.63-70.
[22] A.U. Khan, B.J. Briscoe, P.F. Luckham (2002) *Journal of the European Ceramic Society* **21** pp.2893–2900

IOP Conf. Series: Journal of Physics: Conf. Series **790** (2017) 012014 doi:10.1088/1742-6596/790/1/012014

Mesoscopic Rhelogical Model for Polymeric Media Flows

K Koshelev, A Kuznetcov, D Merzlikina, G Pyshnograi, I Pyshnograi and M Y Tolstykh

[1]Department of Mathematical Analysis and Applyed Mathematics, Altai State Pedagogical University, Barnaul, Russia
[2]Institute for Water and Environmental Problems, Barnaul, Russia
[3]Department of Mathematics, Altai State Technical University, Barnaul, Russia

E-mail: pyshnograi@mail.ru

Abstract. The paper compares hydrodynamic properties of three-dimensional flows of polymer melts. A modified Vinogradov and Pokrovskii rheological model is used for the mathematical description of nonlinear viscoelastic fluid flows in a plane-parallel channel with a sudden convergence. Discrete analogs for partial differential equations were obtained via the control volume method separating physical processes. The numerical implementation is carried out using the GPU-based parallel computing technology. Velocity and pressure fields have been calculated for two samples of polyethylene melts and the circulating flow at the entrance of the slit channel is noticeable. It is shown that the size of the vortex zone depends significantly on melt rheology.

1. Introduction

Mathematical study polymer fluids flows involves two interrelated phases. Firstly, it is selection and justification of the rheological model. Secondly, it is the algorithmic implementation of the obtained equations and the numerical experiment. Regarding the first step, there are plenty of rheological models of various complexity [1,2,5-12]). All these models describe the main effects observed in viscometric experiments: gradient dependence of shear and elongation viscosity, the first and second normal stress difference and non-monotonic establishment of time dependence of shear and elongation stresses.

All viscometric flows have a fairly simple structure, since the velocity gradient tensor is known, and all the models show similar accuracy in the description of these flows. Therefore, the issue of adequacy of the rheological model is solved via calculation of two and three-dimensional flows in areas with complex geometry.

In this paper, we apply a modified Vinogradov-Pokrovskii rheological model to solve the problem of mathematical modeling of three-dimensional nonlinear viscoelastic fluid flows in a plane-parallel channel with a sudden contraction. The method of control volume with the division of physical processes is used upon obtaining discrete analogs. The numerical implementation is carried out using GPU-based parallel computing technology CUDA.

2. The mathematical model

It is well-known that linear polymer melts are non-linear viscoelastic media. Nowadays the equations that take into account essential features of polymeric liquids behavior are commonly used to describe their flows [5-9]. The priority in choosing a mathematical model should be given to models which to some extent take into account the structure of polymer molecules. It is quite complicated so the most popular models are those based on the mesoscopic approach. In this case, the behavior of polymer macromolecules is replaced by the behavior of one or more relaxation oscillators, and the transition to macroscopic description is carried out via methods of statistical mechanics [5,8,9,12,13]. One such model is a modified Vinogradov and Pokrovskii rheological model [9]

$$\sigma_{ik} = -\hat{p}\delta_{ik} + 3\frac{\eta_0}{\tau_0}a_{ik},$$
$$\frac{d}{dt}a_{ik} - \nu_{ij}a_{jk} - \nu_{kj}a_{ji} + \frac{1+(\kappa-\beta)I}{\tau_0}a_{ik} = \frac{2}{3}\gamma_{ik} - \frac{3\beta}{\tau_0}a_{ij}a_{jk}. \quad (1)$$

Where a_{ik} – stress tensor of the polymer system; a_{ik} – dimensionless extra stress tensor; p – hydrostatic pressure; no – initial value of shear viscosity, t_0 – initial relaxation time; κ and – scalar coefficients of anisotropy taking into account shape and size of macromolecular coil; v_{ik} – velocity gradient tensor; y_{ik} – symmetrized velocity gradient tensor; I – trace of tensor a_{ik}.

In paper [10] the two- and three-dimensional flows under the constant pressure gradient are considered in channels with a rectangular cross-section. The pressure field gradient has been known. In paper [15] an axisymmetric circular channel was used to model the input flows. However, an attempt to use the numerical method [15] for calculating flat channels failed. For calculating real flows on the basis of this model the following equations of momentum conservation and mass should be added to (1)

$$\rho\Big(\frac{\partial}{\partial t}v_i + v_k\frac{\partial}{\partial x_k}v_i\Big) = \frac{\partial}{\partial x_k}\sigma_{ik}, \qquad \frac{\partial}{\partial x_k}v_k = 0. \quad (2)$$

Here p - the density of the polymer; v_i - i-th component of the velocity vector.

The system of equations (1, 2) is closed with respect to variables a_{ik}, v_i, p. To solve the problem initial and boundary conditions are required. We first discuss the computational domain to establish these conditions.

The computational domain involves two 3D boxes. The first reservoir is channel 14 x 14(mm) with a square section, the second a slotted channel with a cross section of 14 x 1(mm). Lengths of the parallelepipeds are selected sufficiently large -80 < x < 80(mm) to eliminate the influence both of the input into the tank and output from the slotted channel. It is expected that within this area the flow will be three-dimensional, i.e. different directions in different axes. In this case, the main flow is directed along axis *Ox;* whereas along axis Oy a strong flow compression will occur. This can cause vortices though along axis *Oz* changes will be immaterial. Therefore, we call axis *Ox* - the flow direction, axis *Oy* - the direction perpendicular to the flow, and axis *Oz* - the direction neutral to the flow.

The main boundary conditions are stick conditions on a solid surface for velocity: $v_i = 0$.

The boundary conditions for the dimensionless stresses are obtained by substituting these conditions in (1) and discarding the corresponding terms.

The boundary conditions $z = 0$ and $y = 0$ are used for terms of symmetry or vanishing of the corresponding partial derivatives. At the entrance to the reservoir at x = -80 for the velocity components we use expressions $v_2(-80,y,z) = v_3(-80,y,z) = 0$ and $v_1(-80,y,z) = 9V(z-7)^2(y-7)^2/5488$(mm/s), where V - volume flow rate.

Flows in a plane-parallel channel with a sudden contraction, are commonly referred to as converging flows or input flows. In order to calculate such flows via model (1, 2) it is necessary to determine the numerical values of the parameters for the rheological model: κ, P, no and To . To do this, we turn to the experimental data on the dependence of steady-state shear viscosity on the shear rate [18] where flows of two samples of polyethylene melts were applied - linear low density polyethylene (LLDPE) and low density polyethylene (LDPE). The model parameters were chosen to give the best agreement between the experimental and theoretical curves. For LLDPE sample we obtained no = 14500(Pa · s), T_0 = 0.2(s), and for the LDPE n_o = 18500(Pa · s), T_0 = 2(s). Anisotropy parameters P = 0,1 and K = 0,12 were the same. Density values were given in [18] and account for 918(kg/m^3) for LDPE and 926(kg/m^3) for LLDPE.

The numerical method is constructed for relatively easy implementation when creating a computer model using technologies GPU NVIDIA and CUDA. The finite-difference approach is employed to find the solution. For the difference approximation the method of control volume was used that does not cause difficulties in a relatively simple geometry of the domain.

3. Mathematical modeling for exit flows

To compare the results of calculations we used the experimental data [18] where flows of two samples of polyethylene melts were applied LLDPE and LDPE. The differences between these two examples is that LLDPE shear viscosity changes less than for LDPE, and LLDPE relaxation time is shorter than that of LDPE. We found out that there is an explicit secondary flow for LDPE flows in the corners of the flow channel, which is not observed for LLDPE. These vortices change their shape in the channel cross-sections parallel to the axis, which characterizes the three-dimensional nature of the flow field. Furthermore, it is noted that within the investigated vortex flow there is a helical stream, which is directed to the tank wall. Particular attention is paid to the distribution of velocity along the symmetry axis of the channel. It was found that LDPE maximum velocity is observed directly in the entrance portion of the channel slot. This effect does not appear for LLDPE under the same conditions.

All these effects are found during the numerical experiment. The calculations imply that the increase in speed in the slit of the channel happens due to the three-dimensional nature of the flow field and the stress caused by the increase in the flow of LDPE compared with a sample of LLDPE. The three-dimensional nature of LDPE flow is confirmed by the presence of the neutral component of velocity in the flow direction, which was not observed for LLDPE flow. This additional component of velocity results in increasing the volumetric flow along the centerline at the entrance to the slit channel.

Thus, the velocity, pressure and stress fields were calculated in a steady flow. Herewith circulating areas were found at the entrance of the slit channel. As seen from Figure 1 for LLDPE sample (T_0 = 0,2(s)), the reverse flow zone is negligible. For LDPE sample (T_0 = 2(s)) this zone, as seen from Figure 2 is much larger. Note that the value of viscous parameter n_o in these samples is approximately the same and the difference in their behavior should be attributed to time parameter T_0.

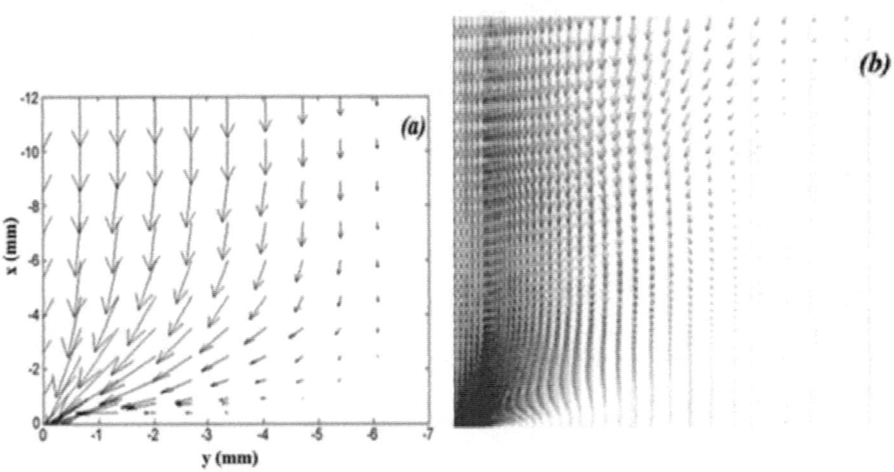

Figure 1. Comparison of experimental (a) and theoretical (b) LLDPE velocity vector fields in a slit entrance channel (V = 0, 20 [cm^3/s])

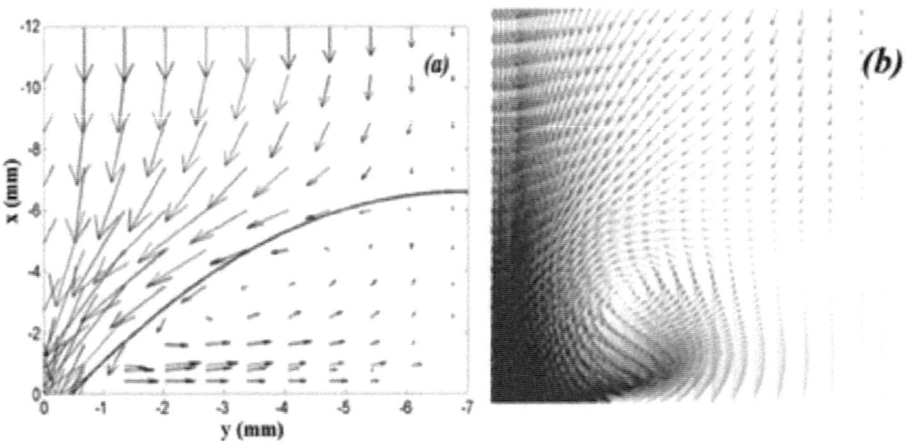

Figure 2. Comparison of experimental (a) and theoretical (b) LDPE velocity vector fields in a slit entrance channel (V = 0,20 [cm^3/s])

We note that the values of the Reynolds number calculated using the formula (Re = $\rho v_0 h/n_0$) for the considered currents in the slot channel do not exceed 0, 02. Such flows are fairly well understood in the classical Newtonian model, and eddy zones in this case are not observed.

Thus, converging flows for some polymer melts may exhibit a substantial three-dimensional picture that is shown by the presence of velocity components in the direction neutral to the flow. This should be taken into account in the design of the experiment, as there are methods that are not able to measure all the components of velocity.

Also one should mention the fact of long-term relaxation of the velocity profile in the gap. Calculations showed that the steady velocity profile is observed at a considerable distance from the entrance slit. This should be considered when conducting measurements in a narrow channel portion.

Previously, it was noted the presence of eddy currents in the corners of the flow channel, the size of which depends strongly on the temperature of the melt and on specific consumption.

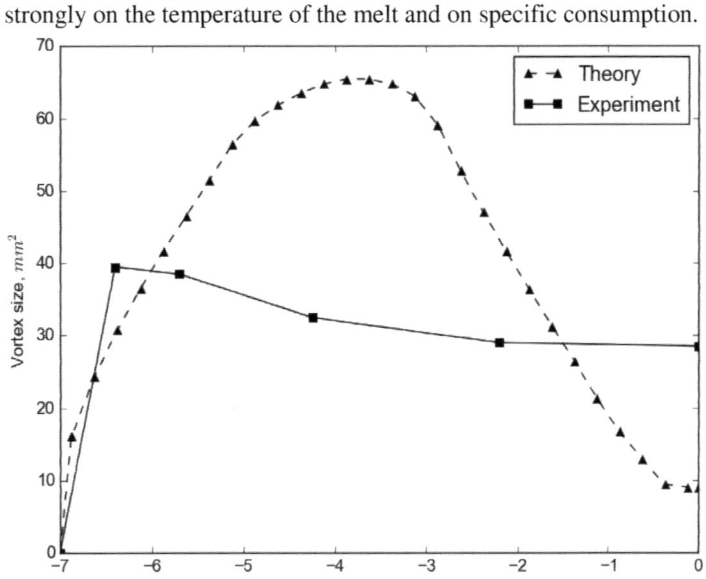

Figure 3. The size of the vortex in different sections

One of the essential characteristics of the test flow is the size of the vortex. We calculated the area of the vortex zones when flow section planes z = *const.* Comparing shown in Figure 3 the calculated and experimental vortex sizes depending on the distance from the axis of the channel may be concluded that an increase in the intensity of the vortex flows away from the axis of the channel. This fact can be explained by Weissenberg effect. From this comparison it can be concluded qualitative agreement of the experimental and the payment data. Note that in the calculations carried out for the Newtonian law of behavior (TO = 0; $P_0 = p_0 = 0$) demonstrate the absence of vortex zones, and calculations for viscoelastic Oldroyd-B ($P_o = p_0 = 0$) show low values of the vortex area and the lack of increase in the intensity of the vortex when you remove the cross section on the axis of the channel.

References

[1] Wilco M.H. Verbeeten, Gerrit W.M. Peters, Frank P.T. 2004 *J. Non-Newtonian Fluid Mech.* **117** 73-84.
[2] Mitsoulis E., Schwetz M, Mnstedt H. 2003 *J. Non-Newtonian Fluid Mech.* **111** 41-61.
[3] H. Munstedt, M. Schmidt, E. Wassner 2000 *J. Rheol.* **44** 413-427.
[4] Afonso A.M., Oliveira P.J., Pinho F.T., Alves M.A. 2011 *J. Fluid Mech.* **677** 272-304.
[5] McLeish T.C.B., Larson R.G. 1998 *J. Rheol.* **42** 81.
[6] Verbeeten W.M.H., Peters G.W.M., Baaijens F.P.T. 2001 *J. Rheol.* **45** 823.
[7] Leonov A.I., Prokunin A.N. 1994 *Chapman and Hall, New York. 1sted* 356-395.
[8] Pyshnograi G.V., Gusev A.S., Pokrovskii V.N. 2009 *J. Non-Newtonian Fluid Mech.* **163** 17-28.
[9] Pokrovskii V.N. 2010 *The Mesoscopic Theory of Polymer Dynamics. 2nd Edition.* (Berlin: Springer) 184.
[10] Kuznetsova Yu.L., Skul'skiy O.I., Pyshnogray G.V. 2010 *Computational Continuum Mechanics.* ***2*** 55-69 (in Russian).
[11] Altukhov Yu.A., Samoylov V.S., Pyshnograi I.G., Pyshnograi G.V. 2012 *Journal on Composite Mechanics* and Design. ***18*** *325-332.*
[12] Altukhov Yu.A., Pyshnograi G.V., Pokrovskii V.N. 2004 *J. Non-Newtonian Fluid Mech.* ***121*** 73-86.
[13] Pokrovskii V.N., Altukhov Y.A., Pyshnograi G.V. 1998 *J. Non-Newtonian Fluid Mech.* ***76*** 153-181.
[14] Altukhov Yu.A., Golovicheva I.E., Pyshnograi G.V. 2000 *Fluid Dynamics.* ***1*** 3.
[15] Altukhov Yu.A., Pyshnograi G.V. 2001 *Journal on Composite Mechanics and Design.* ***1*** 16-23.
[16] Koshelev K.B., Pyshnograi G.V., Tolstykh M.Yu. 2015 *Fluid Dynamics* **50** 315-321.
[17] Hertel D., Mnstedt H. 2008 *J. Non-Newtonian Fluid Mech.* ***152*** 73-81.
[18] Hertel D., Valette R., Mnstedt H. 2008 *J. Non-Newtonian Fluid Mech.* **153** 82-94.

IOP Conf. Series: Journal of Physics: Conf. Series **790** (2017) 012015 doi:10.1088/1742-6596/790/1/012015

Zirconia-Based Powders Produced by Plasma-Spray Pyrolisys and Properties of Sintered Ceramics

S N Kulkov[1], S Buyakova[1], L A Gömze[2]

[1]Institute of Strength Physics and Materials Science SB RAS, Tomsk, Russia
National research Tomsk State University, avenue Lenin, 36, 634050, Tomsk, Russia
[2]Department of Ceramics and Silicate Engineering, University of Miskolc, Miskolc, Hungary

E-mail: kulkov@ms.tsc.ru

Abstract. It have been studied zirconia-based powders and sintered ceramic. It was shown that in the porous structure of zirconia-based ceramics there is a critical value of porosity the material divides into two sub-systems, being variously deformable under external loading. It have been shown that *m*-phase in ZrO_2 is formed due to increase in the microdistortion level which destabilizes the nanocrystalline *t* phase. It has been found out the correlation between the sizes of crystallites and porosity, which associated with transition of the isolated porous structure to the continuous one and the porosity of 20%, corresponds to the first percolation threshold.

1. Introduction

Nanocrystalline powders obtained by plasma-chemistry methods are a promising material for production of zirconia-based ceramics. Powders of this sort exhibit a more uniform constituent element distribution over particles as compared to those prepared by conventional processes [1], and a larger amount of excess energy is stored in these powders than in their coarse-grained cousins. In this connection, an investigation of the structure and phase composition in the powders and sintered materials under consideration assumes particular importance, because it extends the possibilities to generate a necessary structure of the materials.

It is known that materials with a stochastic pore structure are attracting great interest because of their wide application as filters for fluid purification and separation, catalyst carriers, heat-insulating coatings, etc. [2]. Ceramic materials have high chemical and corrosion resistance and, therefore, have advantages over metals and high-molecular compounds when used in an aggressive environment or at elevated temperatures. Most ceramics are chemically inactive in biological media, so that associated porous products can be implanted into the human organism (bone tissue endoprotheses and medicinal preparation dispensers).

The synthesis of porous materials with desired property call for the detailed investigation of a correlation between the physics-mechanical properties of the material and its structure. This is especially true for ceramic materials, because porosity, as a rule, leads to a catastrophic decreasing of the strength. Porous ceramics are a heterogeneous material. Depending on its applications, the pore size and the porosity may vary in a very wide range. Therefore, their mechanical behavior is also of scientific interest.

Zirconia is among the materials that are promising for porous structures. It has high strength, fracture toughness, and corrosion resistance. Also, zirconia is stable in aggressive media and does not react with human organism tissues. This ceramic, containing a controllable pore concentration, retains strength properties due to transformation toughening [3].

The aim of this work is to study changes in the features of the structure and phase composition of nanocrystalline zirconia-based powders and the effect of pore morphology on the mechanical behavior of partially stabilized zirconia subjected to high compressive deformation.

2. Materials and Experimental Technique

We have studied powders produced by plasma spray synthesis of nitrate salts of Zr, Y, and Al in a high-frequency discharge. During the experiments, a drop of the aqueous solution used turned into a solid powder particle (within < 1s) that cooled down very quickly. The powders were of the following chemical composition: ZrO_2 + 3 mol % Y_2O_3 (ZrO_2 (3Y), called a binary system), and ZrO_2 (3Y) + 20 wt % Al_2O_3 (ZrO_2 (ZrO_2 (3Y) + 20 wt % Al_2O_3, called a ternary system).

X-ray analysis was performed by means of a DRON-UM1 diffractometer with filtered CuK_a or FeK_a radiation. Point-by-point measurements were taken with a step of 0.1 or 0.05^0. X-ray diffraction lines of the tetragonal phase were indexed on the basis of a face-centered tetragonal unit cell. The lattice parameters were determined from all reflections at angles lying in the range $20 < 2\theta < 120$. Overlapping x-ray diffraction peaks were separated by means of a computer code based on a minimum deviation of the total approximation profile from the experimental one. The phase abundances in the zirconia powder were calculated from the integrated line intensity ratio for the tetragonal (111) and monoclinic ($\bar{1}$ 11) and (111) phases, using a standard formula [6]. The phase composition was calculated to within 5 %.The lattice strains and coherently diffracting domain (CDD) size were found from x-ray line broadening (FWHM). The contributions associated with the effects of microdistortions and range of crystallite dimensions were separated by the approximation method, using the Cauchy-type functions. The physical broadening β served as a basis for construction of a plot on the ($\beta\cos\theta/\lambda$)–($\sin\theta/\lambda$) coordinates. The slope $4\langle\varepsilon\rangle$ and the quantity $1/D$ plotted and extrapolated on the ordinate (ε is the rms lattice microdistortions and D is the CDD average size) were estimated by the least-squares technique [4]. Coarse-grained α-quartz was used as a standard for determining the instrumental broadening.

The grain structure was examined under a transmission electron microscope (TESLA BS-500) at an accelerating voltage of 90 kV, using carbon replicas or foils. The latter were obtained by ion-beam milling of thin powder-compact plates.

The porosity of ceramic samples ranged between 2 and 70%, and the mean pore size was comparable to or much larger (by several orders of magnitude) than the grain size. The samples were prepared by the power metallurgy method (pressing with the subsequent sintering of the ultra-fine powder at temperatures of 1200-1600°C). In this way, samples with a pore size comparable to the grain size were obtained. Larger pores were produced by adding granulated admixtures that are easy to burn out, such as rosin or paraffin, into the powder. Compression tests were carried out on an INSTRON-1185 machine with a constant loading rate of 3×10^{-4} s^{-1}. From stress-strain curves, it have been evaluated the strain and the ultimate strength of the material.

3. Results and Discussion

3.1. Electron microscopy of the structure of powders produced by plasma spray synthesis

Data obtained by transmission electron microscopy (TEM) show that the morphology of the ZrO_2 (3Y) powder particles is typical for plasma spray synthesis. The major fraction of the particles is hollow spheres, semispheres or films. Figure 1 shows typical electron micrographs of ZrO_2 (3Y) and ZrO_2 (3Y) + 20wt. % Al_2O_3 powders. The particles are seen to be polycrystals, with their electron diffraction patterns having a ring structure typical for the nanocrystalline state. The system of the Debye rings seen in the electron diffraction patterns of the majority of particles corresponds to high-temperature tetragonal (*t*) or cubic (*c*) ZrO_2 modifications. Moreover, reflections of the monoclinic (*m*) zirconia modification are observed in the electron diffraction patterns of the structure characterized by larger grains. Dark-field microscopy of the nanostructure of zirconia powder particles suggests that the major fraction of crystallites of size $D < 50$ nm is single-domain particles. X-ray spectral microanalysis of individual powder particles shows no difference in Y_2O_3 concentration between particles made up of coarse and fine crystallites. The average amount of Y_2O_3 is ~ 3 mol %, with the standard deviation of the Y_2O_3 content for different powder particles being ≤ 1 mol %. This implies no correlation between the occurrence of the *m* modification and the changing composition. More likely the reflections of *m*-ZrO_2 seen in the electron diffraction patterns of coarse-grained particles suggest a crystallite size of several tens of nanometers for the powder under study. In the case where the crystallite size is above the critical value, the *m* modification is formed. According to the data obtained from dark-field microscopy, up to three defocusing levels of polycrystalline structure images are revealed for the majority of particles. This bears witness to a multiple-grain layered structure of the particles.

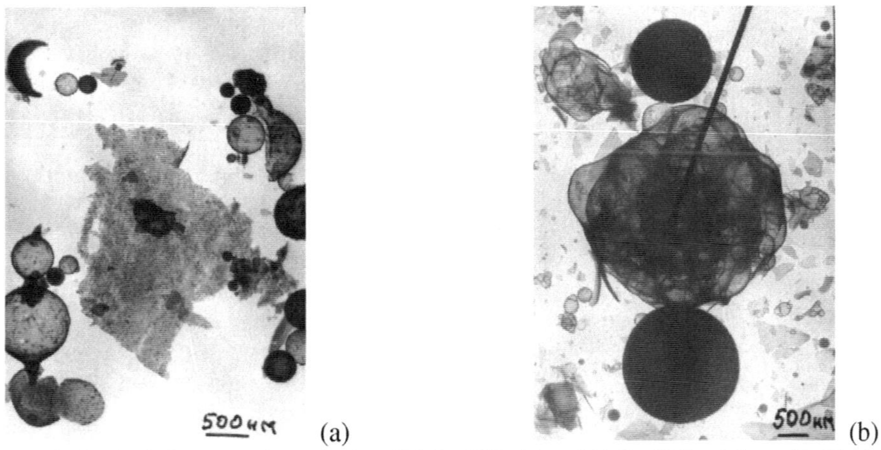

Figure 1. Typical electron micrographs of ZrO_2 (3Y) (a) and ZrO_2 (3Y) + 20 wt. % Al_2O_3 (*b*)

Figure 2 shows the crystallite particle size distribution among nanocrystalline particles in the powders under review transparent to the incident electron beam. The average crystallite size for ZrO_2 (3Y) powder is 20 nm (Fig. 2*a*), approaching an average size of 22 nm found in other zirconia powders produced by plasma spray synthesis [5,6]. ZrO_2 (3Y) powder exhibits a "broad" crystallite size distribution with variance σ = 11.1. The largest crystallite size (D > 30 nm) may run into 80 nm. However, their contribution is moderate (≤ 10 %). The zirconia crystallite size distribution in the nanocrystalline particles of ZrO_2 (3Y) + 20 % Al_2O_3 is illustrated in Fig.2*b*. The structure of the transparent powder particles under study is seen to consist of smaller grains than those found in the binary system. The average zirconia crystallite size revealed by dark-field microscopy in the (111) reflection of high-temperature ZrO_2 phases is 9 nm with σ = 5.2.

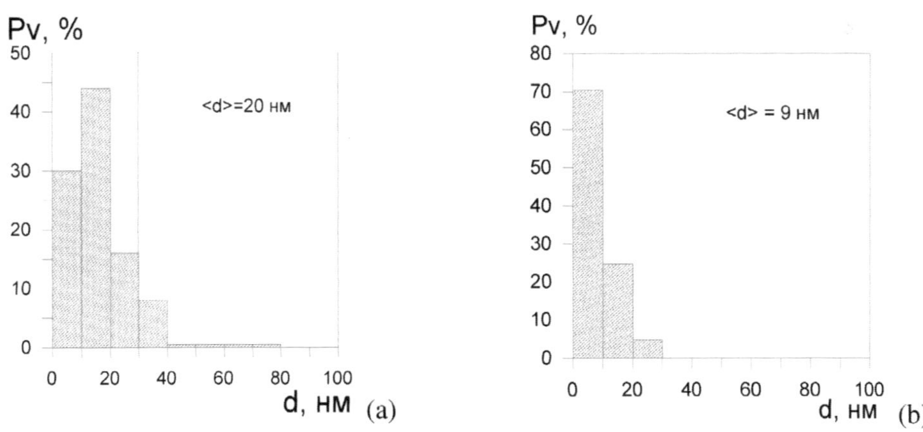

Figure 2. Zirconia crystallite size distributions in initial ZrO_2 (3Y) and in ZrO_2 (3Y) + 20 wt. % Al_2O_2 powders (*a* and *b*, respectively)

The crystallite size for all nanocrystalline powder particles of the ternary system under investigation is ≤ 30 nm, which is below the critical value. This means that for a phase composition of zirconia in the particles being consistent with the critical-size concepts, no reflections of the *m* modification may be observed in the electron diffraction patterns. The foregoing electron diffraction data substantiate this conclusion and are evidence of the fact that the critical zirconia crystallite size in the nanocrystalline particles is no less than 30 nm.

IOP Conf. Series: Journal of Physics: Conf. Series **790** (2017) 012015 doi:10.1088/1742-6596/790/1/012015

Thus, the TEM data suggest that ZrO_2 (3Y) particles transparent to the incident electron beam have a nanocrystalline structure made up of zirconia grains. The majority of crystallites in ZrO_2 (3Y) are single-domain particles of high-temperature modifications. Reflections of the *m* modifications are seen solely in the electron diffraction patterns of particles made up of coarse grains.

In ZrO_2 (3Y) + 20wt. % Al_2O_3, there are both crystalline and quasi-amorphous particles. X-ray spectral microanalysis shows the presence of alumina in particles of the two kinds, with the alumina content in the particles approaching its total amount in the powder (20 wt %). The structure of the nanocrystalline particles in this powder is similar to that found in the binary system with nearly the same crystallite size. Judging by the electron diffraction data and those obtained by x-ray spectral microanalysis, the crystalline structure of the quasi-amorphous particles appears to correspond to a nonequilibrium solid solution based on high–temperature zirconia modifications with alumina addition.

3.2. Variations of the t-lattice microdistortions and coherently diffracting domain size in ZrO_2 (3Y)

Treatment and annealing of ZrO_2 (3Y) produced by plasma spray synthesis enable us to form a large number of *t* states with varying CDD average size and varying microdistortion level. Figure 3 is a plot of the microdistortion level against the CDD average size constructed with the use of data on the initial state and those obtained for the powder subjected to different type of treatment and annealing. For the CDD size lying in the range from 20 to 40 nm, the microdistortions increase with decrease in *D*. In the case where the CDD size is above 40 nm, the microdistortion level remains virtually constant ($\varepsilon_0 = 0.7 \cdot 10^{-3}$). Also shown in the figure is the microdistortion level ε_1 for ZrO_2 (3Y) in the initial state.

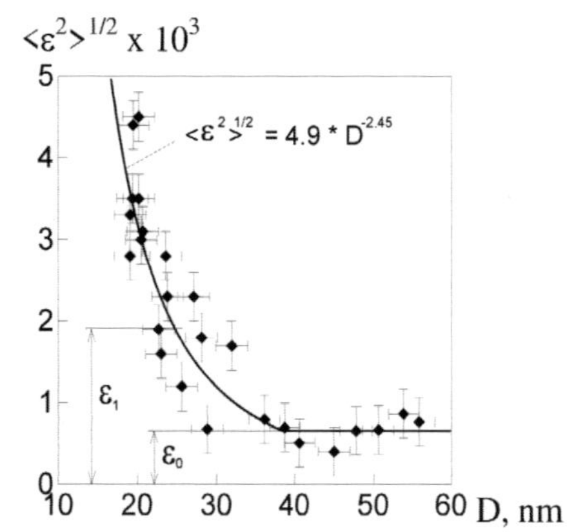

Figure 3. Lattice microdistortions versus CDD average size for nanocrystalline *t*-ZrO_2 (3Y) powder

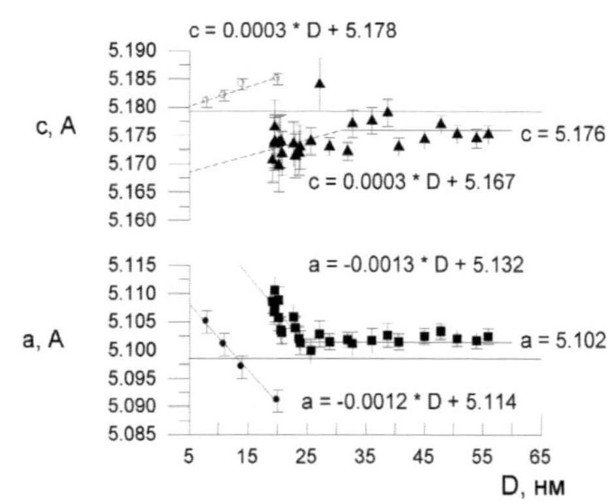

Figure 4. Tetragonal lattice parameters versus CDD average size for nanocrystalline *t*-ZrO_2 (3Y)

It is probable that the residual microdistortion level observed in *t*-ZrO_2 (3Y) for $D > 40$ nm (i.e., in the region where the grain boundary contribution is at a minimum) is primarily associated with defects of the inner regions of the crystal lattice. These are most likely to be point defects due to annealing like Y^{3+} ions of a larger ionic crystal radius than that of Zr^{4+} ions and oxygen lattice vacancies due to charge neutralization.

The increase in the microdistortion level with reduction of the CDD size down to < 40 nm testifies that the CDD size is limited by defects concentrated at the CDD boundaries. This is also demonstrated by the fact that the CDD average size in the powder material under study is in good agreement with the crystallite size, as revealed by TEM.

In the case where all microdistortions are due to interfaces, the following relation should be imposed: $\varepsilon \sim D^{-1}$. As is seen from Fig.3, the microdistortion level in the material under study for a CDD size of < 40 nm is increased in proportion to $D^{-2.45}$, i.e., the microdistortion accumulation rate is much higher in this case. Notably, the data for this relationship correspond for the most part to the *t*-

phase state after treatment. This behavior of the material appears to demonstrate that the increase in the microdistortion level occurs not only due to distortions of the subsurface crystallite layer, but due to an accumulation of lattice defects in the bulk of the crystallites as well. Hence, defects may be cation vacancies and interstitial Zr^{4+} and O^{2-} ions.

The *t* phase in ZrO_2 (Y) produced by plasma spray sintering is characterized by reduced tetragonality as compared to the coarse-grained material. A set of the *t*-phase states with varying CDD size obtained as the result of treatment and annealing enables us to examine the CDD size dependence of the lattice parameters. Figure 4 depicts *t*-lattice parameters as functions of the CDD average size for the powders under consideration. The *t*-lattice parameters reported in [7,8] for unstabilized powders of ZrO_2 produced by plasmochemical techniques are given for comparison. The dotted horizontal lines represent lattice parameters for a coarse-grained material of the ZrO_2 (Y) composition [9]. The lattice parameters are seen to exhibit regular changes for a CDD size of < 25 nm as compared to the unstabilized material. This appears to be due to a dissolved yttria effect. It is evident from the figure that in the ZrO_2 (3Y) powders under consideration, the lattice parameters approach each other for a CDD size of < 25 nm. For the CDDs of size > 25 nm, the two *t*-phase parameters in the stabilized material are no longer varied.

The fact that the ZrO_2 (3 Y) powders under study have close *t*-phase parameters for a CDD size of < 25 nm correlates well with the changes in the lattice parameters in unstabilized ZrO_2 as the CDD size is reduced from 20 down to 8 nm (Fig.4). This means that the lattice parameters of the powders produced by plasma spray synthesis for our experiments do change due to the CDD size effect. The slopes Δa and Δc of the curves representing the linear relationships between the parameters *a* and *c* for ZrO_2 (3Y) used in our investigations approach those observed in the unstabilized material. This suggests that similar mechanisms responsible for the *t*-lattice parameter variations with CDD size are involved in the two types of materials.

It should be noted that while the lattice parameters of the powders under consideration do vary with increase in the CDD size to > 25 nm, they do not reach the level characteristic of the coarse grained material (Fig.4). For D > 25 nm, their values correspond to those found in as-annealed powders. The behavior of the nanocrystalline *t*-phase parameters observed for D > 25 nm may be attributed to the presence of residual stresses in the crystallite structure. In this case, even annealing fails to provide complete stress relaxation. The residual stresses appear to favor the *m* to *t* transformation on annealing.

The set of the *t*-phase states of the powders also enables us to examine the correlation between the amount of the *m* modification in the material, on the one hand, and the average CDD size and lattice microdistortion level, on the other. Figure 5 shows variations of the *m*-phase abundance in ZrO_2 (3Y) with the CDD average size (*a*) and with *t*-lattice microdistortion level (*b*). Normally, the *m*-phase abundance is increased with increase in the average crystallite size. As is seen in Fig.5, this relation breaks down for D = 20–60 nm in the powders under consideration: as the CDD average size is increased, the *m*-phase abundance is decreased down to the point that the reverse (*m*–*t*) transformation takes place. The decrease in the *m* content with increase in the CDD average size for D > 23 nm testifies that the critical size approaches the characteristic values for sintered material (500–1000 nm). This fact substantiates our conclusion that grain boundaries become more perfect on annealing (i.e., annealing decreases the number of defects).

Inspection of Fig.5*b* discloses that the *m*-phase abundance, on the contrary, correlates closely with lattice microdistortions. This means that the increase in the *m* content for small values of D (Fig.5*a*) is really due to the microdistortion effect on the critical size, thus lending credence to the proposition that the phase composition of nanocrystalline zirconia is governed by *t*-lattice microdistortions.

Thus, the *t* phase in ZrO_2 (3Y) produced by plasma spray sintering exhibits variations of the lattice parameters due to the crystallite size. Unlike the coarse-grained material, the variations of the *t*-ZrO_2 (3Y) content are observed for a CDD average size of < 25 nm, with the *t*-phase structure approaching that of the *c* phase. For a CDD average size of < 40 nm, elastic straining of crystallite regions in the vicinity of grain boundaries was observed as microdistortions in diffraction experiments. The phase composition of the material under study is really controlled by the microdistortion level.

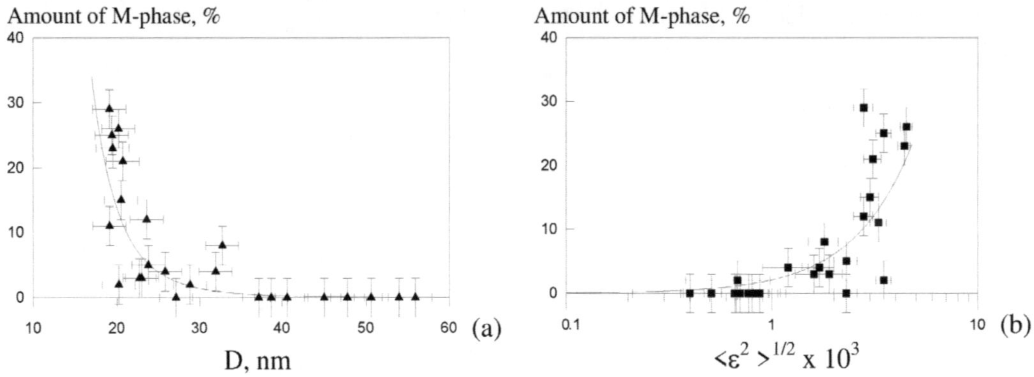

Figure 5. Variations of the amount of the *m* modification with CDD average size (*a*) and lattice microdistortions (*b*) in nanocrystalline *t*-ZrO_2 (3Y)

3. Porous structure and phase composition of ceramics

A porous structure of ceramics with porosity less than 10% mainly are the isolated pores. There are two types of porosity in ceramic with porosity 10 - 60%, those are isolated and continuous one, or the isolated pores and the porous clasters. The porous structure of ceramic with porosity more than 50% is a system of the continuous pores. One can observe the unimodal sizes distribution of pore, Fig. 6, the mean pore size being commensurable with the grain size. There appears the second peak in distribution in ceramic sintered with porophores: the first peak is defined by small pores, the presence of which is owing to packing of the initial powder particles, the second one is conditioned by the pores with configuration of porophores particles (Fig. 7), the greater portion of continuous porosity the greater mean pore size. As one can see in Fig. 8, the increasing volume of porosity decreases tetragonal phase content almost by 2 times. It is stipulated by the growing of crystallites and absence of compressive stresses in high-porous media, which stabilize the tetragonal phase.

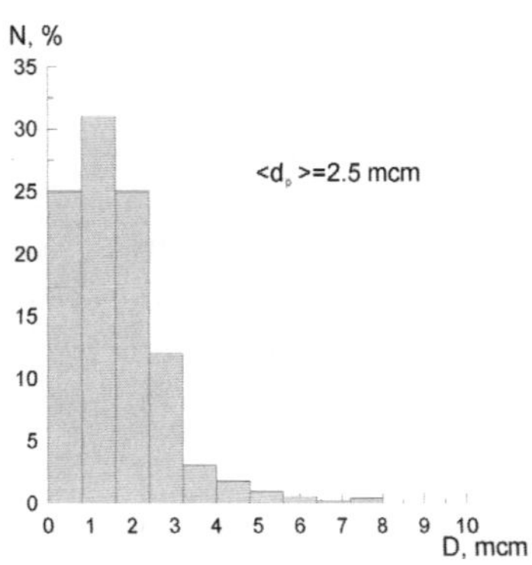

Figure 6. The size distribution of pores in ceramics with the mean size being commensurable with the grain size

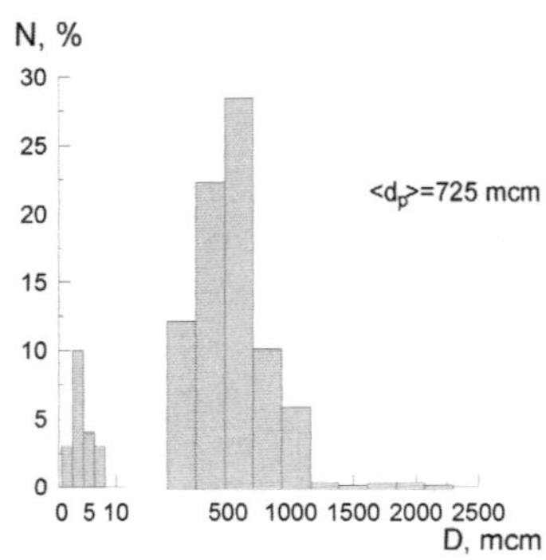

Figure 7. The size distribution of pores in ceramics with the mean size exceeding the grain size

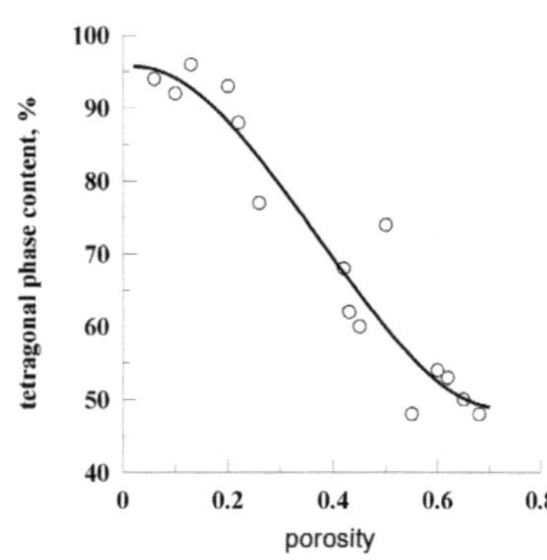

Fig ure8. The content of tetragonal phase in $ZrO_2(Y_2O_3)$ vs. porosity

4. Deformation behavior and mechanical properties of porous ceramics

The stress-strain curves for the ceramics with different porosity are shown that the material with porosity less than 10% has linear stress-strain curves with small deviation from linearity at the tip of a curve being apparently owing to t-m phase transition occurring under stresses, which is characteristic for partially stabilized ZrO_2 [3]. Ceramic with the higher porosity has the more complex curves. One can separate out some parts in curves. It is seen that the increase of the pore volume causes multiple microdamages during the deformation, which are the greater, the higher the porosity. In the curves, this process shows up as sharp stress drops due to microcracking. After the microcracks have terminated at the pores, the material can elastically deform again. As the porosity grows the range of microcracking shifts toward higher strains and expands. When the porosity exceeds 20%, the stress-strain curves become concave upward, which is totally untypical of the stress-strain curves for sintered materials.

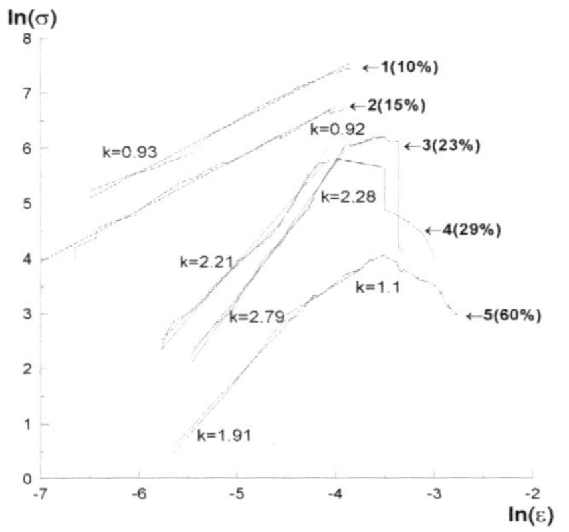

Figure 9. Stress-strain curves in log-log coordinates. Numbers is porosity

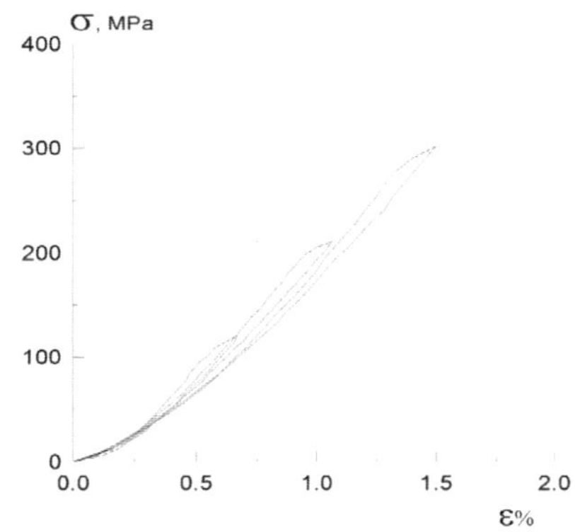

Figure 10 Stress-strain curves of $ZrO_2(Y_2O_3)$ under cycle

At the stage of high deformation (prior to microcracking), the general slope of the curves a σ=**f**(ε) varies with the porosity. However, the slope varies also locally (within one curve), depending on the porosity type and pore size. Such dependencies can be described by a power function like $Y=bX^k$ for both deformation and pressing-related processes (the latter are also possible in this system). Here, the exponent ***k*** depends on which of the processes, pressing or plastic deformation is governing. For purely elastic deformation, $k=1$; for plastic deformation, $k < 1$; and for pressing, $k > 1$. In the most general case it is possible superposition of all the above mechanisms.

Figure 9 demonstrates the curves σ=**f**(ε) in the "log-log" coordinates. For the ceramic of porosity higher than 20%, the curves take the form of several linear segments, which obviously have various slopes *k*. The higher the porosity, the larger the number of the linear segments and all the values of $k>1$, with the highest values achieved for a number of materials, up to 2.5. So, it should be expected above all, there will be revealed pressing process, specifically under active deformation. In general, it is wonderful that there are $k > 1$ under deformation to small degrees, since it is impossible to find here the traces of fracture. For this purpose the special metallographic investigations has been carried out with specimens being exposed to deformation to small degrees, up to the very beginning of visual traces of fracture, and to cycle loading.

The examination of the sample surface after the deformation indicates no signs of local pressing - related material displacement before microcracking. Moreover, at this stage, the stress-strain curves are reversible, fig. 10, that is, the deformation process is purely elastic, as demonstrated by direct metallographic studies: there is no effect of irreversible displacement of the fragments into the porous space.

It is possible to determine the effective elastic modulus of system by the initial parts of the obtained deformation curves and the yield strength – by their maximum values. The value of the effective compressive modulus is known to depend on the porous space volume but it is insensible to pore configuration, whereas strength must depend on morphological structure of porosity owing to different curvature of its inner surface, for example. The dependencies of the yield strength and the effective compressive modulus on the porous space volume are illustrated in Figs. 11 and 12. As should be expected, the experimental results on strength are divided into two systems of points that is for high and low-level porosity, being arranged higher than in the case of microporous ceramics with the mean level porosity highly exceeding the grain size. The difference almost disappears with the porosity of ~ 60%. Modulus dependence on porosity is common for all the porosity types.

These strength and modulus dependencies on porosity are described by the equations as follows $Y=Y_o\exp(bX)$, which permit to determine strength and modulus of non-porous material. Their values well agree with the literature data, the coefficient **b** also correlating with the well-known values for similar materials.

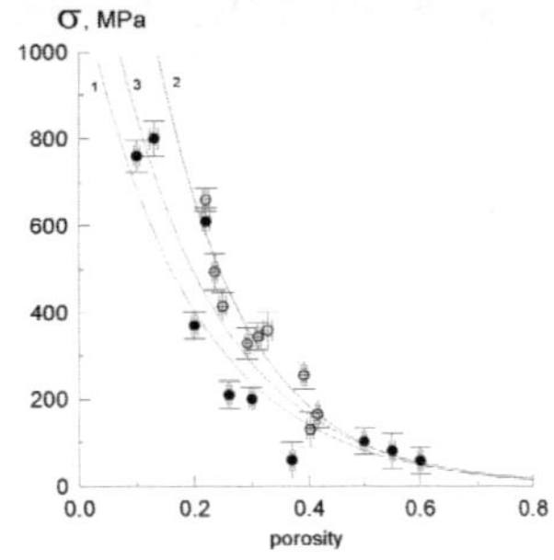

Figure 11. Strength of ZrO_2 (Y_2O_3) vs. porosity:
1. ceramics with pore size is comparable to the grain size, Y=1200exp(-5X);
2. ceramics pore size is much larger than the grain size, Y=2400exp(-6X);
3. averaging for all values, Y=1500exp (-5.5X)

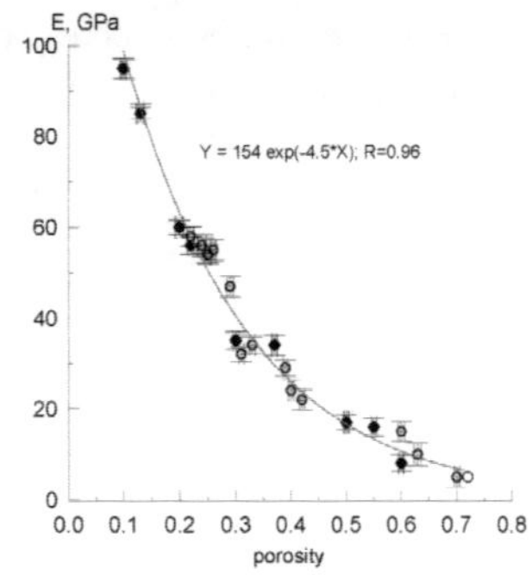

Figure 12. Effective elasticity modulus of ZrO_2 (Y_2O_3) vs. porosity

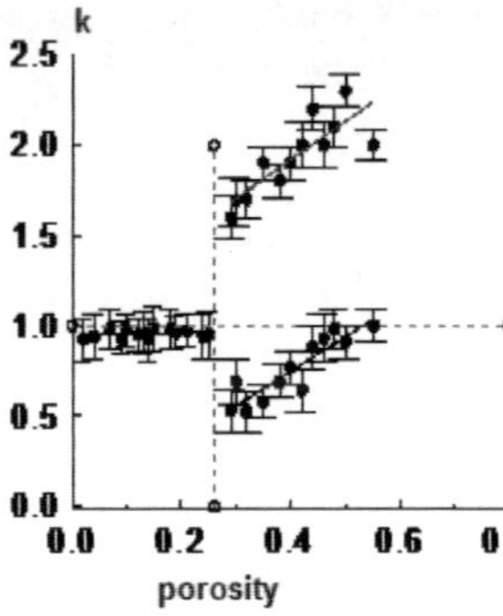

Figure 13. The parabolic factor **k** of the deformation equation vs. porous space volume of ZrO_2 (Y_2O_3) with pore size is comparable to the grain size

In Fig. 13, the exponents ***k*** measured from the slopes of the linear segments are plotted against porosity for the ceramic where the mean pore diameter is comparable to the mean grain size. The data points for ***k*** are well fitted by three straight lines. There exists some critical porosity value at which the deformation of the porous material radically changes: the second exponent of the power function (much larger than that in the initial state) appears. Most likely, this is related to a change in the pore distribution: individual pores give way to the continuous pore distribution. In essence, the material is split into two subsystems, which deform in different ways under stressing. It may be associated with change of pore distribution character from the isolated ones to the continuous porous structure.

As it was shown in [10], in this systems one can forms a linear or plate-like structures which admitting a model representation in the form of a rod, can exist in high-porosily ceramics with a limited area of contact between grains. Owing to a small size of the grains and the significant size of the contact area compared with the grain size and strong contact between grains formed during sintering, the bending stresses arising as a result of the loss of stability such rods do not exceed the ultimate strength of the ceramic material (which is virtually free of defects within the limits of the grain). That is there occurs the micromechanical instability of rod-like structures, obtained after sintering of porous body, with their considerable macro-deformation as structural elements, which may be realized in the elastic area as one can see from Fig.10.

5. Percolation transitions in porous ceramics

It should be assumed that qualitative changes occurring in ceramic macrostructure, namely the transition from the isolated pores to communicating porous structure, would be affected on the microstructural parameters. For this purpose, the coherent diffracting domain (CDD) sizes obtained from X–ray patterns have been plotted with the volume of ceramic porous space. These results are illustrated in Fig.14. As one can see, there takes place the increasing of CDD sizes with the increasing of ceramic porosity in the very beginning of a pattern; those being decreasing with the further increasing of porosity. With the increasing volume of porous space in the initial range (up to 15%) the increasing CDD size depend on diffusion growth of crystallites, which will be then controlled by stresses.

It is possible to find approximating by the dependencies like $F \sim (q-q_c)^m$, here m is the critical power index [11] and q_c is 0.15-0.17, that on the right and on the left from q_c both of the dependencies are put to the following form: $F_- \sim x^{5.3}$ and $F_+ \sim x^{5.9}$, which consequently, there are strong grounds to think that it is the correct interpretation of the obtained results in the terms of "the first percolation threshold" [11].

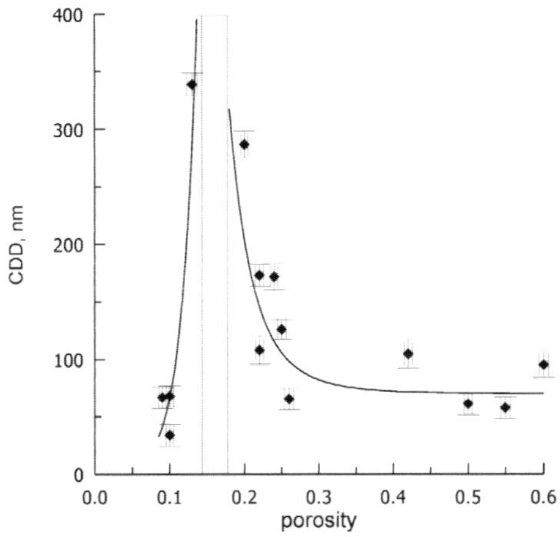

Figure 14. CDD-sizes vs. porosity

That is, in case of achieving some critical level of porosity, the basic changing character of the CDD dependence on porosity is associated with the percolation threshold in ceramics from the isolated pores to the continuous porous structure. The isolated pores do not qualitatively affect on the microstructural parameters, but in turn, the existing stresses restrict growth of the crystallites in the continuous porous ceramic structure.

4. Conclusion

1. Yttria and/or alumina-doped zirconia are in nanocrystalline and quasi-amorphous structural states. The nanocrystalline part of the material is in the *t*-phase state, while its quasi-amorphous part is in the *t*- or *c*-phase state. The crystal structure of the nanocrystalline and quasi-amorphous parts of the material is a nonequilibrium solid solution of ZrO_2(Y, Al). In the powder yttria stabilized with alumina addition, the *m* phase in ZrO_2 is formed due to increase in the microdistortion level which destabilizes the nanocrystalline *t* phase.
2. It was shown that in the porous structure of zirconia ceramics there is a critical value of porosity the material divides into two sub-systems, being variously deformable under external loading. There is no effect of irreversible displacement of the fragments into porous space.
3. In ceramics with the porosity up to 70% there occurs the micromechanical instability of rod-like structures, obtained after sintering of porous body, with their considerable macro-deformation as structural elements, which may be realized in the elastic area.
4. It has been found out the correlation between the sizes of crystallites and porosity, which associated with transition of the isolated porous structure to the continuous one and the porosity of 20%, corresponds to the first percolation threshold.

Acknowledgement
This work was partial financial supported by Tomsk State University Competitiveness Improvement Program and Siberian Branch Program #III.23.2.3.

5. References

[1] G.L.Messing, Shi-Chang Zang, and G.V.J-Ayanti, J.Amer.Ceram.Soc., **76**, No.11, 2707–2726 (1993)
[2] Porous Structural Ceramics. Ed. By Yu.Krasulin. (Metallurgia, Moscow, 1980)
[3] Evans A.G. J. Amer. Ceram. Soc, 1990. V.**73**, #2. Pp.187-206.
[4] G.K.Williamson and W.H.Hall, Acta Metall., **1**, 22–31 (1953).
[5] S.N.Kulkov, P.V.Korolev, A.G.Melnikov, *et al.*, Russ. Phys. J., **38**, No.1, 42–45 (1995).
[6] Yu.F.Ivanov, V.V.Lopatin, and V.S.Dedov, Russ. Phys. J., **37**, No.1, 95–101 (1994).
[7] V.F.Petrunin, Zhur. Vsesoyuz Khimich. Obshch. Im. D.I.Mendeleev, **36**, No.2, 146–150 (1991).
[8] V.F.Petrunin, A.G.Ermolaev, A.V.Burkhanov, *et al.*, Poroshk.Metallurg., No.**3**, 47–52 (1989).
[9] D. –J.Kim, J. Amer. Ceram. Soc., **73**, No.1, 115–120 (1990).
[10] Kulkov S., Maslovskii V. et al.Technical Physics, v.**47**, #3, 2002, pp.320-324.
[11] Sahimi M. Application of Percolation Theory. (L.: Taylor &Francis. 1994)

Rheological behaviour of egg white and egg yolk from different poultry specimen

V Kumbár[1], Š Nedomová[2], J Votava[1] and J Buchar[1]

[1]Department of Technology and Automobile Transport, Mendel University in Brno, Brno, Czech Republic
[2]Department of Food Technology, Mendel University in Brno, Brno, Czech Republic

E-mail: vojtech.kumbar@mendelu.cz

Abstract. The main goal of this study is differences in rheological behaviour of hen (ISA BROWN), goose (*Anser anser f. domestica*) and Japanese quail (*Coturnix japonica*) egg white and egg yolk.

The rheological behaviour of egg white and egg yolk was studied using a concentric cylinder viscometer. Rheological behaviour was pseudoplastic and flow curves were fitted by the Herschel–Bulkley model and Ostwald–de Waele model with high values of coeficients of determination R^2.

The meaning of rheological parameters on friction factors during flow of egg white and egg yolk in real tube has been shown. Preliminary information on time–dependent behaviour of tested liquids has been also obtained.

Keywords: egg white, egg yolk, rheology, viscosity, shear stress, shear strain rate, fluid dynamic, thixotropy, time-dependent behaviour

1. Introduction

Rigorous knowledge ofthe rheological properties of food products is essential for the product development, quality control, sensory evaluation and design, and evaluation of the process equipment. The flow behaviour of a fluid can be varied from Newtonian to time dependent non-Newtonian in nature depending on its origin, composition and structure behaviour and previous history [1]. The knowledge of this behaviour is also very important for egg liquids owing to an increasing demand on the processed egg products. The term "egg products" refers to eggs that are removed from their shells for processing and convenience forms of eggs for commercial, foodservice, and home use. These products can be classified as refrigerated liquid, frozen, and dried products. Traditionally, eggs are marketed as shell eggs, but in recent years, egg consumption in the form of egg liquid products has increased [2]. Several researchers [3-10] studied the rheological characteristics of egg yolk, white and liquid whole egg and reported Newtonian as well as time-dependent non-Newtonian flow behaviour of egg. These works are summarized in [11]. Most of these studies have been usually carried out with only egg yolk. There is a lack of information about the rheological behaviour of liquid egg white at different temperature range [12]. Nearly no information are available of the rheological behaviour of liquid products of another eggs as e.g. Japanese quail eggs, which are more and more used in food industry.

Considering this lack of published information on fluid dynamics of egg yolk and egg white, the main purpose of this work was to determine rheological properties of these products for eggs of three domestic fowls: hen (ISA BROWN), goose (*Anseranser f. domestica*) and Japanese quail (*Coturnix japonica*). The meaning of these data for the calculation of friction factors for tube flow is discussed in details.

2. Materials and methods

2.1. Materials

Fresh shell eggs hen (ISA BROWN), Japanese quails (*Coturnix japonica*) and goose (*Anseranser f. domestica*), no older than 3 days, were purchased from local farms (region South Moravia). Egg shells were washed with deionized water and hand broken properly. All liquids were filtered in order to separate impurities (chalaza, membranes). Samples, with volume 200 ml, for egg yolk and egg white were prepared and stored at 4 °C before measurement. From many physical parameters which have been measured only the densities of the tested liquids are presented in the Table 1.

Table 1. Density of liquid egg products (data are displayed as mean value ± standard deviation, n=20)

Product	Poultry	Density ρ (kgm^{-3})
Egg white	Japanese quail	1020.4±1.8
Egg yolk		1027.6±2.9
Egg white	Hen	1034.1±1.9
Egg yolk		1027.0±2.4
Egg white	Goose	1035.5±2.8
Egg yolk		1031.9±3.1

2.2. Rheological measurement

Rheological measurements were carried out using an Anton Paar DV3-P rotary viscometer, equipped with a coaxial cylinder sensor system. Rotational speed ranged between 0.3 (0.279 s^{-1}) and 200 rpm (186 s^{-1}). Apparent viscosity η which is the ratio of shear stress σ and shear strain rate$\dot{\gamma}$. By [13] it has been evaluated as:

$$\eta = \frac{\sigma}{\dot{\gamma}} [Pas] \tag{1}$$

All measurements were performed at the room temperature 18°C.

3. Results and discussion

In the Figures 1 and 2 the flow curves, i.e. shear stress vs. shear strain rate, are shown. These curves can be fitted using of Herschel–Bulkley model [14]:

$$\sigma = \sigma_0 + K\dot{\gamma}^n \tag{2}$$

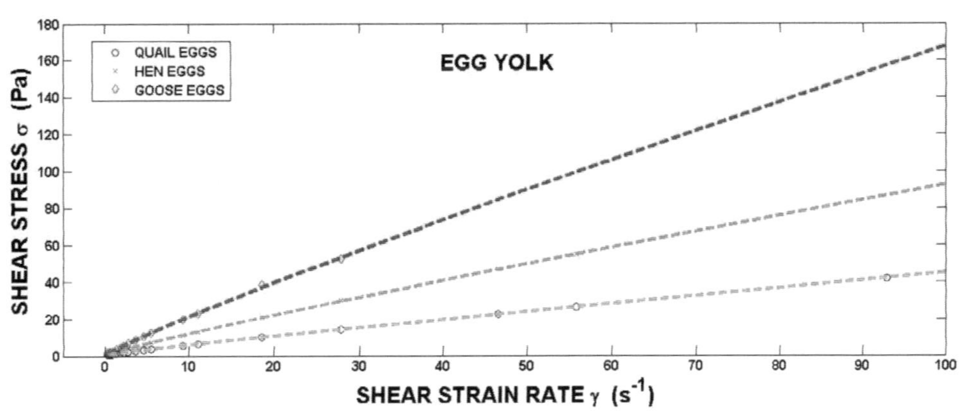

Figure 1. Effect of shear strain rate on the shear stress – egg yolk

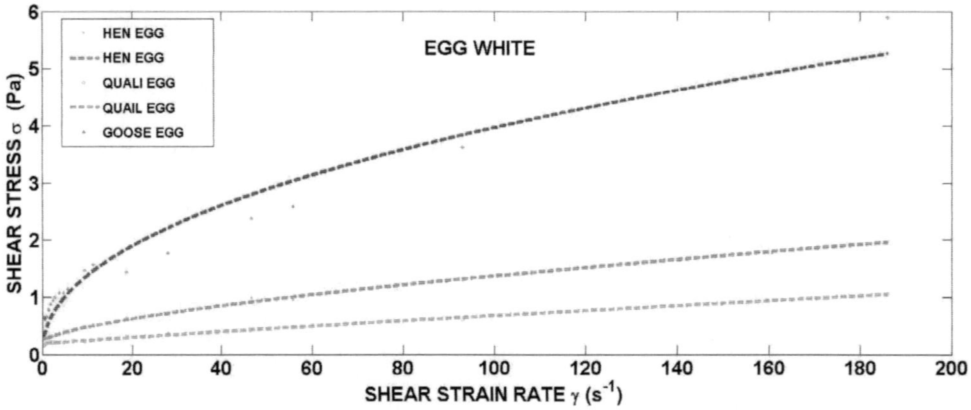

Figure 2. Effect of shear strain rate on the shear stress – egg white

In Eq. (2), σ is the shear stress, $\dot{\gamma}$is the shear strain rate, K is the consistency index, n is the flow behaviour index and σ_o is the yield stress.

For egg yolk and egg white of goose eggs this model reduces to the Ostwald–De Waele model, given by Eq.(3), also is known as the power-law model [15]:

$$\sigma = K\dot{\gamma}^n \tag{3}$$

Eqs.(2) and (3) can be used for both Newtonian and power law fluids. For Newtonian fluids n equals 1, and K equals η or $\eta+\sigma_0$, respectively.

The Eq.(2) can be used for the egg yolk and egg white of hen and quail. Parameters of Eq.(2) are given in the Tables 2–3 (R^2 is the coefficient of determination).

Table 2. Parameters of Herschel–Bulkley model for egg yolk

EGG	σ_0 (Pa)	K (Pas^n)	n (1)	R^2
HEN	0.1466	1.499	0.8951	0.9999
GOOSE	0	2.678	0.8982	0.9989
QUAIL	0.4115	0.6533	0.9164	0.9999

Table 3. Parameters of Herschel–Bulkley model for egg white

EGG	σ_0 (Pa)	K (Pas^n)	n (1)	R^2
HEN	0.2372	0.05134	0.672	0.9954
GOOSE	0	0.4831	0.457	0.9415
QUAIL	0.1904	0.007022	0.9192	0.9923

The apparent viscosity is given using the Eq. (1). The viscosity of tested egg liquids is shown in the Figures 3–4.

The egg liquids exhibit shear thinning behaviour. The main differences between liquids of different eggs can be observed for the egg yolks. The highest values of the apparent viscosity were achieved for the goose eggs, following by the hen eggs and the minimum values exhibited mail eggs. The differences between egg white of different eggs are not too significant. The differences in rheological properties of egg liquids are caused by chemical composition of eggs from hens, geese and quail (differences in water and lipid content).

In order to study of the time on the liquid egg products these liquids were sheared at constant shear rates (46.5 s^{-1}) for one hour (3600 seconds) and changes of apparent viscosity with time was considered as time dependence, see Figures 5–6.

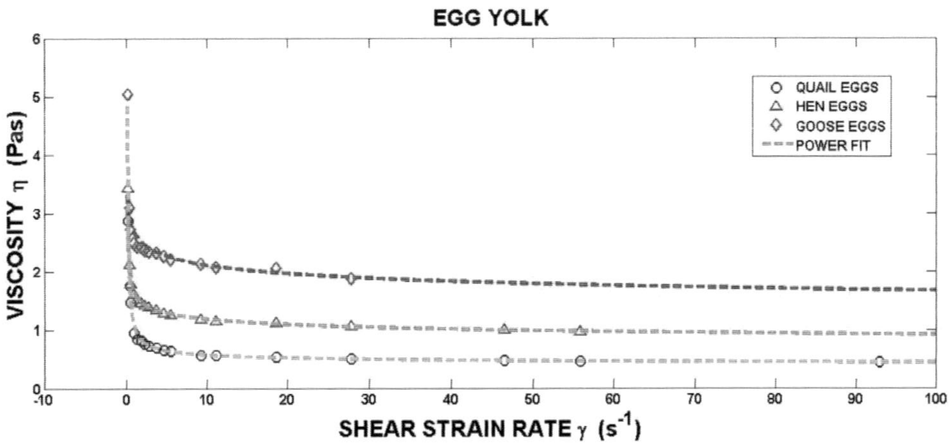

Figure 3. Viscosity of the egg yolk

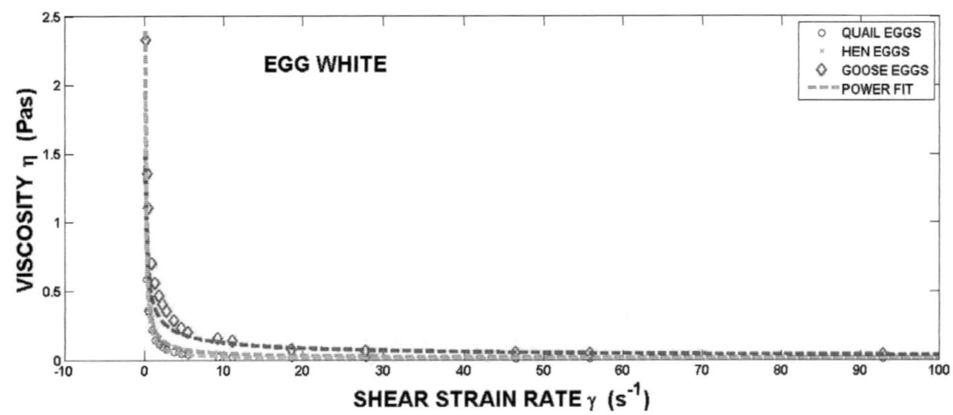

Figure 4. Viscosity of the egg white

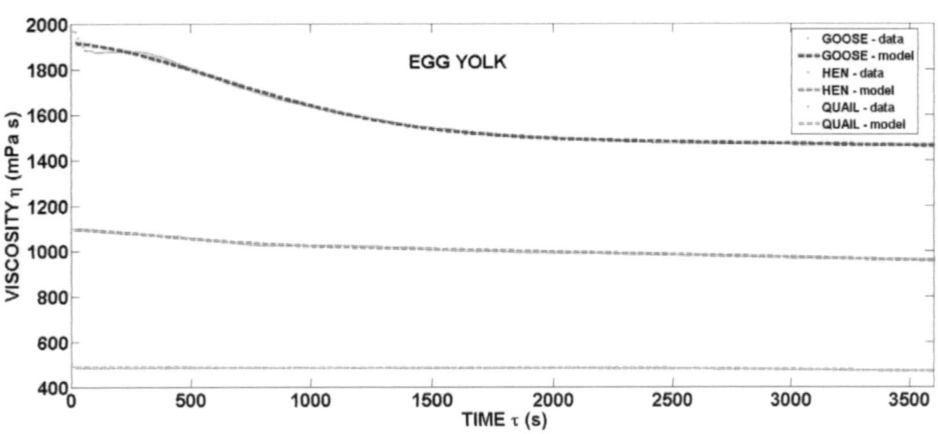

Figure 5. Time dependence if apparent viscosity of the egg yolk

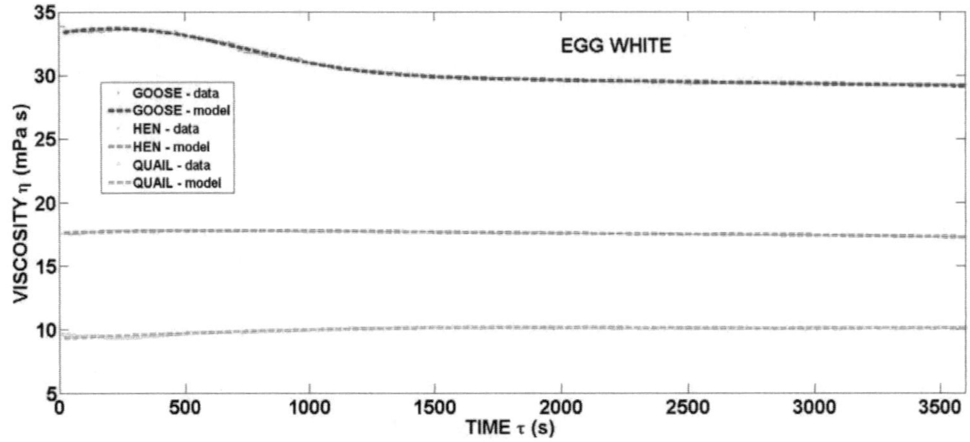

Figure 6. Time dependence if apparent viscosity of the egg white

The egg yolk exhibits typical thixotropic behaviour as is described in [4]. The egg white (especially hen and quail) is shown time–depend behaviour of apparent viscosity [16].

The experimental data were fitted by Gaussian function [17]:

$$\eta = a_1 e^{-\left(\frac{\tau - b_1}{c_1}\right)^2} + a_2 e^{-\left(\frac{\tau - b_2}{c_2}\right)^2}, \tag{4}$$

where η is apparent viscosity, τ is time and a_i, b_i, c_i are parameters. These parameters are given in the Table 3.

Table 3. Parameters of Gaussian model for egg yolk and egg white

PARAMETERS:		a_1	b_1	c_1	a_2	b_2	c_2	R^2
	GOOSE	410.3	–148.5	1098	5.029e7	–2.057e6	6.375e5	0.9969
EGG YOLK	HEN	51.25	–40.28	573.9	1.378e4	–2.124e5	1.322e5	0.9949
	QUAIL	489	–1776	2.621e4	6.706	2464	1050	0.9945
	GOOSE	3.52	241	694.8	6.067e12	–5.343e6	1.047e6	0.9966
EGG WHITE	HEN	2.321	–526.7	2052	17.29	3009	8982	0.9693
	QUAIL	10.18	1863	6306	0.7469	3853	829.8	0.9264

The obtained rheological parameters have great meaning in many problems of industry. For example the design of piping and pumping systems requires knowledge of the pressure drop due to flow in straight pipe segments and through valves and fittings. Friction losses caused by the presence of valves and fittings usually result from disturbances of the flow, which is forced to change direction abruptly to overcome path obstructions and to adapt itself to sudden or gradual changes in the cross section or shape of the duct. This problem is described e.g. in [18]. The pressure drop is calculated using the friction factor, f. The friction factor is defined as [19]:

$$f = \frac{2\sigma_w}{\rho v^2}, \tag{5}$$

where ρ is fluid density, v is the average flow velocity, and σ_w is the stress in the wall, given by

$$\sigma_w = \frac{D\Delta P}{4L}. \tag{6}$$

In Eq. (6), D is the tube diameter and ΔP is the pressure drop observed in a length L of the tube. For laminar flow, the friction factor can be obtained from a simple function of the generalized Reynolds number, which is identical to the dimensionless form of the Hagen–Poiseuille equation [20]:

$$f = \frac{16}{R_{eg}}, \tag{7}$$

in which

$$R_{eg} = \frac{D^n v^{2-n} \rho}{8^{n-1} K} \left(\frac{4n}{1+3n} \right)^n. \tag{8}$$

Eqs.(7) and (8) can be used for both Newtonian and power law fluids. For Newtonian fluids n equals 1 and Kequalsη. So that the generalized Reynolds number (Eq.(8)) is reduced to well-known number$Re = Dv\rho/\eta$.

The values of generalized Reynolds number for egg liquids tested in this article are given in the Table 4. Under turbulent flow conditions, the existing correlations to estimate the friction factor are semi-empirical. For power law fluids, probably the best-known correlation is that presented by [21]:

$$\frac{1}{\sqrt{f}} = \frac{4}{n^{0.75}} ln\left(R_{eg} f^{1-\frac{n}{2}}\right) - \frac{0.4}{n^{1.2}}. \tag{9}$$

Let us consider a cylindrical tube of diameter D = 0.01 m and an average flow velocity v = 1 m/s. This type of tube is usually used in pasteurize process of egg products. The values of generalized Reynolds numbers are given in the Table 4.

Table 4. Reynolds numbers given by the Eq. (8)

Poultry	R_{eg} (Egg yolk)	R_{eg} (Egg white)
Japanese quail	27	2444
Hen	13	1670
Goose	7	718

The Reynolds number describes namely the transitive from laminar to turbulent flow. One can see that the highest R_{eg} of yolk and white exhibits the quail eggs and the smallest one the goose eggs.

Laminar flow of a power law fluid exists in the tube when:

$$R_{eg} \leq \left(R_{eg}\right)_{critical}. \tag{10}$$

The critical value of the power law Reynolds number depends on the value of the flow index behaviour n according to [22]:

$$\left(R_{eg}\right)_{critical} = 2100 + 875(1-n). \tag{11}$$

Values of critical Reynolds number vary from 2888 at n = 0.1 to the familiar value 2100 for Newton liquids (n = 1). In all cases the flows of liquid egg products are laminar.

The next application of the rheological properties is connected with continuous thermal processing system. Such system generally involves a heat exchanger in form of a tube. A length of this tube is known as a „hold tube", must be sufficient in order to achieve sufficient fluid residence time. Because the hold tube is a critical part of the system, understanding velocity profiles found in tube flow is important for the numerical simulation of thermal process.

For power law fluid in laminar flow the velocity $v(x)$ is function of the distance x from the centre of the pipe:

$$v(x) = \left(\frac{\Delta P}{2KL}\right)^{\frac{1}{n}} \left(\frac{n}{n+1}\right) \left(R^{\frac{n+1}{n}} - x^{\frac{n+1}{n}}\right), \tag{12}$$

whereΔP denotes the driving over pressure, L is the tube length and R is its radius. For the illustration the values of R = 0.05 m, L = 1 m and ΔP = 300 Pa were chosen. This values are usually used in pasteurize process of egg products. Results are shown in the Figures 7–8.

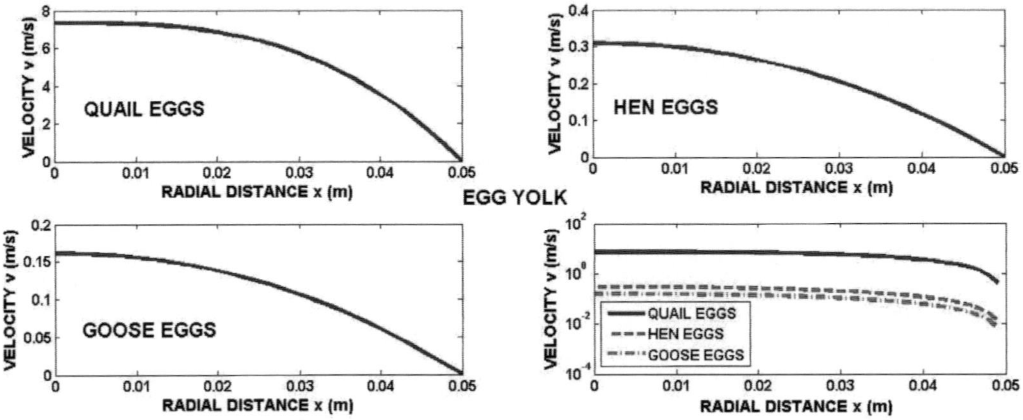

Figure 7. Velocity profiles of egg yolk flow

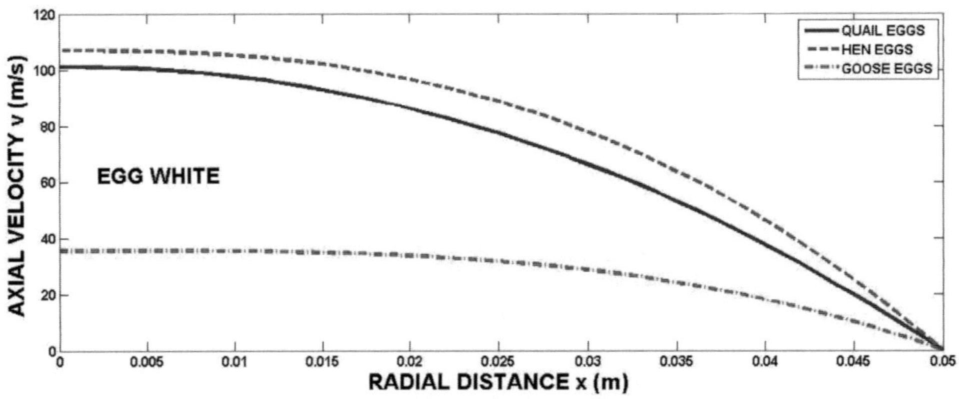

Figure 8. Velocity profile of egg white flow

The higher values of flow velocities are observed for the flow of egg white. The lowest one was achieved for the flow of egg yolk. Velocities are different for different eggs. Liquids of the goose eggs exhibited the lowest values. The highest values of egg white were observed for the hen eggs. The maximum of the flow velocity of the egg yolk was observed for the quail eggs.

The velocity equation given above is valid for fully developer undisturbed flow in straight, horizontal tubes. Reals processing systems contain many elements like valves, tees, elbows, etc. that cause fluid mixing during flow [23-26]. In addition pipe vibration caused by energy inputs from pumps may contribute to mixing. It means that the equation given above represents only general guidelines in examining velocity profiles during tube flow.

4. Conclusion

Rheological properties of liquid egg yolk and egg white were studied for the eggs of three poultry species – hen (Isa Brown), Japanese quail (Coturnix japonica) and goose (Anseranser f. domestica). Experimental data were successfully fitted to Ostwald-De Waele and Herschel-Bulkley model. The egg liquids exhibit shear thinning behaviour. The main differences between liquids of different eggs can be observed for the egg yolk. The highest values of the apparent viscosity were achieved for the goose eggs, following by the hen eggs and the minimum values exhibited mail eggs. The differences in rheological properties of egg yolk are caused by chemical composition of yolks from hens, geese and quail (differences in water and lipid content). In the publication [27] is described that the goose eggs (especially yolk) contained the highest contents of saturated and monounsaturated fatty acids but the lowest content of polyunsaturated fatty acids (PUFA). The ω-3 PUFA content and the ω-6/ω-3 ratio were higher in the yolks of goose. It can cause the higher values of apparent viscosity of goose yolk. The differences between egg white of different eggs are not too significant.

The practical importance of knowledge of rheological parameters was outlined. These parameters can be used in much software dealing with a numerical simulation of flow problems.

Acknowledgement
This work was supported by the project TP 6/2015 „Impact loading of agricultural products and foodstuffs“, financed by IGAFA MENDELU.

References

[1] Migliori M, Gabriele D, Baldino N, Lupi FR, de Cindio B: *Journal of Food Process Engineering* **34** (2011) 1266-1281.
[2] de Jesus MN, Zanqui AB, Valderrama P, Tanamati A, Maruyama SA, de Souza NE, Matsushita M *Food Science and Technology* **33** (2013) 549-554.
[3] Jones DR: *International Journal of Poultry Science* **6** (2007) 157-162.
[4] Atilgan MR, Unluturk S: *International Journal of Food Properties* **11** (2008) 296–309.
[5] Takeuchi J, Nagashima T: *Food Science and Technology Research* **16** (2010) 149-156.
[6] Toyosaki T: *Poultry Science* **89** (2010) 1009-1014.
[7] Alamprese C, Casiraghi E, Rossi M: *International Journal of Food Science and Technology* **47** (2012) 2503-2509.
[8] de Souza PM, Fernández A *Food Hydrocolloids* **31** (2013) 127-134.
[9] Toyosaki T, Sakane Y: *Advance Journal of Food Science and Technology* **5** (2013) 1351-1354.
[10] Navidghasemizad S, Temelli F, Wu J: *LWT - Food Science and Technology* **55** (2014) 170-175.
[11] Singh J, Sharma HK, Premi M, Kumari K: *Journal of Food Science and Technology* **51** (2014) 543-550.
[12] Telis-Romero J, Thomaz CEP, Bernardi M, Telis VRN, Gabas AL: *Journal of Food Engineering* **74** (2006) 191-197.
[13] Steffe JF: Introduction to rheology. In Rheological methods in food process engineering, Freeman Press, East Lansing (1996).
[14] Travnicek P, Krcalova E, Vitez T: *Polish Journal of Environmental Studies*, **22** (2013) 1499-1504.
[15] Antunes SA, Lanza M, Hense H:. *Industrial Crops and Products* **46** (2013) 111-116.
[16] Riscardo MA, Moros JE, Franco JM, Gallegos C: *European Food Research and Technology*, **220** (2005) 380-388.
[17] Kumbár V, Nedomová Š, Strnková J, Buchar J: *Journal of Food Engineering*, **156** (2015) 45-54.
[18] Cabral RAF, Telis VRN, Parkb KJ, Telis-Romero J: *Food and bioproducts processing* **89** (2011) 375-382.
[19] Garcia EJ, Steffe JF: *Journal of Food Process and Engineering* **9** (1987) 93-120.
[20] Tao J, Chen S, Su W: *Mechanics and Astronomy* **56** (2013) 263-269.
[21] Bahrani SA, Nouar C: *Journal of Applied Fluid Mechanics* **7** (2014) 1-6.
[22] Steffe JF, Daubert CR: Bioprocessing Pipelines: Rheology and Analysis. Freeman Press, Michigan (2006).
[23] Telis-Romero J, Polizelli MA, Gabas AL, Telis VRN: *Canadian Journal of Chemical Engineering* **83** (2005) 181–187.
[24] Girón-Palomares B, Hernández-Guerrero A, Romero-Méndez R, Oviedo-Tolentino F: *International Journal of Heat and Fluid Flow*, **30** (2009) 158-171.
[25] Wu B: *Water Research*, **44** (2010) 1507-1519.
[26] Courtois P, Ettelaie R, Chen J: *Appl. Rheol.* **16** (2006) 275-286.
[27] Wang Q, Jin G, Jin Y, Ma M, Wang N, Liu C, He L: *European Journal of Lipid Science and Technology*, **116** (2014) 1044-1053.

Rheological Characterisation of the Flow Behaviour of Wood Plastic Composites in Consideration of Different Volume Fractions of Wood

N Laufer[1], H Hansmann[1,2], M Koch[3]

[1]Institute for Polymer Technologies e. V., Wismar, Germany
[2]Department of Mechanical Engineering, Hochschule Wismar University of Applied Sciences, Technology, Business and Design, Wismar, Germany
[3]Department of Plastics Technologies, Ilmenau University of Technology, Ilmenau, Germany

E-mail: laufer@ipt-wismar.de, h.hansmann@ipt-wismar.de, michael.koch@tu-ilmenau.de

Abstract. In this study, the rheological properties of wood plastic composites (WPC) with different polymeric matrices (LDPE, low-density polyethylene and PP, polypropylene) and with different types of wood filler (hardwood flour and softwood flour) have been investigated by means of high pressure capillary rheometry. The volume fraction of wood was varied between 0 and 60 %. The shear thinning behaviour of the WPC melts can be well described by the Ostwald - de Waele power law relationship. The flow consistency index K of the power law shows a good correlation with the volume fraction of wood. Interparticular interaction effects of wood particles can be mathematically taken into account by implementation of an interaction exponent (defined as the ratio between flow exponent of WPC and flow exponent of polymeric matrix). The interaction exponent shows a good correlation with the flow consistency index. On the basis of these relationships the concept of shear-stress-equivalent inner shear rate has been modified. Thus, the flow behaviour of the investigated wood filled polymer melts could be well described mathematically by the modified concept of shear-stress-equivalent inner shear rate. On this basis, the shear thinning behaviour of WPC can now be estimated with good accuracy, taking into account the volume fraction of wood.

Keywords: Wood Plastic Composites, High Pressure Capillary Rheometry, Rheology, Interaction Exponent, Flow Function, Interparticular Interaction Effects

1. Introduction

Flow and viscosity functions of wood filled polymer melts are shifted to higher values in comparison to those of the unfilled polymer melts (figure 1). The offset depends on the volume fraction of wood as well as the applied shear stress.

In general, the offset of the vertical shift of flow functions of suspensions can be well described by the concept of shear-stress-equivalent inner shear rate according to W. Gleißle and M. K. Baloch.

According to this concept (figure 2), the hydraulic width of the gap between the plates is reduced, due to the volume content of the filler (solid particles idealized as rigid plates). This leads to an increasing inner shear rate of the pure matrix while the external shear rate remains constant [1], [2],

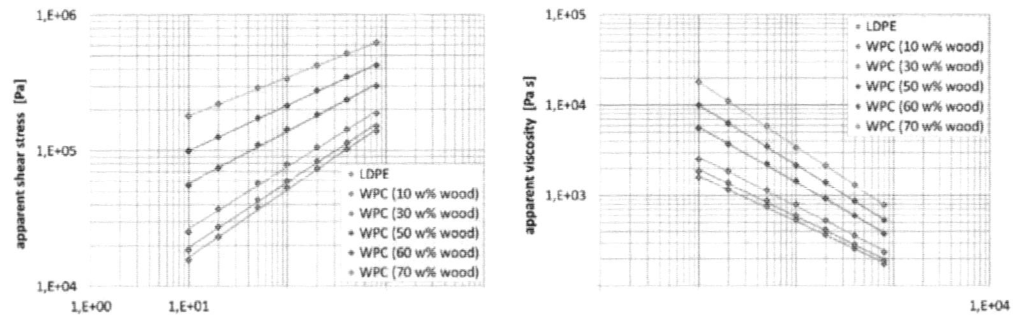

Figure 1. Flow and viscosity functions of LDPE-based WPC with varying wood content (T=190°C)

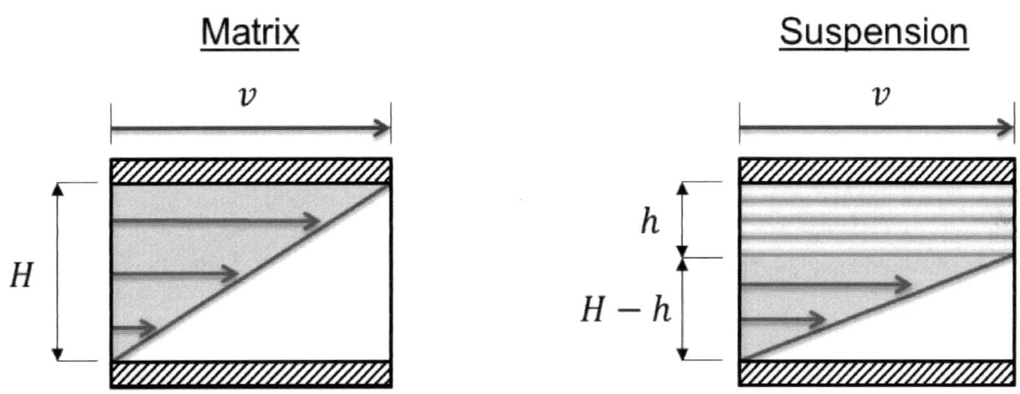

Figure 2. Idealization of flow processes in suspensions according to the concept of the shear-stress-equivalent inner shear rate

The external shear $\dot{\gamma}_0$ rate is calculated as:

$$\dot{\gamma}_0 = \frac{v}{H} \tag{1}$$

The inner shear $\dot{\gamma}_c$ rate is given by:

$$\dot{\gamma}_c = \frac{v}{H-h} \tag{2}$$

The relative increase of inner shear rate is expressed by the shift factor B:

$$B = \frac{\dot{\gamma}_c}{\dot{\gamma}_0} = \frac{H}{H-h} \tag{3}$$

The flow behavior of a shear thinning fluid, such as polymer melt, is generally described by the power law of Ostwald / de Waele (figure 3). According to the concept of shear-stress-equivalent inner shear rate, the relative increase of inner shear rate is attributed only to the volumetric content of the filler particles. Interaction effects of any kind need to be negligibly small ($n_c = n_0 = n$). On these conditions, the higher values for shear stresses of a shear thinning suspension can be calculated in accordance with shift factor as follows:

$$\tau_0 = K_0 \cdot \dot{\gamma}_c^{\,n_0} = \left(\frac{\dot{\gamma}_c}{\dot{\gamma}_0}\right) \cdot \tau_0 = B^{n_0} \cdot \tau_0 \tag{4}$$

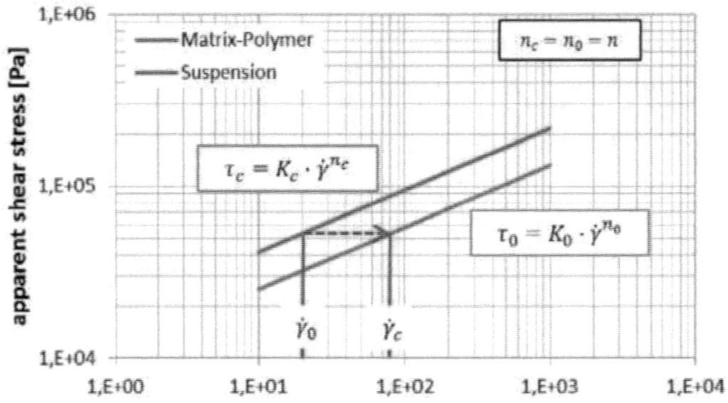

Figure 3. Comparison of flow functions (matrix-polymer and suspension) when no interactions effects

2. Materials and methods

2.1. Materials

In this study two different types of matrix polymers were used. On one hand the low-density polyethylene (LDEP, Lupolen 2420 K), which is highly branched and on the other hand the polypropylene (PP, Moplen HP 501L), which is a homopolymer with linear structure.

Softwood flour (Lignocel® BK 40/90) as well as hardwood flour (Lignocel® HB 120) were used as filler particles in this study (figure 4 and figure 5). The wood filler differ in particle size and compressibility.

Figure 4. Lignocel® BK 40/90 magnification: left: 50x, right: 100x

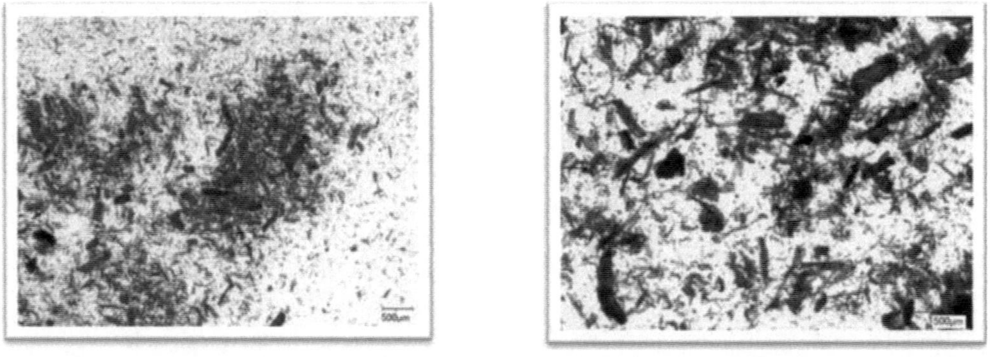

Figure 5. Lignocel® HB 120 magnification: left: 50x, right: 100x

IOP Conf. Series: Journal of Physics: Conf. Series **790** (2017) 012017 doi:10.1088/1742-6596/790/1/012017

2.2. Rheometry

All rheological measurements were performed on a high pressure capillary rheometer (CEAST Smart RHEO 5000, Instron®) using a round die with a diameter of 2 mm and a length of 40 mm. Before the measurements begun, the WPC was dried in an oven at 105°C for at least 3 hours.

The measured data were analysed on the basis of the apparent shear stress and shear rates. The apparent shear stress is calculated as:

$$\tau_{ap} = \frac{p \cdot r}{2 \cdot L} \tag{5}$$

with:

τ_{ap}	apparent shear stress	[Pa]
p	measured pressure	[Pa]
r	radius capillary	[mm]
L	length of capillary	[mm]

The apparent shear rate is given by:

$$\dot{\gamma}_{ap} = \frac{4 \cdot \dot{V}}{\pi \cdot r^3} = 4 \cdot \frac{R^2}{r^3} \cdot v \tag{6}$$

with:

$\dot{\gamma}_{ap}$	apparent shear rate	[l/s]
R	radius of cylinder/piston	[mm]
v	piston velocity	[mm]

3. Results and discussion

3.1. LDPE filled with softwood flour Lignocel® BK 40/90

For the unfilled LDPE and the LDPE filled with different content of softwood flour BK 40/90 the flow functions are displayed in figure 6. All flow functions show strongly pronounced shear thinning flow behaviour. The corresponding power law model parameters are reported in table 1.

It seems obvious that the fibrous structure of the wood particles has a great impact on the flow behaviour of WPC melts. Due to their eometric shape, these particles have a great tendency to interact with one another.

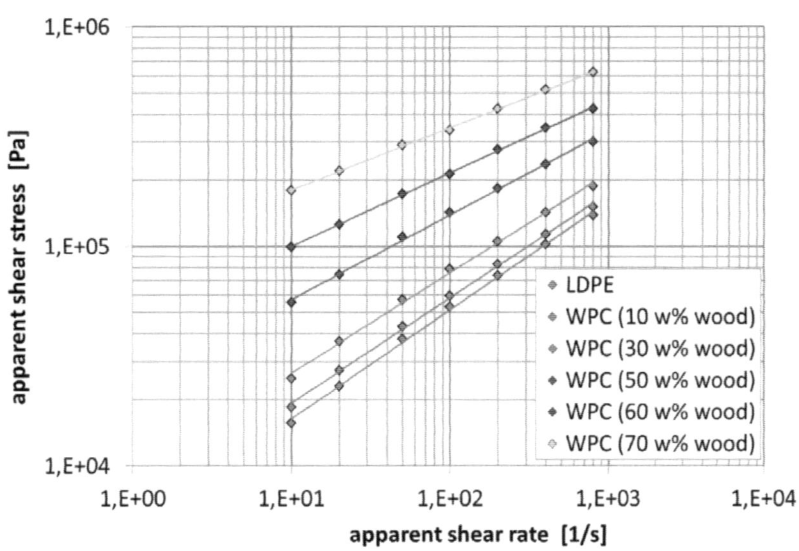

Figure 6. Flow functions of LDPE-WPC with varying wood contents (T = 190°C)

IOP Conf. Series: Journal of Physics: Conf. Series **790** (2017) 012017 doi:10.1088/1742-6596/790/1/012017

Table 1. Power law model parameters for LDPE filled with softwood flour BK 40/90

LDPE-WPC	softwood flour volumetric content [%]	K [Pa s^n]	n [-]
LDPE-BK0	0	5231	0.496
LDPE-BK10	5.9	6426	0.479
LDPE-BK30	19.4	9262	0.457
LDPE-BK50	36.0	23627	0.385
LDPE-BK60	45.7	46619	0.333
LDPE-BK70	56.7	93922	0.284

The decreasing flow index (n) with increasing wood content shows that interactions effects are by no means negligibly small. Therefore, the shift factor B, which represents the relative increase of the inner shear rate, becomes dependent on both the volumetric filler concentration and the applied shear stress. [3]

$$B(\tau)^{n_0} = \frac{K_c}{K_0^{\frac{n_0}{n_c}}} \cdot \tau^{\left(\frac{n_0}{n_c}-1\right)} \quad (7)$$

The term in the exponent of equation (7) expresses the intensity of interparticular interactions. For that reason, it can be referred to as interaction exponent χ.

$$B(\tau)^{n_0} = \frac{K_c}{K_0^{\chi}} \cdot \tau^{(\chi-1)} \quad (8)$$

Table 1 shows that the consistency (K) is dependent on the volumetric content of the wood filler. In order to describe the quantitative impact of the wood content on the consistency a mathematically correlation has been derived (figure 7).

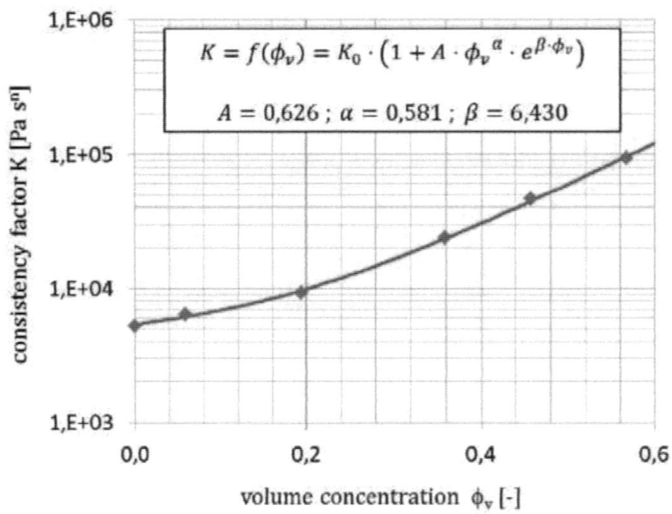

Figure 7. Dependency between consistency factor and wood content of LDPE filled with softwood flour BK 40/90 (T = 190°C)

In addition, a logarithmic correlation between interaction exponent *(x)* and consistency factor (K) was found (figure 8).

On the basis of these relationships, the influence of the volumetric filler concentration and the applied shear stress on the shift factor B can now be described mathematically for LDPE filled with softwood flour BK 40/90 with good accuracy (figure 9).

This approach is of great practical importance due to the good consistency to the measured values. Based on this approach the flow curves of WPC formulations based on LDPE, which is filled with softwood flour BK 40/90 can now be estimated in consideration of different volumetric contents of wood and the applied shear stress and shear rate, respectively (figure 10).

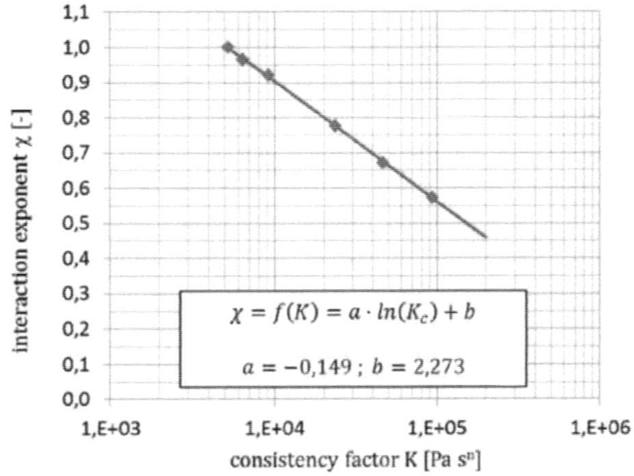

Figure 8. Dependency between interaction exponent and consistency factor for LDPE filled with softwood flour BK 40/90 (T = 190 °C)

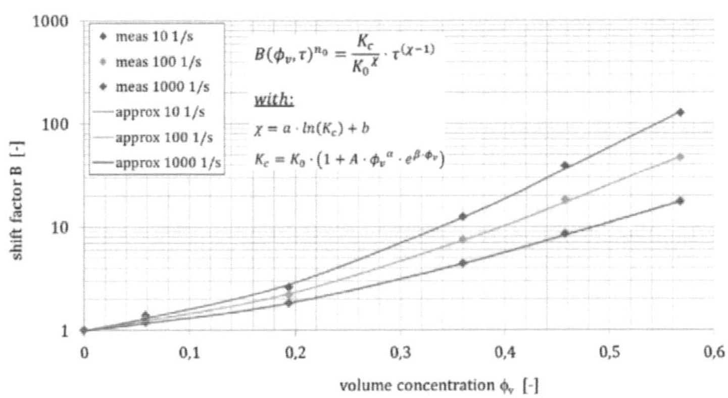

Figure 9. Shift factor B as a function of wood content and applied shear rate for LDPE filled with softwood flour BK 40/90 (T = 190 °C

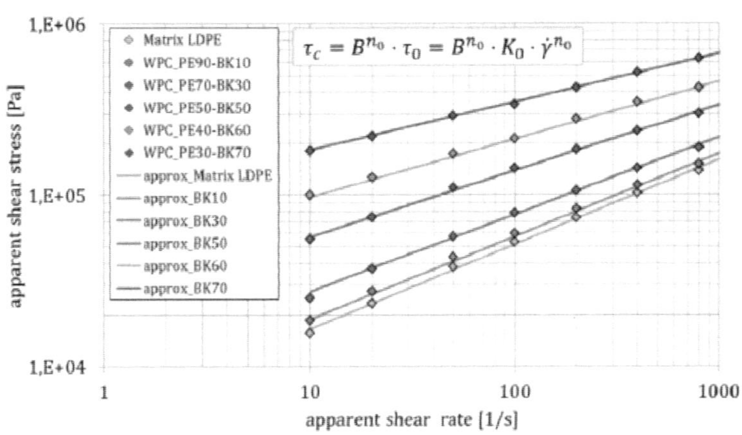

Figure 10. Comparison of estimated and measured flow functions of LDPE filled with softwood flour BK 40/90 (T = 190°C)

3.2. LDPE filled with hardwood flour Lignocel® HB 120

To investigate the applicability of this approach to wood particles with different particle sizes and compressibility, WPC formulations were prepared based on LDPE filled with different contents of hardwood flour HB 120.

Table 2 shows the corresponding power law model parameters of these WPC formulations.

Table 2. Power law model parameters for LDPE filled with hardwood flour HB 120

LDPE-WPC	hardwood flour volumetric content [%]	K [Pa s^n]	n [-]
LDPE-HB0	0	5231	0.496
LDPE-HB10	6.1	6396	0.481
LDPE-HB30	20.1	9553	0.454
LDPE-HB50	37.0	20205	0.405
LDPE-HB60	46.9	38596	0.345
LDPE-HB70	57.9	71810	0.30

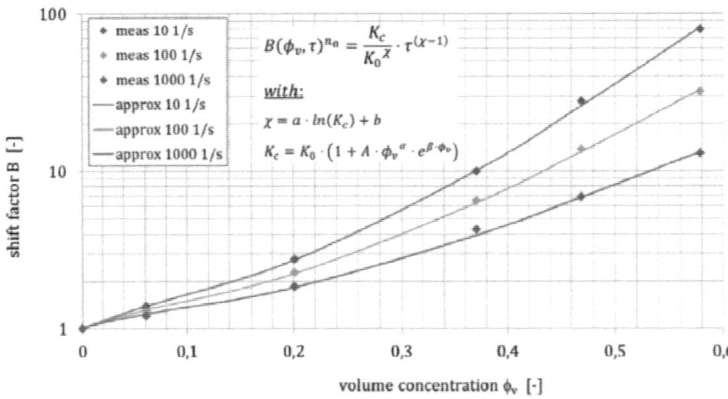

Figure 11. Shift factor B as a function of wood content and applied shear rate for LDPE filled with hardwood flour HB 120 (T = 190°C)

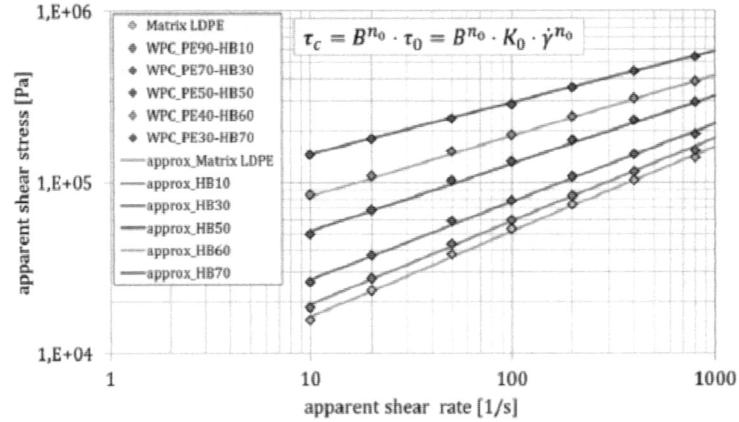

Figure 12. Comparison of estimated and measured flow functions of LDPE filled with hardwood flour HB 120 (T = 190°C)

It can be clearly seen in figure 11 and 12 that this approach is also appropriate to describe the influence of volumetric filler concentration and applied shear stress on the shift factor B for WPC formulations based on LDPE filled with hardwood flour HB 120. The estimated values correlate to the measured values with good consistency.

3.3. PP filled with softwood flour Lignocel® BK 40/90

To investigate the applicability of this approach to another matrix polymer, with different molecular structure and shear thinning flow behaviour, WPC formulations were prepared based on PP filled with different contents of softwood flour BK 40/90.

Table 3 shows the corresponding power law model parameters of these WPC formulations.

Table 3. Power law model parameters for PP filled with softwood flour BK 40/90

PP-WPC	softwood flour volumetric content [%]	K [Pa s^n]	n [-]
PP-BK0	0	12963	0.340
PP-BK10	5.9	14360	0.330
PP-BK30	19.5	21624	0.300
PP-BK50	36.1	49641	0.230
PP-BK60	45.9	70301	0.200
PP-BK70	56.9	109602	0.166

As a result of this investigation it can be concluded that this approach is also appropriate to calculate shift factor B for WPC formulations based on PP filled with softwood flour BK 40/90, by taking into consideration different volumetric filler concentrations and applied shear stresses (figure 13, 14).

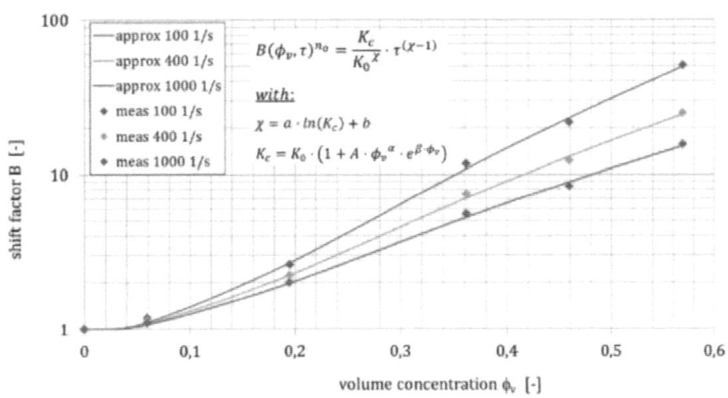

Figure 13. Shift factor B as a function of wood content and applied shear rate for PP filled with softwood flour BK 40/90 (T=200°C)

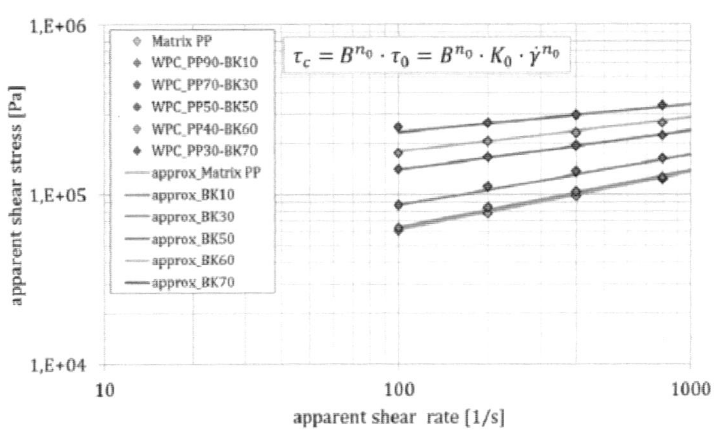

Figure 14. Comparison of estimated and measured flow functions of PP filled with softwood flour BK 40/90 (T = 200°C)

4. Conclusion

In this study, the flow behaviour of the investigated WPC formulations could be mathematically described by a modified approach, which is based on the concept of shear-stress-equivalent inner shear rate. It has been proven, that this approach is able to take account of interaction effects that typically occur in highly filled WPC formulations. On this basis, flow functions have been estimated with good consistency to the measured values, independently of the molecular structure of the matrix polymer (linear or highly branched structure) or the particle size and compressibility of the wood particles.

References

[1] M. Pahl, W. Geißle und H.-M. Laun, Praktische Rheologie der Kunststoffe und Elastomere, Düsseldorf: VDI-Verlag GmbH, 1995
[2] B. Hochstein, Rheologie von Kugel- und Fasersuspensionen mit viskoelastischen Matrixflüssigkeiten, Universität Fridericiana Karlsruhe (Technische Hochschule), 1997
[3] N. Laufer, H. Hansmann, M. Koch, High Pressure Capillary Rheometer on Wood Plastic Composites with Variation of Wood Content and Matrix Polymer, 58th Internationale Wissenschaftliche Kolloquium, Ilmenau 2014

IOP Conf. Series: Journal of Physics: Conf. Series **790** (2017) 012018 doi:10.1088/1742-6596/790/1/012018

Investigation of the stress distribution around a mode 1 crack with a novel strain gradient theory

M Lederer and G Khatibi

Institute of Chemical Technologies and Analytics, Vienna University of Technology, Christian Doppler Laboratory for Lifetime and Reliability of Interfaces in Complex Multi-Material Electronics, Wien, Austria

E-mail: martin.lederer@tuwien.ac.at

Abstract. Stress concentrations at the tip of a sharp crack have extensively been investigated in the past century. According to the calculations of Inglis, the stress ahead of a mode 1 crack shows the characteristics of a singularity. This solution is exact in the framework of linear elastic fracture mechanics (LEFM). From the viewpoint of multiscale modelling, however, it is evident that the stress at the tip of a stable crack cannot be infinite, because the strengths of atomic bonds are finite. In order to prevent the problem of this singularity, a new version of strain gradient elasticity is employed here. This theory is implemented in the commercial FEM code ABAQUS through user subroutine UEL. Convergence of the model is proved through consecutive mesh refinement. In consequence, the stresses ahead of a mode 1 crack become finite. Furthermore, the model predicts a size effect in the sense "smaller is stronger".

1. Introduction

It is now more than 100 years ago that Inglis [1] presented his celebrated calculation of the stress field of an elliptical hole which takes on the shape of a sharp crack when the minor axis decreases relative to the major axis. At the tip of a sharp crack, the stress concentration factor led to a stress singularity. The solution of Inglis is exact within linear elastic fracture mechanics (LEFM). Or, to put it in other words, the solution is restricted to the case where Hooke's law is generally valid.

However, even in the elastic deformation regime deviations from Hooke's law are nowadays accepted. In the 1960s, theories of couple stresses and strain gradient elasticity were introduced [2-6]. In fact, the appearance of strain gradients is related to a bending of the microstructure. But since the microstructure tries to preserve its integrity, the material reacts with stiffness against bending. In consequence, couple-stresses are generated which appear in addition to the stresses following from Hooke's law. Thereby, the couple-stresses are connected to their corresponding components of the strain gradient tensor. The versions of strain gradient elasticity mentioned all have in common that the magnitude of stress concentrations is in general reduced. For the case of a circular hole in a field of simple tension Mindlin [5] obtained stress concentration factors depending on Poisson ratio and size of the hole. In the macroscopic limit for large holes, the result of strain gradient theory coincides with the classical result of Inglis. But for small holes, the stress concentration is reduced.

When the strain gradient theory of Toupin and Mindlin is applied to a sharp crack in a stress field, however, the stress singularity at the crack tip is still present. In spite of the general tendency to reduce stress concentrations, the strain gradient effects seem to be incapable of reducing the stress concentration at the tip of a sharp crack to a finite value [8]. Nevertheless, the aim of removing stress singularities seems to be an important issue of elasticity theory, and several researchers are still convinced that singularities of strain should somehow be removed at least for cases of static equilibrium [9, 10].

In the present study, we therefore employ a new version of strain gradient elasticity which consequently leads to a removal of the stress singularity at the tip of a mode 1 crack. The main difference of the present approach compared to the older versions of strain gradient elasticity is that we consequently use a symmetric stress tensor throughout the model calculations, while Toupin and Mindlin postulated an asymmetric stress tensor.

2. Constitutive equation of isotropic strain gradient elasticity

The fundamental equation of our theory is an equation for the internal energy, whereby the deformation regime is restricted to small deformations. The model is reversible elastic. Thus, the internal energy writes as [11, 12]

$$W = \frac{1}{2}\int_V \left(\sigma_{ij}\epsilon_{ij} + \mu_{ijk}\eta_{ijk}\right) dV \tag{1}$$

where summation is carried out over repeated indices. σ_{iJ} is the Cauchy stress and ε_{iJ} is the linear strain. η_{ijk} are the strain gradients $\delta^2 u_i/\delta x_J \delta x_k$, where u_i are the displacements and x_i are the coordinates. μ_{ijk} are higher order stresses related to the strain gradients. In stable equilibrium, the energy of the system takes a minimum. Per definition, our theory is linear. The relation between higher order stresses μ_{ijk} and strain gradients η_{ijk} reads as

$$\mu_{ijk} = B\eta_{ijk} \tag{2}$$

where B is called the bending modulus. It should be noticed here that the energy density

$$W^{sg} = \frac{1}{2}B{\eta_{ijk}}^2 \tag{3}$$

related to the strain gradients is invariant with respect to rotations. Hence, the requirements for an isotropic material model are fulfilled. An even more general version of this theory was suggested by the present authors in reference [13], where B was consequently treated as sixth rank tensor leading to five independent bending moduli. The present version is a special case of the general theory. This means, we will here demonstrate that the consideration of one bending modulus is already sufficient to suppress the stress singularities at the tip of a mode 1 crack.

3. Finite Element implementation in ABAQUS through User subroutine UEL

The Finite Element implementation of this theory has already been explained elsewhere [12].

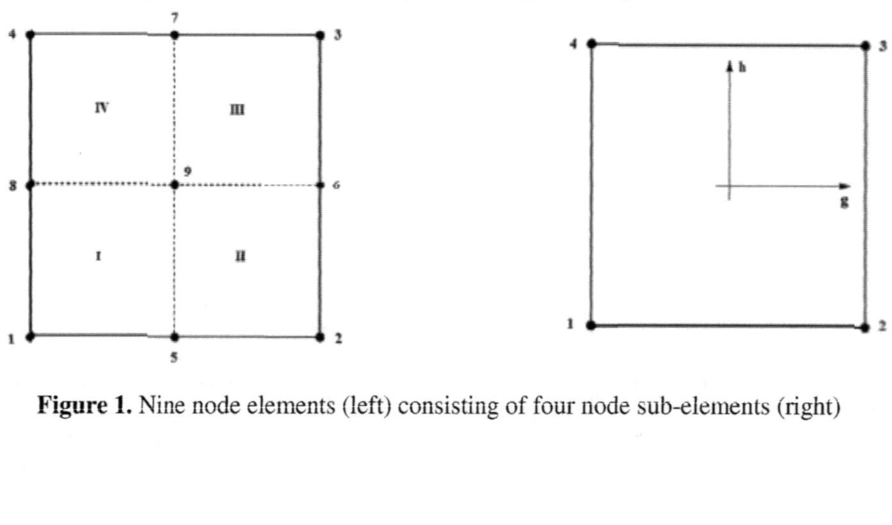

Figure 1. Nine node elements (left) consisting of four node sub-elements (right)

Therefore, only the fundamental principles of this approach are explained here. We have defined square nine node elements consisting of isoparametric four node sub-elements as depicted in figure 1. The values for the nodal forces are derived from a principle of virtual work which requires that the work done by external forces equals the energy stored in the volume element.

In strain gradient elasticity neighbouring elements must fulfill boundary conditions in order to achieve C_1 continuity. Otherwise strain gradients would occur mainly at the element boundaries, where their energy would be neglected. In order to avoid this problem, we have introduced an overlapping mesh technique, which is considered as the most precise approach to handle this issue.

4. Definition of an intrinsic material length scale

Owing to the fact that Young's modulus E and bending modulus B have different dimension, a size effect of strain gradient elasticity is expected. This behavior may be described by the definition of an intrinsic material length l_0 [6]. Samples or structures with size larger than l_0 will behave similar as bulk material, and Hooke's law will provide a reasonable good approximation of the material properties. On the other hand, structures with size of l_0 or below are dominated by a size effect in the sense smaller is stronger. In order to define the material length l_0, we compare the contributions to the elastic energy arising either from strains or from strain gradients. Let us therefore assume a quadratic volume element with constant strain gradient $\wedge_{112}$ = ^m. The strain in the middle of this volume element is assumed to be 0. The elastic energy of this square with side length l may be calculated using equations (1) and(2). Thus, one gets for the elastic energy of the strains

$$W^{strain} = \int_{-1/2}^{1/2}\int_{-1/2}^{1/2}\left(\frac{1}{2}\sigma_{xx}\varepsilon_{xx} + \frac{1}{2}\sigma_{xx}\gamma_{xx}\right)dxdy \quad (4)$$

where

$$\varepsilon_{xx} = y\ \eta_{112} \quad (4a)$$

and the engineering shea

$$\gamma_{xy} = x\ \eta_{121}. \quad (4b)$$

The stresses occurring in equations (4a, b) are obtained from the condition $\varepsilon_{yy} = 0$ in combination with the plane stress approximation. Taking into account $\eta_{112} = \eta_{21} = \eta$, one obtains

$$W^{strain} = \frac{E\eta^2 l^4}{48}\left(\frac{2}{1-\nu^2} + \frac{1}{1+\nu}\right) \quad (5)$$

where ν is the Poisson ratio. For the same volume element the elastic energy related to the strain gradients is

$$W^{sg} = \frac{1}{2}B\eta_{112}{}^2 l^2 + \frac{1}{2}B\eta_{121}{}^2 l^2 = \frac{1}{2}B\eta^2 l^2 \quad (6)$$

The total energy of this volume element is W^{trran} + W^{g}. Now we are in the position to define the intrinsic material length l_0 by the condition that the energies W^{strain} and W^{sg} shall be equal for the square with side length l_0. This condition yields

$$l_0 = \sqrt{\frac{48B}{E\left(\frac{2}{1-\nu^2} + \frac{1}{1+\nu}\right)}} \quad (7)$$

5. Results

Now, the theory is applied to a crack in a transverse field of uniform uniaxial tension. According to the definition of a mode 1 crack, one should consider an infinite sample. But since the simulation of an infinite sample leads to numerical difficulties, we will here present a series of simulations where length and width of the sample are twice the crack length. Afterwards, it is demonstrated how an increase of sample size relative to the crack length influences the results for the stress at the crack tip.

We here assume a Young's modulus E of 68 GPa and a Poisson ratio ν of 0.33. These are values for aluminum, which seems to be a good example for a solid with nearly isotropic elasticity. Unfortunately, an experimental value for the bending modulus B of aluminum is not known. But when the material length l_0 is used as unit of the length scale, the value of the bending modulus may be obtained from equation (7). Therefrom, we get B = 4.245 [GPa l_0^2].

In figure 2, one can see a plot of the von Mises stress for a sample exposed to a tensile stress of 1 MPa with a crack length equal to l_0. For this simulation a rather coarse mesh consisting of 40500 nodes was used. The maximum of the von Mises stress appeared at the crack tip where a value of 1.2728 MPa was found.

In order to check the mesh convergence of the solution another two simulations were performed with refined meshes using 251001 and 1002001 nodes for a quarter of the sample, respectively. Due to the mesh refinement, the von Mises stress at the crack tip was increased to 1.2867 MPa in the second and to 1.2885 MPa in the third simulation. This is only a small increase in stress compared to a reduction of the element length to $^1/_5$ and $^1/_{10}$ of the first mesh, respectively. The convergence norm reflecting the largest imbalance of nodal forces divided by the average of nodal forces was better than 10^{-6} for each of the 3 simulations.

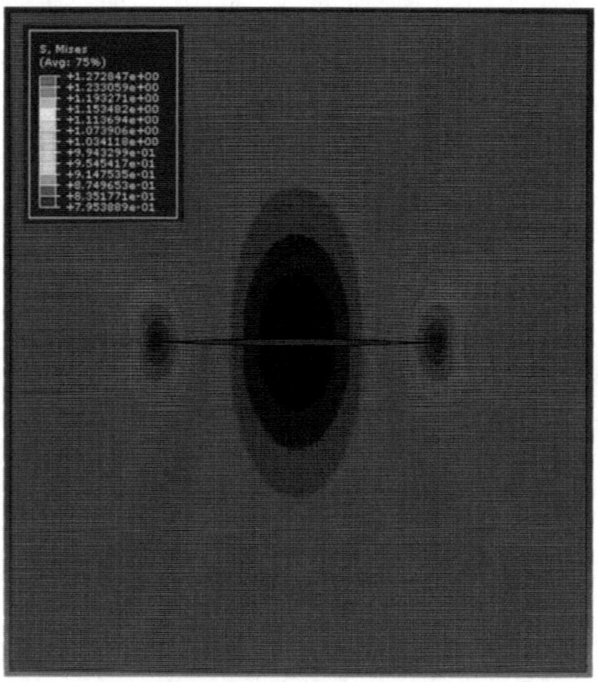

Figure 2. Plot of the von Mises stress [MPa] for a tensile load of 1 MPa and a crack length equal to the intrinsic material length l_0

A plot of the stress distribution using the finest mesh is depicted in figure 3. For comparison, a simulation of the model without strain gradient effects is shown in figure 4. Here, the maximum of von Mises stress is 23.42 MPa. This plot shows the typical characteristics of a stress singularity, where the maximum value for the stress observed in the plot is governed by the element size.

Let us now investigate how the crack length influences the results for the stress distribution within strain gradient elasticity. Simulations were performed for crack lengths between $^1/10$ and 100 l, whereby the proportions of crack length to sample length and width remained the same, respectively.

IOP Conf. Series: Journal of Physics: Conf. Series **790** (2017) 012018 doi:10.1088/1742-6596/790/1/012018

The results of this comparison are shown in table 1. They indicate a size effect in the sense smaller is stronger. In the case of short micro-cracks stress concentrations are drastically reduced, and in the limit to infinitesimal cracks they disappear completely.

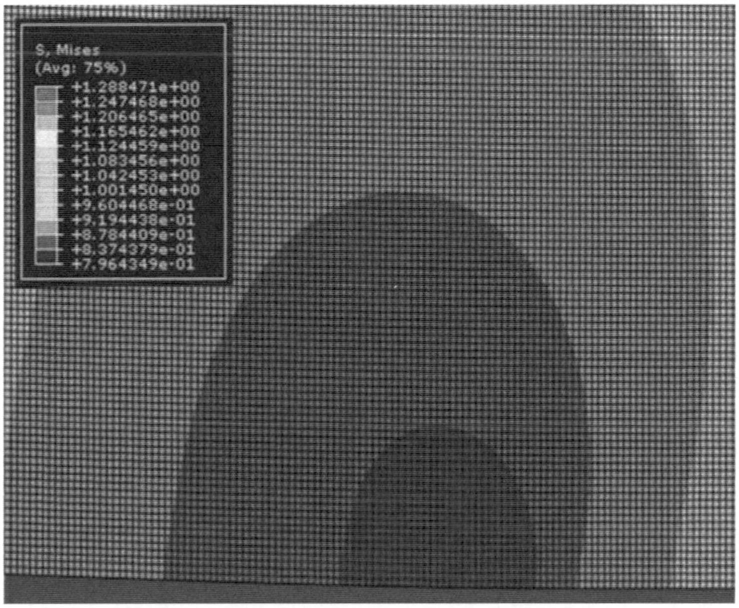

Figure 3. Plot of the von Mises stress [MPa] obtained for % of the sample with refined mesh of 1002001 nodes. Tensile stress: 1 MPa. Crack length = l_0. The region around the crack tip is shown

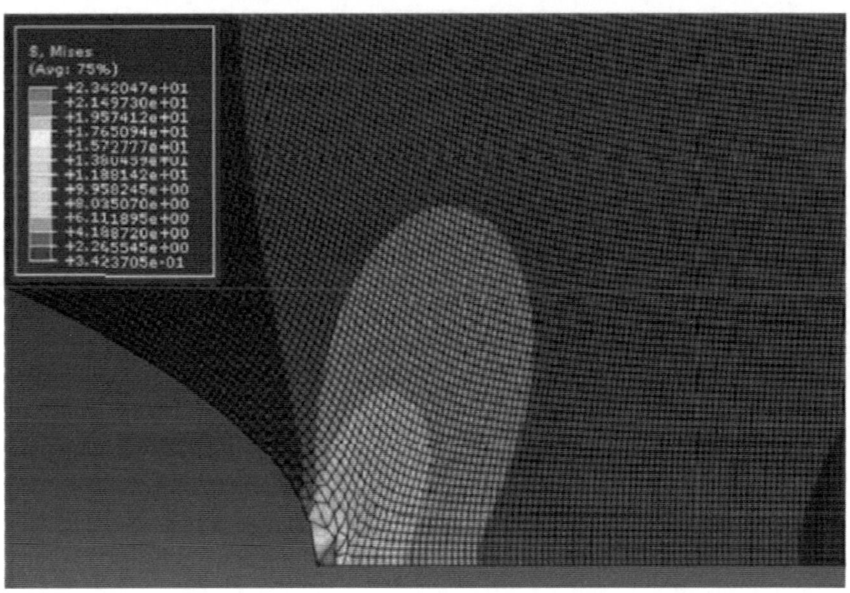

Figure 4. Plot of the von Mises stress [MPa] obtained from a simulation without strain gradient effects. Tensile stress, crack length and mesh size are the same as in figure 3.

Table 1. Results for the von Mises stress [MPa] at the crack tip for crack lengths from ¡$_0$/10 to 100 ¡, whereby the proportions for crack length relative to length and width of the sample remained the same. Tensile stress: 1 MPa. Number of nodes in the mesh: always 1002001.

Crack length	**von Mises stress at crack tip [MPa]**
lfl/10	1.003312
lo	1.288471
10 l_0	4.925857
100 *l0*	14.55232

In the next series of simulations, it was investigated how the stress values change when sample length and width are increased relative to the crack length. The results are summarized in table 2.

Table 2. Results for the von Mises stress [MPa] at the crack tip for a crack length of l_0 and various sample dimensions. For this comparison the element size used for the meshes was kept constant. This led to meshes with 564001, 1002001 and 1565001 nodes for sample widths of $3l_0$, $4l_0$ and $5l_0$, respectively.

Sample length and width	**von Mises stress at crack tip [MPa]**
3 l_0	1.271981
4 l_0	1.265688
5 l_0	1.262201

In general, an increase of the sample size relative to the crack length slightly reduces the stress value at the crack tip, since the additional material increases the stiffness of the structure.

6. Summary and conclusions

A new approach of strain gradient elasticity was employed to study the stress distribution around a mode 1 crack. Thereby, the stress singularity at the crack tip has been removed, and a pronounced size effect has been found. The stress values at the tips of short cracks are lower compared to stresses observed at longer cracks.

Acknowledgments
The financial support by the Austrian Federal Ministry of Science, Research and Economy and the National Foundation for Research, Technology and Development is gratefully acknowledged.

References

[1] Inglis C E, 1913, *Transactions of the Institute of Naval Architects* **55**, 219-241.
[2] Toupin R, 1962, *Arch. Rat. Mech. Anal.* **11**, 385-414.
[3] Toupin R, 1964, *Arch. Rat. Mech. Anal.* **17**, 85-112.
[4] Mindlin R D and Tiersten H, 1962, *Arch. Rat. Mech. Anal.* **11**, 415-448.
[5] Mindlin R D, 1963, *Experimental Mechanics* **3**, 1-7.
[6] Mindlin R D, 1964, *Arch. Rat. Mech. Anal.* **16**, 51-78.
[7] Mindlin and Eshel, 1968, *Int. J. Solids Structures* **4**, 109-124.
[8] Sternberg E and Muki R, 1967, *Int. J. Solids Structures* **3**, 69-95.
[9] Aifantis E, 1992, *Int. J. Eng. Sci.* **30**, 1279-1299.
[10] Askes Hand Aifantis E, 2011, *Int. J. Solids Structures* **48**, 1962-1990.
[11] Lederer M, Khatibi G, Thuiner D and Danninger H, 2014, *Key. Eng. Mat.* **592-593**,pp 246-249
[12] Lederer M, Magnien J, Khatibi G and Weiss B, 2015,J. of Phys.: Conf.Series **602**, 012020
[13] Lederer M and Khatibi G, 2015, *WSEAS Trans. Appl.Theor.Mech.*, **10**, 213-219.

On the relation between the energy of a distorted crystal lattice and the bending modulus of strain gradient elasticity

M Lederer and G Khatibi

Institute of Chemical Technologies and Analytics, Vienna University of Technology, Christian Doppler Laboratory for Lifetime and Reliability of Interfaces in Complex Multi-Material Electronics, Wien, Austria

E-mail: martin.lederer@tuwien.ac.at

Abstract. It is a well known fact that linear elastic fracture mechanics (LEFM) predicts stress singularities at the tips of sharp cracks, at sharp edges, at corners and at the surface of material transitions. However, from the viewpoint of the strengths of atomic bonds it is clear that only finite forces may be present at the tip of a stable crack. Therefore, theories of strain gradient elasticity were developed which reduce the values of stress concentrations. Within these theories a bending modulus is postulated which introduces an increased stiffness of the crystal lattice against bending. In the present study, the value of this bending modulus is evaluated on the basis of the electrostatic energy of a bent crystal lattice. This is done for the face centred cubic structure of NaCl. In fact, results for the bending modulus could be obtained although they depend on the crystal size.

1. Introduction

Nowadays, stress singularities are commonplace in linear elastic fracture mechanics. Inglis [1] was the first to evaluate the stress ahead of a sharp crack in a transverse stress field. This calculation of Inglis is said to be exact within linear elastic fracture mechanics (LEFM). Since then singularities of stress were found by other authors on numerous occasions. On the other hand, there are others who believe that singularities of stress or strain should be avoided [2-5] within continuum mechanics. But this can only be achieved through a change of the physical assumptions made in Hooke's law. One approach of this kind was proposed by Mindlin [6] who assumed an elastic bending stiffness in addition to the elastic response described in Hooke's law. Thereby, the bending stiffness was related to a bending modulus which has the dimension of a force. In fact, the theory of Mindlin could reduce the values of stress concentrations in some cases. But unfortunately, the theory could not reduce the stresses around sharp cracks to finite values. Nevertheless, the stress singularities can successfully be removed after a modification of Mindlin's theory, which was recently proposed by the authors of the present article [7, 8]. The main idea behind this approach is that the asymmetric stress tensor used by Mindlin is replaced by a symmetric stress tensor. This version of the theory, which was implemented in the commercial FEM code ABAQUS through user subroutine UEL, has successfully been applied to several cases of stress concentrations [7-9]. Thereby, stresses singularities were consequently removed.

However, one disadvantage of strain gradient elasticity seems to be that reliable values for the bending modulus are still not known. In this context it should be said that nowadays there exist several versions of strain gradient elasticity which can include different numbers of bending moduli. In the

IOP Conf. Series: Journal of Physics: Conf. Series **790** (2017) 012019 doi:10.1088/1742-6596/790/1/012019

case of isotropic materials, the most general version includes 5 independent bending moduli, while simplified versions may be formulated using only one bending modulus. In the anisotropic case the calculations become very complicated. Mindlin [10] reported that in his version of the theory there are all together 903 independent material constants including the usual 21 elastic constant of the triclinic system. The large number of constants makes it difficult to measure their values. Until today, precise values for the bending moduli are not known.

Therefore, the present article is devoted to a numerical computation of the bending modulus. The computation is performed for the NaCl crystal structure, because the electrostatic energy of this crystal lattice may be computed in straightforward manner. The computations reveal that the value of the bending modulus is governed by long range electrostatic interactions.

2. Assumptions at the level of continuum theory

The constitutive equation for the energy density w of the strain gradient theory used in the present investigation writes as [7]

$$w = 1/2\,(\sigma_{ij}\varepsilon_{ij} + \mu_{ijk}\boldsymbol{\eta}_{ijk}) \tag{1}$$

where summation is carried out over repeated indices. c_{iJ} is the Cauchy stress and s_{iJ} is the linear strain. η_{ijk} are the strain gradients $\delta^2 u_i/\delta x_J \delta x_k$, where u_i are the displacements and x_i are the coordinates. μ_{ijk} are higher order stresses related to the strain gradients. The relation between higher order stresses μ_{ijk} and strain gradients η_{ijk} reads as

$$\mu_{ijk} = B\,\eta_{ijk} \tag{2}$$

where B is called the bending modulus. Further, the contribution of higher order stresses to the internal energy may be rewritten as

$$W^{sg} = 1/2\,B\eta_{ijk}{}^2 \tag{3}$$

It should be mentioned here that also more general types of strain gradient elasticity have been developed. In reference [8] an energy density was used which may be considered as the most general isotropic invariant expression. Therefrom, a sixth rank bending tensor was derived which has 5 independent components. The model used in the present investigation is a special case of this general theory. Aside from the isotropic models also anisotropic versions of strain gradient elasticity have been developed [10]. But for sake of simplicity, we will here concentrate on a model with only one bending modulus.

3. Energy of the deformed crystal lattice

There are several contributions to the internal energy of a bent crystal lattice. First, one has to consider the cohesive energy of the material, second there are the ordinary strains and third we get the contribution of the additional bending stiffness. All these contributions are a result of the electrostatic interactions between the atoms together with repulsive forces originating from quantum mechanical principles or from the kinetic motion of atoms.

3.1 Inter-atomic potential of Na+ and Cl⁻ ions and the electrostatic energy

For alkali halides the inter-atomic potential of two atoms may be written in the form [11]

$$\phi_{ij}(r_{ij}) = \frac{q_1 q_j}{4\pi\varepsilon_0 |r_{ij}|} + A_{ij} exp\left(\frac{-r_{ij}}{B_{ij}}\right) - \frac{C_{ij}}{r_{ij}{}^6} - \frac{D_{ij}}{r_{ij}{}^8} \tag{4}$$

where we can distinguish between short range and long range interactions. When the coefficients of this equation are known for some material, then one may determine the value of the elastic constants through molecular dynamic simulations. However, in the present study we are more interested in the value for the bending modulus. In this respect long range interactions play an important role. We will therefore try to extract the value of the bending modulus from the electrostatic energy

$$E^{electrostatic} = \frac{1}{2}\sum_{i\neq j}^{N}\sum_{j=1}^{N}\frac{q_i \cdot q_j}{4\pi\varepsilon_0 \cdot |r_{ij}|} \quad (5)$$

of a bent crystal lattice. Thereby, the cohesive energy of the undeformed crystal and the energy of the ordinary strains are to be subtracted from the overall energy in order to obtain the energy contribution of bending. But in contrast to molecular dynamic simulations, we will here neglect the influence of the atomic motion. We claim that the repulsive forces caused by the motion of atoms are very similar in homogeneously strained and in bent crystals. Therefore, we here concentrate on the potential energy instead of simulating the atomic motion. Thereby, the amount of computation time is drastically reduced.

3.2 Comparison of homogeneously strained and bent crystals

When the energies of homogeneously strained and bent crystals are compared, a size effect must be taken into account. Therefore, only samples with equal numbers of atoms should be compared. Moreover, the strain of a bent crystal can only be represented by an ensemble of homogeneously strained samples. Consequently, the bent crystal is divided into fibers of different lengths, and for each length a homogeneously strained crystal is constructed which has the same length as the fiber. Thereby, the cross contraction is considered in either case: in the bent crystal as well as in the homogeneously strained samples. Thus, the electrostatic energy of the homogeneously strained crystal corresponding to the strain of the bent crystal is obtained from the ensemble average $\bar{E}_{strained}^{electrostatic}$ representing all fiber lengths.

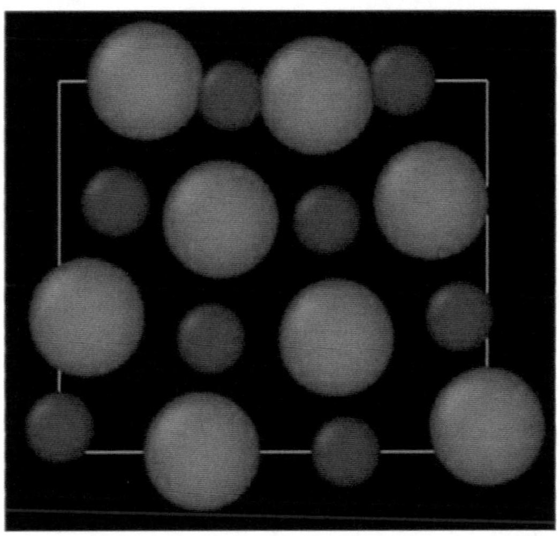

Figure 1. The atomic structure of a bent crystal, schematically.

The atomic structure of the bent crystal is depicted in figure 1, schematically. A neutral fiber of the same length as the undeformed sample is assumed in the middle of the cross section. Cross contractions were considered according to a Poisson ratio of v = 0.252. The lattice constant of NaCl is a = 0.564 nm. Computations were performed for a two-dimensional and for a three-dimensional model. But due to a size effect, only the three-dimensional model seems to be relevant for the real material properties. One derives the contribution of strain gradients to the internal energy through the equations

$$\boldsymbol{E}_{\mathbf{strained}}^{\mathbf{electrostatic}} - \bar{\mathbf{E}}_{\mathbf{strained}}^{\mathbf{electrostatic}} = \boldsymbol{E}^{sg} = \frac{1}{2}\int_V \sum_{\eta_{ijk}} \boldsymbol{B} \cdot \boldsymbol{\eta}_{ijk}{}^2 \boldsymbol{dV} \quad (6)$$

where the strain gradients j_{ijk} are inserted according to their geometrical values used in the computation. This means, we have used the equation

IOP Conf. Series: Journal of Physics: Conf. Series **790** (2017) 012019 doi:10.1088/1742-6596/790/1/012019

$$|K| = |\eta_{112}| \tag{7a}$$

for the curvature K in the case where the radius of curvature

$$r = \frac{1}{K} \tag{7b}$$

points exactly along the direction of the y-axis. Since the derivatives with respect to $\delta x\delta y$ and $\delta y\delta x$ must be equal, one gets $\eta_{112} = \eta_{121}$. Furthermore, we have assumed that $\eta_{222} = v\ \eta_{112}$, because of the cross contraction. The remaining components of the strain gradient tensor were assumed to be zero. During the computation it has been verified that the values which we get from equation (6) for the bending modulus are independent from the radius of curvature chosen for the samples. But since our theory of strain gradient elasticity is restricted to small deformations, the radii of curvature in the computation were always much larger than the sample size.

The computation was repeated for different sample sizes. The samples always had cubic shape, and the largest sample included 64000 atoms. It was found that the value obtained for the bending modulus strongly depends on the sample size.

4. Results

Figure 2 shows the results for the energy E^{sg} per atom in units of $e^2/\pi\varepsilon_0 a$, where e is the elementary charge, s_0 is the vacuum permittivity and a is the lattice constant. The lattice constant of NaCl is twice the atomic distance between neighbouring ions of Na and Cl. In this computation the radius of curvature for the neutral fibre was chosen to be 1000 times the lattice constant. The charges of the Na and Cl ions were assumed to be e and -e, respectively. The maximum number of atoms considered in the model corresponds to a side length of 40 atoms. Obviously, the computation shows a pronounced size effect. Unfortunately, the results obtained from the numerical calculation are only available for microscopic sample sizes due to the large computation time of the algorithm. In this context it should be noticed that the electrostatic energy of the homogeneously strained crystal $\bar{E}^{electrostatic}_{strained}$ was derived through an ensemble average over up to 40 replicas of the model representing the different strain values of the bent crystal. However, we are here interested in the model results which may be expected for mesoscopic sample sizes.

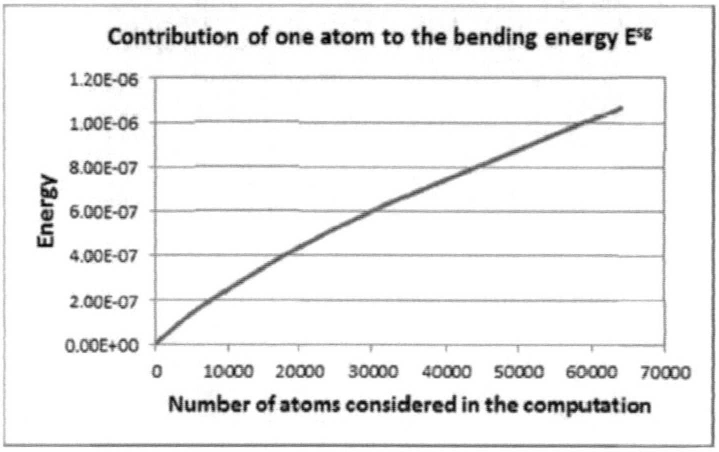

Figure 2. Energy per atom in units of e^2 / $4ns_0a$ versus number of atoms considered in thecomputation. The results show a strong dependence on the size of the model due to long range interactions.

We therefore extrapolate the results in order to estimate the bending modulus of a crystal with side length of 10 μm. Hence, we replot figure 2 to derive an estimate for larger samples: In fact, figure 3 permits an estimate regarding the extrapolation. If the energy per atom is divided by the square of the number of atoms along the side length, then numerical saturation is observed in the plot. In conclusion, the plot in figure 3 saturates numerically at a value of approximately $7 \cdot 10^{-10}$. On the basis of this estimate one obtains a value for the bending modulus of about 4.94 [mN] for a sample with side length of 10 μm. This side length of 10 μm could be a typical value for the grain size of a polycrystal. But it should be considered here that possible lattice defects are not included in this computation.

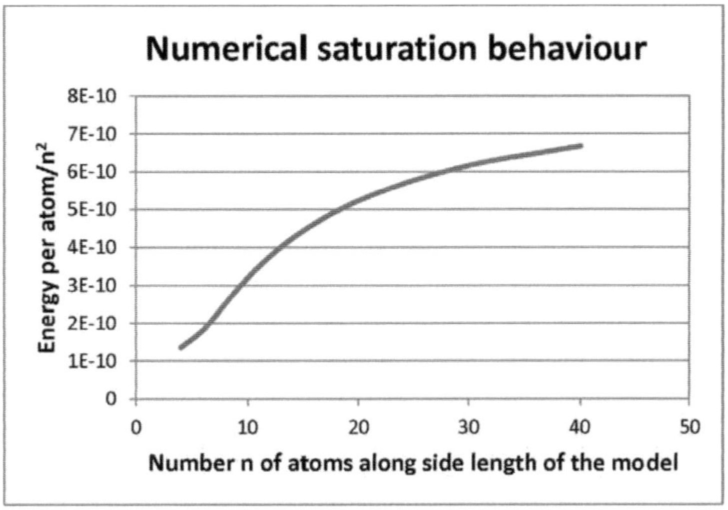

Figure 3. Numerical saturation behaviour. On the basis of this graph an extrapolation of the results to mesoscopic sample sizes was achieved

5. Discussion

It must be admitted that the extrapolation of the calculated value for the bending modulus to mesoscopic scale is somewhat speculative, because we have evaluated only a fraction of the desired sample size. On the other hand, it must be said that the computation is very time consuming due to the large number of summands occurring in equation (5). Within molecular dynamic simulations such problems are usually handled by introducing a cutoff radius for long range interactions. But such a method cannot be used here, since the value of the bending modulus mainly depends on long range forces.

Nevertheless, the computation led to the result that there exists some bending stiffness in excess of that stiffness what would be expected from Hooke's law alone. The main purpose of the inclusion of this bending stiffness in strain gradient elasticity is the removal of stress singularities at sharp edges, corners, material transitions and crack tips [7-9]. In this context it is sufficient that there exists some bending modulus $B > 0$. Insofar, the present investigation may be viewed as justification of the strain gradient approach even though a value for the bending modulus on mesoscopic level could only be estimated.

However, the size effect predicted by strain gradient elasticity needs to be reconsidered. So far, it has been assumed that a size independent bending modulus will introduce a strong size effect in the sense smaller is stronger. On the other hand, the present study shows that the bending modulus itself is size dependent. This may lead to competing interactions with various types of size effects in different materials.

The geometrical reason for the size effect observed in the computation seems to be that crystallographic layers which are further away from the middle show a larger missorientation of crystallographic directions due to bending of the structure. But for crystals with high dislocation density or high concentration of other lattice defects the present calculations might not be accurate.

6. Summary and conclusions

We have here elaborated a relationship between the theory of strain gradient elasticity and the electrostatic energy of a distorted crystal lattice. The results obtained here may be seen as a justification for strain gradient theory. Moreover, a quantitative value for the bending modulus could be estimated. The evaluations led to the conclusion that this value is size dependent or at least grain size dependent. This discovery came unexpected, since earlier publications assumed bending moduli which are size independent constants [12]. However, the result concerning size dependence is only valid if the concentration of crystallographic defects is not too high. If the periodicity of the lattice is disturbed, then long range interactions will occur randomly distributed. In consequence, the value for the bending modulus could saturate already at a microscopic length scale. Nevertheless, the existence of a bending modulus $B > 0$ leads to the consequence that the stress singularities of continuum mechanics are suppressed.

In this investigation we have used a simple version of strain gradient elasticity which uses only one bending modulus. For the future it will be interesting to extend such evaluations to more general versions of this theory including more independent bending moduli and also anisotropic behaviour. Furthermore, it seems eligible to perform such calculations for larger samples.

Acknowledgments

The financial support by the Austrian Federal Ministry of Science, Research and Economy and the National Foundation for Research, Technology and Development is gratefully acknowledged.

References

[1] Inglis C E, 1913, *Transactions of the Institute of Naval Architects* **55**, 219-241.
[2] Aifantis E, 1992, *Int. J. Eng. Sci.* **30**, 1279-1299.
[3] Altan S and Aifantis E, 1992, *Scripta Metall. Mater.* **26**, 319-324
[4] Ru C and Aifantis E, 1993, *Acta Mech.* **101**, 59-68
[5] Askes H and Aifantis E, 2011, *Int. J. Solids Structures* **48**, 1962-1990.
[6] Mindlin R D, 1963, *Experimental Mechanics* **3**, 1-7.
[7] Lederer M, Magnien J, Khatibi G and Weiss B, 2015, J. of Phys.: Conf. Series **602**, 012020
[8] Lederer M and Khatibi G, 2015, *WSEAS Trans. Appl. Theor. Mech.*, **10**, 213-219.
[9] Lederer M, Khatibi G, Thuiner D and Danninger H, 2014, *Key.Eng. Mat.* **592-593**,pp 246-249
[10] Mindlin R D, 1964, *Arch. Rat. Mech. Anal.* **16**, 51-78.
[11] Nevins D and Spera F J, 2007, *Molecular Simulation* Vol 33, **15**, 1261-1266
[12] see for instance: Maranganti R and Sharma P, 2007, Phys. Rev. Lett. **98**, 195504, 1-4

IOP Conf. Series: Journal of Physics: Conf. Series **790** (2017) 012020 doi:10.1088/1742-6596/790/1/012020

Concentration fields of the ternary systems and trajectory of phases in T-x-y Diagrams

V Lutsyk[1,2] **and A Zelenaya**[1]

[1] Institute of Physical Materials Science (Siberian Branch of Russian Academy of Sciences), Ulan-Ude, Russian Federation
[2] Buryat State University, Ulan-Ude, Russian Federation

E-mail: vluts@ipms.bscnet.ru

Abstract. The possibility of computer models of T-x-y diagrams for analysis two-, one- and zero-dimensional concentration fields are demonstrated using T-x-y diagram of systems CaO-SiO_2-Al_2O_3 and MgO-SiO_2-Al_2O_3. The characteristics of processes proceeding in the concentration fields can be analyzed with the help of diagrams of vertical mass balances, which show the increase or decrease of phases portions for each phase region. Calculation of crystallization paths was made.

1. Introduction

Construction of computer models of ternary systems significantly extends the possibilities of their investigation [1] by the calculation of crystallization paths and the analysis of concentration fields in order to obtain the information about the crystallization stages and the microconstituents. The schema of mono- and invariant equilibria is created on a basis of the data for invariant processes in the binary and ternary systems taking into account the existence and type of binary and ternary compounds. At the first stage of construction of 3D model for T-x-y diagrams, the invariant horizontal complexes at the temperatures of ternary points are reconstructed. Then they are completed by the ruled and unruled surfaces, and the phase regions are formed [2-4]. At the description of surfaces having a complex geometric structure (e.g. the large quantity of points on surface's contour, the presence of extremes or the gap of smoothness) we used the kinematic method based on the description of surfaces by means of the directing and forming curves [2-3, 5].

Systems CaO-SiO_2-Al_2O_3 and MgO-SiO_2-Al_2O_3 having a great practical importance are the objects of investigation. These systems are used at the description of properties of building materials [6-7] as well as at the characterization of geological objects [8-9].

2. Model of T-x-y diagrams of system CaO-SiO_2-Al_2O_3

The complexity of elaboration of T-x-y diagram model for system CaO-SiO_2-Al_2O_3 is associated with the ambiguous descriptions for binary systems, as well as the liquidus surfaces. There are different descriptions of types binary and ternary points, the character of binary compounds. The model of T-x-y diagram is based on the experimental data from [11]. It includes 16 liquidus surfaces and liquid immiscibility surface (fig. 1a), 128 ruled surfaces, 16 horizontal complexes at the temperatures of invariant points, 33 two-phase regions and 46 three-phase regions [4, 11-13].

Notations: incongruently melting compounds R_1 – $3CaO{\cdot}SiO_2$, R_3 – $3CaO{\cdot}2SiO_2$, R_6 – $3CaO{\cdot}Al_2O_3$, R_{10} – $CaO{\cdot}6Al_2O_3$; and congruently melting compounds R_2 – $2CaO{\cdot}SiO_2$, R_4 – $CaO{\cdot}SiO_2$, R_5 – $3Al_2O_3{\cdot}2SiO_2$, R_7 – $5CaO{\cdot}3Al_2O_3$, R_8 – $CaO{\cdot}Al_2O_3$, R_9 – $CaO{\cdot}2Al_2O_3$, R_{11} – $2CaO{\cdot}Al_2O_3{\cdot}SiO_2$, R_{12} – $CaO{\cdot}Al_2O_3{\cdot}2SiO_2$.

The most complex geometrical structure have the liquidus surfaces corresponding R_{11} and R_{12} with the contours $Q_4Q_5(Q_5,Q_6)Q_6E_6E_3(E_2,E_3)E_2Q_9$, $E_3(E_3,E_4)E_4(E_4,E_5)E_5Q_8(Q_7,Q_8)Q_7E_6$. There are maximum points on the monovariant liquidus lines E_2E_3, E_3E_4, E_4E_5, E_6E_3, Q_4Q_9, Q_5Q_6, Q_7Q_8 and inside surfaces due to the presence of ternary congruently melting compounds R_{11} and R_{12} (Fig. 1b-d). Such surfaces are constructed from several fragments with regards to the smoothness on boundaries of fragments [3, 5].

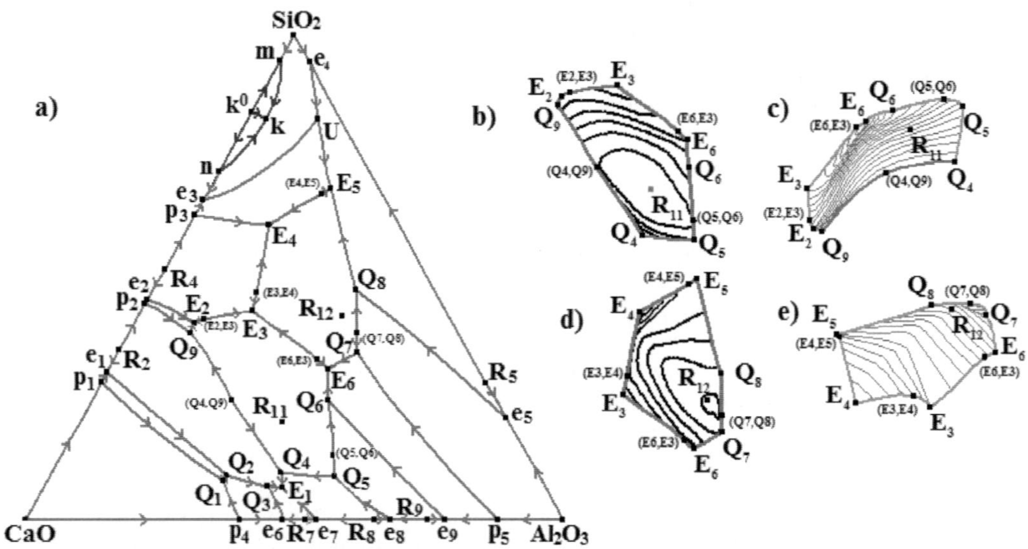

Figure 1. XY projection of liquidus surfaces (a), isolines and 3D models of liquidus surfaces R_{11} (b-c) and R_{12} (d-e)

Let's consider the analysis technology of concentration fields using liquidus field Al_2O_3 ($Al_2O_3p_5Q_7(Q_7,Q_8)Q_8e_5$) as an example. At the projection of considered fragment of phase diagram is divided into 46 concentration fields, including 12 two-dimensional, 23 one-dimensional and 11 zero-dimensional fields (fig. 2a). The diagrams of vertical mass balance have been calculated for each concentration field. They permit to obtain the lists of intersected phase regions and the crystallization schemes for given mass center over the entire temperature range. As a result, it was revealed that two fields coincide with adjacent fields of higher dimension by crystallization stage and microconstituents: R_{10}-2$\in R_{10}$-2-3 and 2$\in$2-3. Furthermore, three concentration fields coincide with adjacent fields by microconstituents, but differ by the crystallization stages: Q_7-5$\in$4-Q_7-5, 6-$Q_8$$\in$$Q_8$-6-7, Q_8-8$\in$$Q_8$-7-8.

Using the diagram of vertical mass balance we can consider in more detail the concentration fields having same microconstituents, but different by crystallization stages, e.g. for fields Q_8-8 and Q_8-7-8. The mass center G_1(0.105; 0.375; 0.52) given in two-dimensional field Q_8-7-8 intersects four-phase region (L+C, L+C+R_5, L+R_5+R_{12}, B+R_5+R_{12}) and two horizontal complexes at the temperatures of invariant points (Q_8 and E_5) (fig. 2b).

IOP Conf. Series: Journal of Physics: Conf. Series **790** (2017) 012020 doi:10.1088/1742-6596/790/1/012020

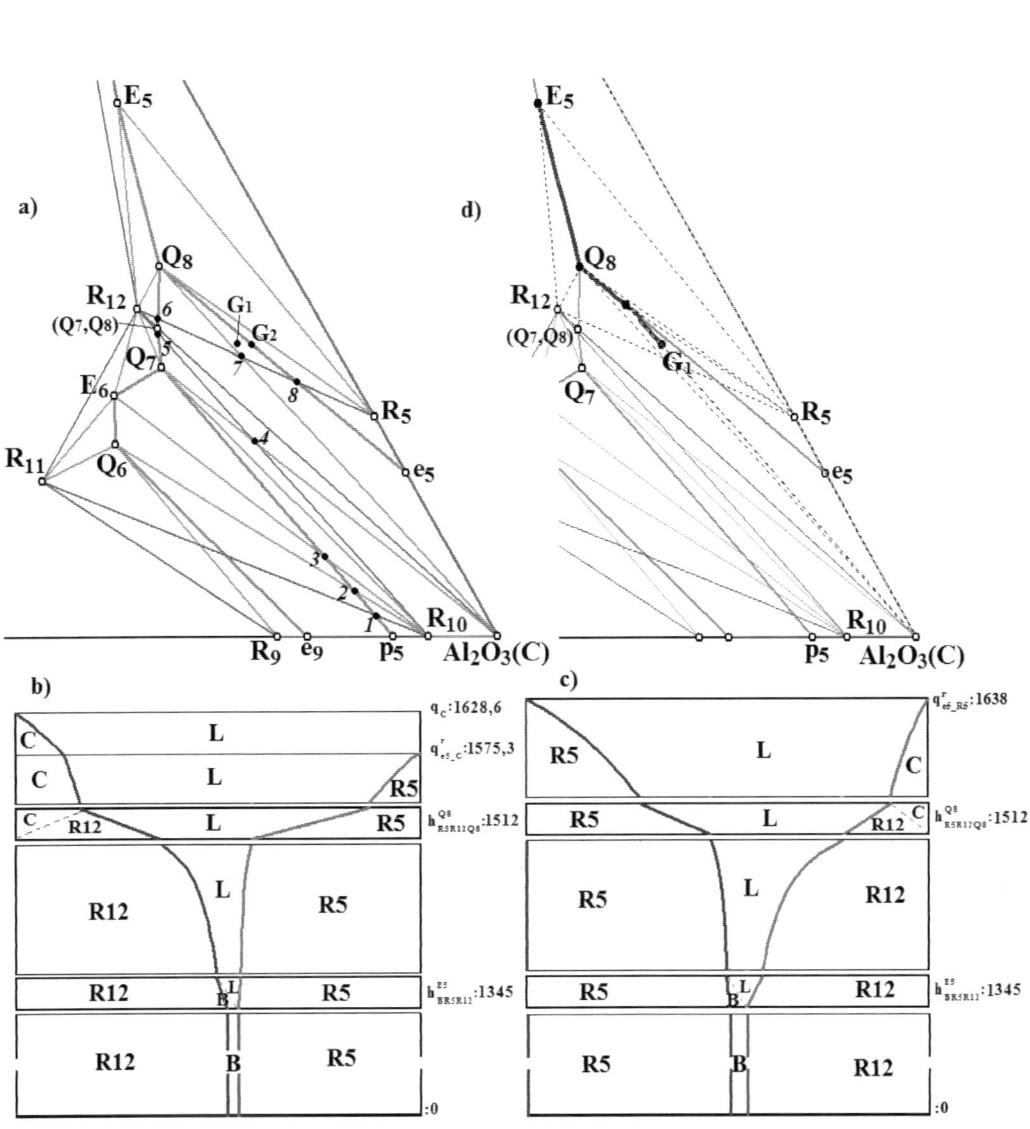

Figure 2. Fragment of projection of T-x-y diagram with dividing of liquidus field Al_2O_3 on the concentration field (a), diagrams of vertical mass balance for compositions G_1 (b) and G_2 (c), crystallization path for G_1 (d)

As can be seen from the diagram of vertical mass balance, there are the increasing of phase portion C and decreasing of phase L in two-phase regions L+C. So, the reaction of primary crystallization takes place $L^1 \rightarrow C^1$. Further, the mass center falls into three-phase L+C+R_5, with the increasing of phases C and R_5, while the phase L continues to decrease. It corresponds the monovariant eutectic reaction $L^e \rightarrow C^{e(R5)}+R_5^{e(C)}$. Next, the invariant quasi-peritectic reaction $L^{Q8}+C^{1,e} \rightarrow R_5^{Q8}+R_{12}^{Q8}$ take place on the horizontal complexed at point Q_8. As a result of this reaction the crystals C is fully expended. In similar way, the postperitectic secondary eutectic reaction $L^{ep} \rightarrow R_5^{ep(R12)}+R_{12}^{ep(R5)}$ occurs in the three-phase region L+R_5+R_{12}. The phase L is fully disappeared as the result of invariant eutectic reaction $L^{E5} \rightarrow B^{E5}+R_5^{E5}+R_{12}^{E5}$ on the plane at the temperature of ternary eutectic points E_5. Below the horizontal complex, the composition gets to the solid-phase region B+R_5+R_{12}. Since the crystals phase

C^1 and C^e are fully expended at the invariant quasi-peritectic reaction, therefore they are not included in the final set of microconstituents. Thus two-dimensional field Q_8-7-8 is characterized by the following set of microconstituents: $R_5^{e5(C)}$, R_5^{Q8}, R_{12}^{Q8}, $R_5^{ep(R12)}$, $R_{12}^{ep(R5)}$, B^{E5}, R_5^{E5}, R_{12}^{E5}. As the field Q_8-8 is a part of monovariant liquidus line e_5Q_8, then it hasn't the reaction of primary crystallization $L^1 \rightarrow C^1$. The mass center G_2(0.092; 0.376; 0.532) for this field at once gets into three-phase region L+C+R_5 (fig. 2c). So, one-dimensional field Q_8-8 has shorter crystallization scheme, but coincides with adjacent two-dimensional field Q_8-7-8 by set of microconstituents.

Figure 2d show the crystallization path for mass center G_1 given in considered two-dimensional field Q_8-7-8. The melt G_1 moves along the segment C-G_1 to the liquidus line e_5Q_8 at the passing through two-phase region L+C. Next it falls into three-phase region L+C+R_5 and the composition of melt moves along the fragment of liquidus line e_5Q_8 to point Q_8. Then melt shifts along the monovariant liquidus line Q_8E_5 to point E_5 at the going through three-phase region L+R_5+R_{12}. The reaction $L^{E5} \rightarrow B^{E5}+R_5^{E5}+R_{12}^{E5}$ is finished on horizontal complex $B_2R_5R_{12}$ and below there are only solid crystals B, R_5,R_{12}.

3. Model of T-x-y diagrams of system MgO-SiO_2-Al_2O_3

The system MgO-SiO_2-Al_2O_3 has a more simple geometric structure. It is includes three binary congruently melting compounds (R_1=2MgO·SiO_2, R_3=3Al_2O_3·2SiO_2, R_4=MgO·Al_2O_3), one binary (R_2=MgO·SiO_2) and two ternary (R_5=4MgO·5Al_2O_3·2SiO_2, R_6=2MgO·2Al_2O_3·5SiO_2) incongruently melting compounds. Systems MgO-SiO_2-Al_2O_3 is characterized by 11 invariant transformations: three eutectic 3 eutectic (E_{1-3}), one peritectic (P), five quasiperitectic (Q_{1-5}) transformations and 2 four-phase regroupings of phases with polymorphous modification of silicon oxide (cristobalite and tridymite) (fig. 3).

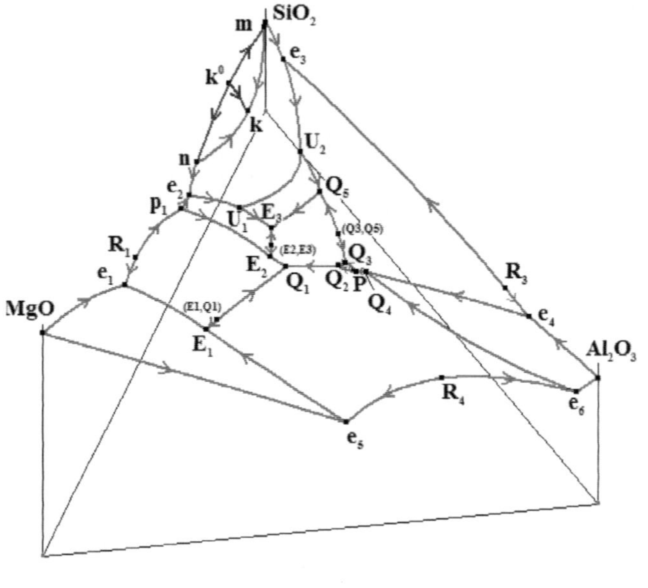

Figure 3. 3D model of liquidus surfaces of system MgO-SiO_2-Al_2O_3

Model of T-x-y diagram includes liquid immiscibility surface, 10 liquidus surfaces, 78 ruled surfaces, 11 horizontal complexes at the temperatures of invariant points, 21 two-phase regions and 29 three-phase regions (fig. 4).

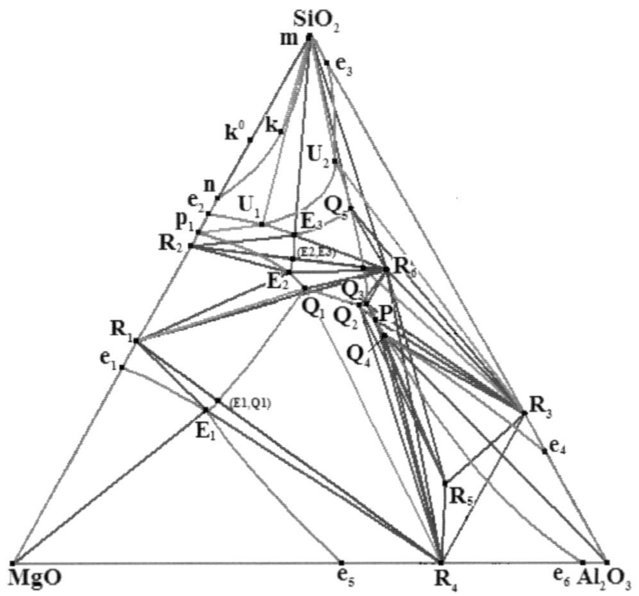

Figure 4. Projection of T-x-y diagram of system $MgO-SiO_2-Al_2O_3$

Ternary points U_1 and U_2 arrange at the same temperature, and the points $B1_{U1}$, $B2_{U1}$, $B1_{U2}$, $B2_{U2}$ coincident into one point on the prism edge. Therefore, the horizontal planes corresponding the four-phase phase regroupings of phases $L_{U1}+B_1 \leftrightarrows R_2+B_2$ and $L_{U2}+B_1 \leftrightarrows R_3+B_2$ have degenerated structure U_1-$R2_{U1}$-$B1_{U1}(B2_{U1})$ and U_2-$R3_{U2}$-$B1_{U2}(B2_{U2})$, correspondingly. Phase region $L+B_1+B_2$ degenerate into the plane and the corresponding monovariant peritectic reaction $L^P+B_1 \rightarrow B_2^P$ occurs at one temperature.

Projection T-x-y diagram is divided into 100 two-dimensional, 170 one- dimensional and 71 zero-dimensional concentration field. As an example, we carry out an analysis of concentration fields located below the immiscibility surface ($nkmk^0$) and the adjacent liquidus field of cristobalite ($e_2U_1U_2e_3$) (fig. 5a). At projection, they are divided into 9 two-dimensional, 16 one-dimensional and 7 zero-dimensional concentration fields. At that four field coincide by crystallization stages and microconstituents: B-3∈B-3-4; B-1∈e_2-n-k-1-B-U_1; k∈nk; 1∈B-1. Additionally 11 concentration fields differ by crystallization stages, by coincide microconstituents (B-m-1, e_2-U_1, n-k, m-1, k-1 (inside immiscibility surface), k-1 (on the contour of immiscibility surface)∈e_2-n-k-1-B-U_1; U_2-3∈B-U_2-3, e_3-4∈B-e_3-4, 4∈B-4, 3∈B-3, U_1∈B-U_1).

A large quantity of concentration fields with coinciding microconstituents due to the presence of phase regions with liquids immiscibility and two polymorphous modifications of component SiO_2. Processes occurring in the phase regions L_1+L_1, $L_1+L_1+B_1$, $L+B_1$ not influence on the final set of microconstituents, because the products of reactions for these fields fully expended. Let's consider two-dimensional field B-U_1-2 as an example. The mass center G (0.1; 0.8; 0.1) given in this field intersects five 5 phase regions $L+B_1$, $L+B_1+B_2$, $L+B_2$, $L+B_2+R_2$, $B_2+R_2+R_6$ (Fig. 5b) with the following phase reactions $L^1 \rightarrow B_1^{\,1}$, $L^P+B_1 \rightarrow B_2^P$, $L^{1p} \rightarrow B_2^{1p}$, $L^{ep} \rightarrow B_2^{R2,ep}+R_2^{B2,ep}$, $L^{E3} \rightarrow B_2^{E3}+R_2^{E3}+R_6^{E3}$ (^{1} – primary crystallization; ^{P} – monovariant peritectic reaction; ^{1p} – postperitectic primary crystallization; ^{ep} – post-peritectic monovariant crystallization; ^{E} – invariant eutectic crystallization). The crystals B_1 are fully disappeared as the result of postperitectic primary reaction $L^{1p} \rightarrow B_2^{1p}$ and they are not included in the set of microconstituents. Thus, the field B-U_1-2 is characterized by the following microconstituents: B_2^P, B_2^{1p}, $B_2^{R2,ep}$, $R_2^{B2,ep}$, B_2^{E3}, R_2^{E3}, R_6^{E3}.

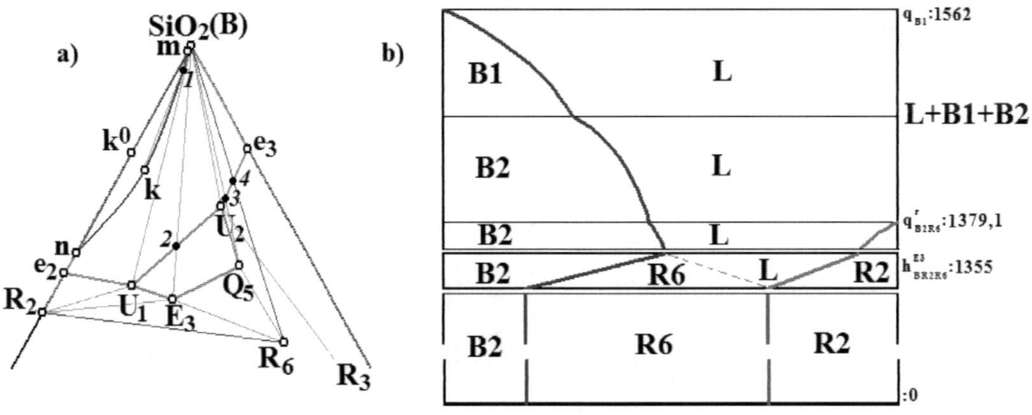

Figure 5. Fragment of phase diagram of system MgO-SiO_2-Al_2O_3 with the dividing into concentration fields

4. Summary

Analysis of two, one and zero-dimensional concentration fields for obtaining the data for the crystallization stages and microconstituents are shown based on the computer models [14] of systems CaO-SiO_2-Al_2O_3 and MgO-SiO_2-Al_2O_3.

Acknowledgement

This work has been performed under the program of fundamental research SB RAS (project 0336-2014-0003) and was partially supported by the Russian Foundation for Basic Research (projects 14-08-00453, 15-43-04304).

References

[1] Lutsyk V I, Vorob'eva V P and Zelenaya A E 2015 *Solid State Phenomena* **230** 51
[2] Lutsyk V I, Zyryanov A M and Zelenaya A E 2008 *Rus. J. Inorg. Chem.* **53** 794
[3] Lutsyk V I, Zelenaya A E and Zyryanov 2009 *Crystallogr. Rep.* **54** 1300
[4] Lutsyk V and Zelenaya 2013 *Journal of Silicate Based and Composite Materials* **2** 34
[5] Lutsyk V I, Zelenaya A E and Zyryanov A M 2008 *Journal of Materials, Methods & Technologies, International Science Publications* **2** 176
[6] Taylor H F W 1997 *Cement Chemistry* (London: Thomas Telford)
[7] Lea F 1998 *Lea's Chemistry of Cement and Concrete* (London: Elsevier Ltd)
[8] Zharikov V A 2005 *Basis of physical geochemistry* (Moscow.: Nauka) (In Russian)
[9] Zavarickij A N and Sobolev V S 1961 *Physico-chemical basis of petrography of eruptive rocks* (Moscow: GOSGEOLTEHIZDAT) (In Russian)
[10] Toropov N A, Bazarkovsky V P, Lapin V V et al. 1972 *Diagrams of Silicate Systems. Vol 3. Ternary Silicate Systems* (Leningrad: Nauka) (In Russian)
[11] Lutsyk V I, Zelenaya A E and Savinov V V 2011 *IOP Conference Series: Materials Science and Engineering* **18** 112005
[12] Lutsyk V and Zelenaya A 2013 *Solid State Phenomena* **200** 73
[13] Lutsyk V I, Zelenaya A E and Savinov V V 2012 *Crystallogr. Rep.* **57** 943
[14] Lutsyk V, Zelena A. 2013 *Építőanyag-JSBCM 65 34* http://dx.doi.org/10.14382/epitoanyag-jsbcm.2013.7

Effect of silicate module of water glass on rheological parameters of poly(sodium acrylate)/ sodium silicate hydrogels

J Mastalska-Popiawska , P Izak

AGH University of Science and Technology, Faculty of Materials Science and Ceramics, Department of Ceramics and Refractories, Krakow, Poland

E-mail: jmast@agh.edu.pl

Abstract. The poly(sodium acrylate)/ sodium silicate hydrogels were synthesized in the presence of sodium thiosulphate and potassium persulphate as the redox initiators and N,N'-methylene-bisacrylamide as the cross-linking monomer. 20 wt% aqueous solution of sodium acrylate was polymerized together with water glass with different silicate modules (M) from 1.74 to 2.29, in three mass ratio of the monomer solution to the water glass 2:1, 1:1 and 1:2. Such obtained hybrid composites were rheologically tested using the oscillation method. It allowed to designate the crossover point during polymerization, as well as to define the viscoelastic properties of the casted hydrogel samples one week after the reaction.

The obtained results of the oscillation measurements showed that cross-linking reaction proceeds very quickly and the lower the silicate module is, the process starts faster. After the completion of the reaction the silicate-polymer hydrogels are strongly elastic materials and the highest elasticity characterizes systems with the mass ratio 1:2, i.e. with the highest water glass *content.*

Keywords: silicate module, sodium water glass, oscillation rheology, silicate-polymer hydrogels, sodium polyacrylate

1. Introduction

Despite the extensive use of water glass in the industry, beginning with production of cosmetic pigments and ending with corrosion protection, this compound is not fully understood in terms of its structure and properties. Due to the fact that water glasses are not very stable systems, i.e. silicic acid undergoes facile polymerization during storage what leads to changes in its structure and composition, a few research centers in the world conduct research in this area. It is also difficult to determine the exact formula of water glass because it is a mixture of cyclic and linear silicate oligomers, which arise from cracking of Si-O-Si bonds in silica to create Si-ONa (K) groups in this place. And for example, sodium water glass, being the aqueous solution of sodium silicates, is described as Na_2O-nSiO_2 [1-3].

Quantity that characterizes composition of the water glass is a molar (silicate) modulus (M) which is defined as it follows (an example for sodium water glass) [4]:

$$M = \frac{X_2}{X_1} \cdot 1.032 \tag{1.1}$$

where: Xi- sodium oxide content [%], X_2- silica content; 1.032- conversion factor of weight units to molar units.

Depending on the molar module and other parameters such as density, viscosity etc., water glass is classified into types. Table 1 shows characteristics of some commonly used in trade sodium water glasses.

Table 1. Characteristics of selected types of sodium water glasses [4]

Requirements	Type of sodium water glass					
	137	**140**	**142**	**145**	**149**	**150**
Molar module	**3.2-3.4**	**2.9-3.1**	**2.6-2.8**	**2.4-2.6**	**2.8-3.0**	**1.9-2.1**
Content of (SiO_2+Na_2O) [%] not less than	35.0	36.0	38.0	39.0	42.5	40.0
Denisty (20°C) [g/cm^3]	1.37-1.40	1.40-1.43	1.42-1.45	1.45-1.48	1.49-1.51	1.50-1.53
Viscosity (Pas) not less than	10	5	10	10	70	10

As it was mentioned before, structure of sodium silicates in aqueous solutions is very complicated. This is due to an atomic structure of silicon itself, which changes the coordination number from 4 to 6 (free 3d orbitals may take the extra pairs of electrons), and the presence of hydroxyl groups. Studies have shown that the colloidal nature of the sodium silicate solutions is the greater the higher the concentration and the molar module are. Thus, in solutions of sodium water glass with a module M=2 there are not only monosilicate groups $[H_3SiO_4]^1$, but also there are dimer silicates $[H_4Si_2O_7]^2$. However, in glasses which have molar module greater than 2, revealed the presence of units containing up to 12 silicon atoms [5].

Due to the fact that water glasses are solution of high pH, above 11, polymerization of mixtures containing them causes certain problems. For example, when we want to introduce an aqueous solution of vinyl monomer such as sodium acrylate with a neutral pH to such system, acrylate precipitates from the solution and it can be only dissolved by the introduction of additional water. We also know that above pH=7 the polymerization rate of acrylates increases, reaching a maximum at pH=10, and then begins to fall in the range of pH = 11-12, what is explained by the decrease in the degree of polymer chain termination caused by the repulsion of polyions of the same charge [6 -10].

In the present work, due to the fact that the polymerization of vinyl monomers takes place at high temperatures and it is an exothermic process, a redox initiators system lowering the temperature of reaction was used [6,11]. As the cross-linking monomer N,N'- methylenebisacrylamide having double vinyl bond was used.

To assess the rate of polymerization and to estimate the effect of the silicate module on the rheological properties of the synthesized polymer-silicate gels method of oscillation measurement, which is based on measuring the deformation caused by sinusoidally variable stress, was used. Oscillation rheometry allows to observe of the gelling process without destroying the resulting gel network, and therefore these measurements can be used both to observe the sol gel transition as well as examination of the gelled samples [12].

Obtained results of rheological tests can be used i. a. to design the composition of polymer-silicate hydrogels used for example in the construction industry for the production of fireproof window panels or waterproofing barriers.

2. Materials and methods

2.1. Materials

As a sodium silicate aqueous solution were used sodium water glasses (WG) with the silicate modulus in the range M=1.74-2.29 (Table 2), purchased from the Rudniki Chemical Plant (Rudniki, Poland). Polymer filler Midafen R-102 (MR-102) came from Lubrina S.A. (Poland). Vinyl monomer solution, i.e. 20 wt% aqueous solution of sodium acrylate, was laboratory synthesized, remembering that due to the exothermic character of the reaction, mixture temperature can not exceed 40 °C. Acrylic acid, produced by Lach-Ner (Czech Republic), was neutralizing to pH=7 with a stoichiometric volume of the 22 wt% aqueous solution of sodium hydroxide (Stanlab SJ, Poland). After that, the resulting solution was diluted to 20 wt% (20%ANa).

The polymerizing agents, i.e. the cross-linking monomer N,N'- methylenebisacrylamide (NNMBA) and system of redox initiators, potassium persulphate (KPS) and sodium thiosulphate (NTS) were delivered by Avantor Performance Materials Poland S.A. All reagents were used as supplied, without further purification.

Table 2. Characteristics of the used in test sodium water glasses

Type of the glass	Silicate module M	Density [g/cm]	pH	Na_2O+SiO_2 [%]
A	1.82	1.51	12.30	39.70
A1	1.74	1.55	12.40	44.60
A2	2.20	1.47	11.90	40.70
A3	2.29	1.52	11.80	43.80

2.2. *Synthesis of hydrogels*

Synthesis of poly(sodium acrylate)/sodium silicate hydrogels was divided into two parts. The first part consisted of the preparation of the so-called base mixture by distributing of 5 wt%, 10 wt% and 20 wt% of Midafen R-102 (MR-102) in particular water glass (WG). Only in the case of 20 wt% of Midafen R-102 with water glass type A1 (M=1.74) base mixture wasn't prepared. Because of the high density of the used water glass this mixture was delaminated.

Table 3. Composition of the synthesized polymer silicate hydrogels

Sample symbol*	MR-102 [g]	WG [g]	20% ANa [g]	NNMBA [g]	KPS [g]	NTS [g]
2:1/5%	0.50	9.50	20.00	0.10	0.05	0.05
2:1/10%	1.00	9.00				
2:1/20%	2.00	8.00				
1:1/5%	0.75	14.25	15.00			
1:1/10%	1.50	13.50				
1:1/20%	3.00	12.00				
1:2/5%	1.00	19.00	10.00			
1:2/10%	2.00	18.00				
1:2/20%	4.00	16.00				

*Sample symbol: for e.g. 1:1/5%- 1:1 means a mass ratio of 20 wt% aqueous solution of sodium acrylate to a base mixture, which consists of sodium water glass A1 and a polymer filler Midafen R-102 (5-20%)

The second step concerning the free radical polymerization of base mixture together with 20 wt% aqueous solution of sodium acrylate (20%ANa) at a mass ratio of 2:1, 1:1 and 1:2 of vinyl monomer solution to the base mixture. After mixing these two main components it was noticed that the higher the silicate module was, the amount of water needed to dissolve the precipitated sodium acrylate was higher (from 1 to 5 ml). The cross-linking monomer (NNMBA) and system of redox initiators

(KPS/NTS) were added to the mixture in an amount of 0.3 wt% to the sample weight (Table 3).

2.3. *Rheological measurements*

The oscillation measurements, i.e. frequency sweep standard tests, were performed with a Physica MCR-301 Anton Paar rheometer at low shear rates in the range of angular frequencies 500-0.05 s^{-1} and in room temperature. As a measuring system was used a parallelplate system where the upper plate (PP50) has a 50 mm diameter.

During the cross-linking reaction, where the sample was immediately transferred to the measuring system after preparation, the gap between plates was set at a distance of 0.10 mm, while during the same oscillation test of the cast samples (7 days after preparation) the gap was set at a distance of 1 mm.

3. Results and discussion

3.1. *Frequency sweep standard test during cross-linking process*

Figures 1 and 2 show the dependence of G′ and G″ moduli as a function of angular frequency during cross-linking process for samples based on sodium water glass (A1) with molar module M=1.74. Because the measurement time was only about 30 minutes, it was hard to determine the plateau region from which we can read the exact values of G′ and G″ moduli. So, to describe the polymer-silicate system during the cross-linking process, we use cross-over points (marked on the graph with colored arrows).

Cross-over point, in other words gel point, the moment at which the phase transition occurs to a viscoelastic solid, occurs quite fast. The fastest for the mass ratio 2:1, then 1:1 and 1:2 at the end. Confirmation of these observation are the values of angular frequency included in Table 4, where we present results for other polymer-silicate samples during cross-linking process.

Analyzing these results, it should be remember that water glass designated as A is not a pure sodium silicate solution, but it contains small amounts of aluminum. Also here we can see that cross-linking process starts faster, if we have more monomer in the solution. Presumably at higher pH, i.e. with higher content of sodium silicate solution, cross-linking monomer and redox initiators decompose more rapidly. Second conclusion concerns the molar module- the higher the content of sodium silicate the process starts later.

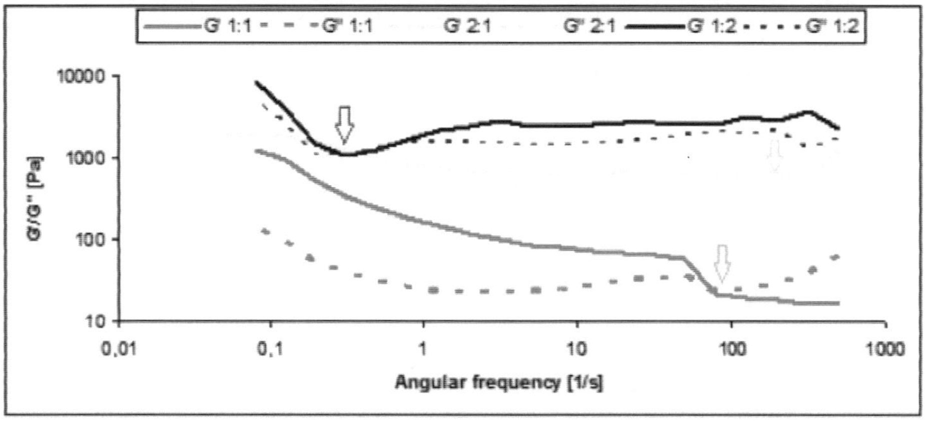

Figure 1. G' and G" moduli as a function of angular frequency during cross-linking process for samples based an Al sodium water glass with 5% of polymer filler

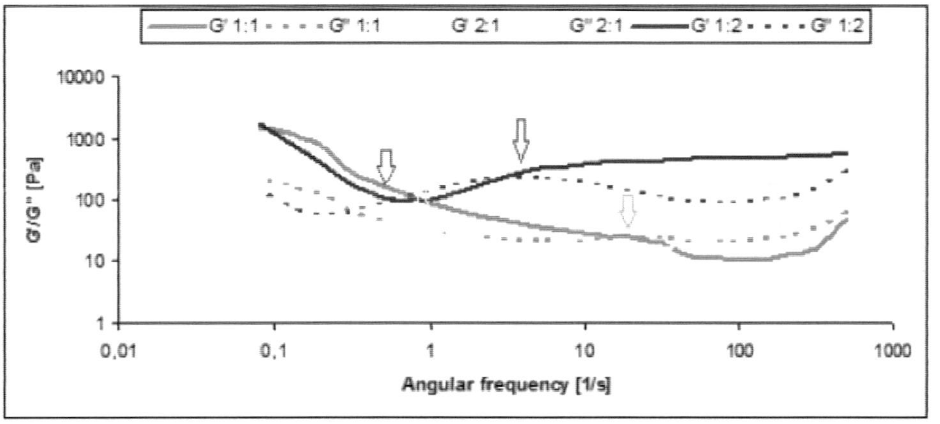

Figure 2. G' and G" moduli as a function of angular frequency during cross-linking process for samples based on A1 sodium water glass with 10% of polymer filler

Table 4. Values of the cross-over angular frequencies for the tested hydrogel samples during cross-linking process

Type of water glass		A	A1	A2	A3
MR-102 [wt%]	**Mass ratio**	**« [rad/s]**			
5	2:1	118.65	360.61	-	287.60
	1:1	105.53	176.29	326.41	4.21
	1:2	0.87	73.19	182.10	2.47
10	2:1	108.21	207.51	216.16	140.53
	1:1	15.48	19.24	65.52	12.30
	1:2	-	2.76	6.68	-
20	2:1	66.52	-	33.89	297.57
	1:1	1.65	-	29.33	10.32
	1:2	-	-	10.41	0.56

3.2. *Frequency sweep standard test- agening tests*

The second frequency sweep standard test was carried out to check the influence of the mass ratio of the poly(sodium acrylate) to the base mixture on the viscoelastic properties of the tested samples after the cross-linking reaction. As it is shown in Figures 3 and 4, the storage (elasticity) modulus G' and the loss (viscous) modulus G" strongly depend on the sample composition and G' is higher than G" which means that we obtain more elastic than viscous hydrogels. Comparing the values of G' and G" with the polymer content, it was found that less poly(sodium acrylate) was present, the values of both moduli were higher.

To asses the viscoelastic properties of the cast samples we calculated phase shift angle, i.e. tangent of the quotient of the G" to G' (Table 5). It was found that for the obtained polymer-silicate hydrogels phase angle is low, below 40°, what indicates that we got highly elastic hydrogels (90° denotes a totally viscous substance and 0° is related to a completely elastic systems) and the most elastic are samples with the higher content of water glass. Comparing the phase angle in terms of molar module we see that this value is the lowest for the highest module 2.29 (A3), i.e. for the water glass with the highest content of sodium silicate.

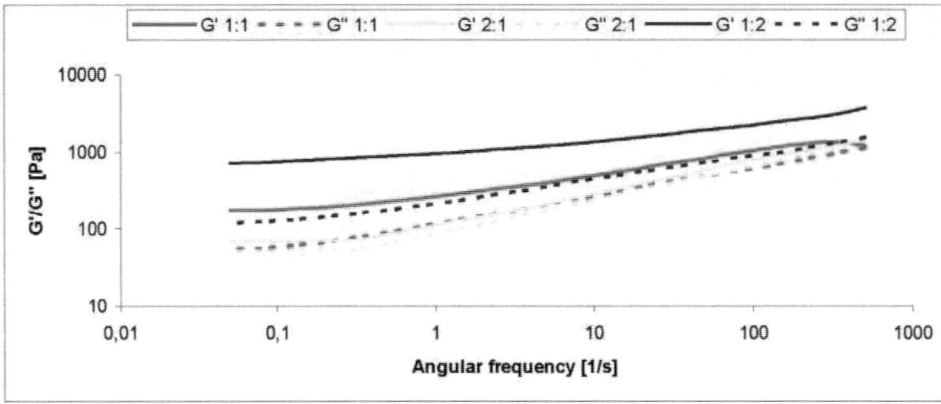

Figure 3. G' and G" modules as a function of angular frequency for cast samples based on A1 sodium water glass with 5 % of polymer filler

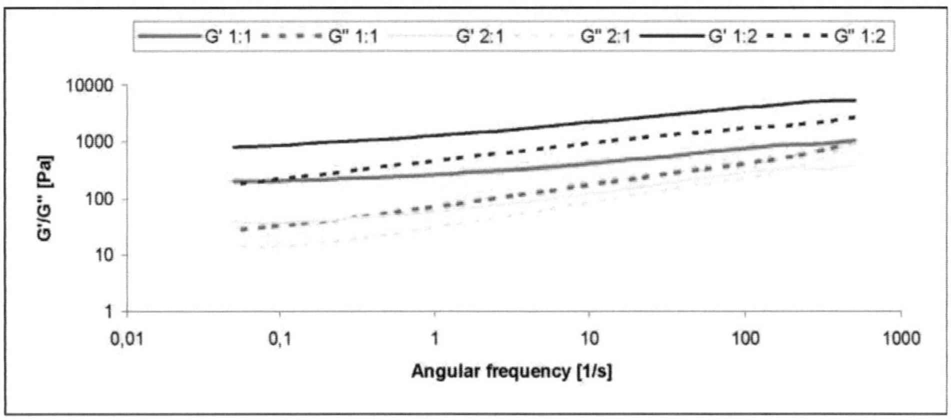

Figure 4. G' and G" modules as a function of angular frequency for cast samples based on Al sodium water glass with 10% of polymer fileler

Table 5. Values of phase shift angle for cast samples of poly(sodium acrylate)/ sodium silicate hydrogel

Type of glass		A	A1	A2	A3
MR-102 [wt%]	**Mass ratio**	**o n**			
5	2:1	-	36.87	-	-
	1:1	31.96	18.83	30.41	15.91
	1:2	24.84	9.98	28.90	-
10	2:1	18.11	21.80	35.22	-
	1:1	11.25	9.20	11.80	38.87
	1:2	5.48	13.60	9.31	22.19
20	2:1	29.60	-	41.70	14.73
	1:1	23.80	-	31.80	10.26
	1:2	13.33	-	12.24	8.87

4. Conclusion

In the present work the oscillation constants of the polymer-silicate system have been investigated. It was noticed that the mass ratio and type of the used sodium water glass has a big influence on the G' and G" values and controlling these two variables more viscous or more elastic gels can be obtained:

- the lower the silicate module is, the amount of water needed to dissolve the precipitated sodium acrylate is smaller;
- the lower the silicate module is, the cross-linking reaction starts faster;
- the obtained polymer-silicate hydrogels are relatively elastic materials; the higher the water glass content the phase shift angle is smaller, i.e. the most elastic hydrogels were obtained for the mass ratio of the sodium acrylate water solution to the base mixture 1:2. There was no dependence between the silicate module value and the phase shift angle.

Summing up, obtained results of rheological tests can be used i. a. to design the composition of polymer-silicate hydrogels used for example in the construction industry for the production of fireproof window panels or waterproofing barriers.

References

[1] Kozlak W 2005 *Wiadomosci Chemiczne* **59** 9-10 (in Polish)
[2] Ming- Chu P et al 2010 *Glass Physics and Chemistry* **36** 110-112
[3] Halasz I, Agarwal M, Li R and Miller L 2007 *Catalysis Today* **126** 196-197
[4] ZN-02/Chemical Plant „Rudniki" S.A. factory standard/257 (in Polish)
[5] Balinski A 2009 About structure of hydrated sodium silicate as a binder for molding masses (Krakow, Foundry Research Institute) (in Polish)
[6] Stempkowska A, Kochanowski A and Izak P 2009 *Ceramic Materials* 61 97-99 (in Polish)
[7] Bortel E 1997 *Synthetic water-soluble polymers, Handbook of Thermoplastics* (Olabis, Marcel Dekker New York)
[8] Blauer G 1960 *Trans. Faraday Soc.* 56 606
[9] Mori H and Muller A 2003 *Prog. Polym. Sci* **28** 1403- 1405
[10] Mayoux C, Dandurand J, Ricard A and Lacabanne C 2000 *J. of App. Polym. Sci.* **77** 2622
[11] Florjanczyk Z and Penczek S 2001 *Chemistry of Polymers vol.1* (Warsaw, Publishing House of Warsaw University of Technology) (in Polish)
[12] Cicha-Szot R and Falkowicz S 2010 *Nafta-Gaz* **12** 1104-1105 (in Polish)

IOP Conf. Series: Journal of Physics: Conf. Series **790** (2017) 012022 doi:10.1088/1742-6596/790/1/012022

Premixes production for synthesis of wear-resistant composite materials

Yu V Kontsevoi, A G Meilakh, A B Shubin, E A Pastukhov, A V Dolmatov and I S Sipatov

Institute of Metallurgy, Ural Branch of Russian Academy of Science, Ekaterinburg, Russia

E-mail: abshubin@gmail.com

Abstract. State of the art line of powder metallurgy is application of initial powders as micro-composites with additional components - premixes. Usage of premixes inhibits segregation of added components and implies the homogeneity of powder charge composition, and finally it has a significant impact on structure formation and properties of end products. The aim of the present work was to design the new production technology of premixes based on iron powder which is layer-by-layer plated by aluminium and copper. We propose to carry out production of Cu-Al-Fe premixes in two stages: cladding of iron powder by aluminium and coating of the obtained composite by copper. The self-developed technique of vibration treatment of iron and aluminium powder mixture was chosen for this purpose. The uniform in thickness and unbroken copper-plating of Fe-Al powders were carried out by chemical technique. Physico-chemical properties and production conditions of premixes-powders were studied, besides optimal parameters of production and further heat-treatment were selected. In the result of the present study the Fe-Al-Cu premixes with laminated structure comprising of iron core, Fe-Al and Cu-Al intermetallide shells were synthesised.

1. Introduction

State-of-the-art problems are development of new multifunctional coatings with required complex of properties, e.g. improved corrosion, abrasive resistance and fatigue resistance, low friction factor, as well as designing of appropriate techniques and machines for production of such materials. From this point of view it appears that the most promising direction for production of composites is powder metallurgy, since its techniques have almost an unlimited number of choices for composition and materials structure. The state of the art line of powder metallurgy is application of input powders as micro-composites with additional components - premixes. Usage of premixes, what inhibits of added components segregation and implies the homogeneity of powder charge composition, has a significant impact on structure formation and properties of end product. For example, the strength and ductility of the steels made of premixes comprised of iron powder plated by Ni-Cu alloys nanocrystals are in 2-4 times higher than steels with the same composition, but which were produced of powder mixture.

Typically, wear-resistant materials have composite structure comprising of plastic matrix and solid inclusions, for instance, copper and intermetallides, respectively. Wear-resistant composites based on copper with inclusion of reinforcing premixes in the form of Cu-C [2], Al-Cu-Fe quasicrystals doped by B and Al_2O_3 [3] granulates and other compositions are generally known.

The aim of the present work was to design the new technology of premixes production based on iron powder which is layer-by-layer plated by aluminium and copper. It was assumed that composites with copper matrix and produced inclusions will have improved mechanical properties owing to formation of Cu-Al premixes and Fe-Al intermetallides in the result of Cu-Al-Fe sintering.

2. Experimental

Aluminium powders PA-4 and PAD-6, PGR 3.160.28 iron and PMS-1 copper powders were used as input materials. Production of Cu-Al-Fe premixes was carried out in two stages: cladding of iron powder by aluminium and coating of the obtained composite by copper. Coppering of Fe-Al premixes was carried out at room temperature by contact technique, namely, copper ions were substituted by iron ones in solution of CuSO4 sulfuric acid which were acidified by 0.3% H2SO4.

Plating of iron powder by copper was performed by the self-developed technique of vibration treatment of iron and aluminium powder mixture [4]. This technique allows us to vary process parameters in the wide range and consequently it provides to obtain predetermined structures and phase composition of cladded powders.

Appearance of apparatus for mechanical plating of powder is shown in Figure 1. The main parts of apparatus are vibrator with crank drive and DC motor, which allow regulating frequency and amplitude of vibration to be made smoothly. Rotating velocity of motor determines the vibration frequency; the value of crank eccentricity is responsible for amplitude. Airtight cylinder of stainless steel with pipes providing the predetermined atmosphere is placed on vibrator. Inductive unit with transformer type VCIN-15 is used for heat treatment of powders after hitting.

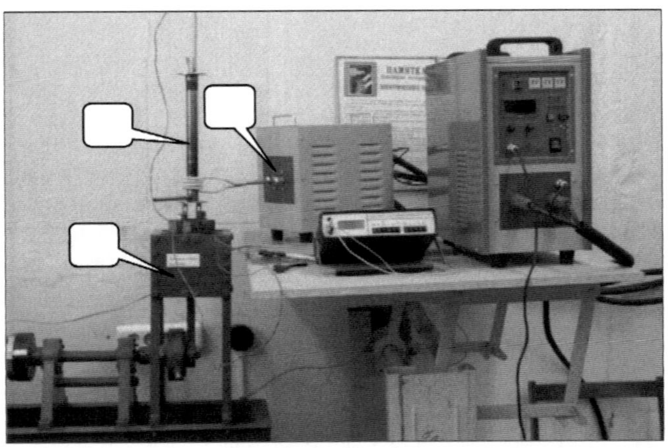

Figure 1. Appearance of apparatus for mechanical cladding of powders: 1 - vibrator unit, 2 - airtight cylinder for powders (plating unit), 3 - inductive heater VCIN-15

In essence the apparatus induces vibratory movements with high amplitude up to 10 mm and operates within the frequency range 30-50 Hz. Reciprocating motion imparts to powder mixture, but the frequencies of plating unit, iron and aluminium powders are different. Whereas the lifting height of the powder particles is up to 200-250 mm. Particles of different composition have various velocities and vibration phases, they have both tangent and central collisions. At the tangent collisions in the result of the tangential stresses between particles the rubbing of softener aluminium on iron powder particles occurs, when in the result of the central collisions the local plastic deformation is taking place, the latter one provides the intensification of diffusion processes [5].

The following instruments and techniques were applied to make our experiments. The frequency of plating unit vibration was determined through rotating velocity of motor which was measured by contactless tachometer CEM-AT-8. Heating temperature of powder mass was determined as the temperature of plating unit walls, and it was measured by Kelvin IR pyrometer with inaccuracy of ±1%. The surface of plated powder was observed on MEIJI EMZ-13TR stereomicroscope. The microstructure observation were conducted using Olympus GX-51optical microscope equipped with the SIAMS system of image analysis, the phase composition analysis of premixes were undertaken using CARL ZEISS EVO 40 scanning electron microscope. The velocity of collided particles was calculated according to the technique described in the literature [6]. The contact time of collided spheres (τ), the closest approach at the impacts (a_m) and the values of plastic and elastic deformations (p_m) were calculated in accordance with equations (1, 2, 3) [7]:

$$\tau = 4{,}53\left[\frac{(\delta_1+\delta_2)m_1m_2}{(m_1+m_2)}\right]^{\frac{2}{5}}\left(\frac{R_1+R_2}{V_0R_1R_2}\right)^{\frac{1}{5}} \quad (1)$$

where $\delta \equiv \frac{1-\mu^2}{E\pi}$; μ is Poisson's ratio; E is Young's modulus; m_1 and m_2 are the masses of collided spheres; R_1 and R_2 are the sphere radii; v_0 is the spheres impact velocity,

$$a_m = \left[\frac{15\pi v_0^2(\delta_1+\delta_2)m_1m_2}{16(m_1+m_2)}\right]^{\frac{2}{5}}\left(\frac{R_1+R_2}{R_1R_2}\right)^{\frac{1}{5}} \quad (2)$$

$$p_m = 0{,}2515\left[\frac{v_0^2}{(\delta_1+\delta_2)^4}\cdot\frac{m_1m_2}{(m_1+m_2)}\cdot\left(\frac{R_1+R_2}{R_1R_2}\right)^3\right]^{\frac{1}{5}} \quad (3)$$

To calculate depth and degree of plastic deformation of every sphere we used the mentioned above data in the process of iteration, besides the growth of the yield strength was taken into account at a given deformation rate.

3. Results and discussion

3.1. Selection of optimal characteristics of plating unit vibration and treatment time

Experiments were carried out within the ranges of vibration frequency and amplitude 20-50 Hz and 2-10 mm, respectively. The airtight quartz tube of the same diameter as plating unit was used to monitor the behaviour of powders. Vibro-boiling process with the separation effect was caused at all applied frequencies when the variation of amplitude was within the range from 2 to 6 mm. It is interesting to note that more dense iron powder was raised, but lesser dense aluminium particles were sunk. When amplitude is higher than 6 mm the vibro-boiling is substituted by vibration motion of powder particles, whereas the increase of vibration frequency is accompanied by the particles lifting height rise. The lifting height is up to 200-250 mm at the amplitude of airtight cylinder about 8-10 mm and frequency about 35-40 Hz, besides at these parameters the vibration phases of powders with diverse densities are changed. The most efficient plating process is performed at such vibration characteristics of airtight cylinder. Further increase of frequency was not tested due to high dynamic loads on part of the apparatus.

The optimal time of treatment in plating unit is 240-300 seconds. During shorter period of treatment some particles remain free from coating, and no positive effect was observed in the result of longer period of particles processing

3.2. Influence of ratio of dispersion and volume of interacted powders on plating efficiency

For plating experiments powders with different particles sizes were used: iron powders – 140÷160 μm, 100÷140 μm, 70÷100 μm, 40÷70 μm; aluminium powder: 140÷160 μm, 100÷140 μm, 70÷100 μm, 40÷70 μm and 1÷10 μm.

The cladding is nonuniform and broken when for treatment iron and aluminium powders had close particles sizes; such characteristic view is shown in figure 2.

The most uniform and unbroken plating of iron powder was obtained at the particles sizes of iron and aluminium within the ranges 100÷140 and 1÷10 μm, respectively.

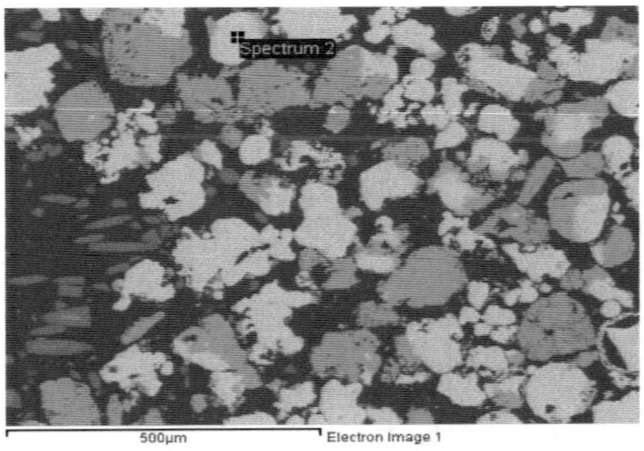

Figure 2. Iron powder with aluminium shell, where fractions of Fe and Al powders were 100÷140 µm and 70÷100 µm, respectively. Aluminium is dark grey.

Figure 3. Iron powder plated by aluminium one, where fractions of Fe and Al powders were 100 – 140 µm and 1÷10 µm, respectively. Aluminium is dark grey.

It is the author's opinion that in this case the iron powder play role of impact tool which provides rubbing and hammering of aluminium powder particles placed between two collided iron particles.

In the result of the present investigation of influence of quantitative volumetric powder ratio on quality and efficiency of plating it was established that volumetric ratio, when powders of 40÷160 µm are used, has no significant effect on cladding process. However, to avoid additional operation of cladded powder separation of excess aluminium it is necessary to maintain powder volumetric ratio Al/Fe=1/4÷1/5.

The effect of 1÷10 µm highly-dispersed aluminium powder is significantly different. When the Al/Fe ratio is higher than 1/4 the cladding process stops, since aluminium powder is tamped down the plating unit and iron powder particles are sunk in it. For this reason the optimal ratio of volumes is Al/Fe=1/5÷1/6 when highly dispersed aluminium powder is used.

3.3. Diffusive interaction of powders

Iron and aluminium powders were sharply heated up in plating unit to provide solid-phase interaction of particles. Earlier it was established [5] that high-rate heating, which is higher than 50 °C per second, leads to increase of real diffusivity. Hence the rate of Fe-Al solid-phase reaction is increased and the treatment time becomes shorter, respectively. Thermal treatment on air leads to iron oxidation; hence the plated powders were heated in hydrogen and nitrogen atmospheres which delay the reactive diffusion. Several parameters were varied in the experiment, namely, the top temperature 650-850°C, the exposure time 30-40 s, heating rates $^{o}Cs^{-1}$. In the result of the experiments it was established that efficiency of solid-phase reactions in hydrogen and nitrogen is almost equal, thus further experiments were carried out in a nitrogen atmosphere.

Characteristic view of iron particles cladded by aluminium after 300 s mechanical treatment, and subsequent 10 s thermal treatment at 700-725°C is shown in Figure 4.

Further increase of both temperature and exposure time leads to rise of diffusion rate and multiphase Fe-Al alloys are obtained, example of this is shown on Figure 5.

The view of ferroaluminide powder cladded by copper and its phase composition are submitted in Figures 6 and 7, respectively.

Copper-plated ferroaluminium powders were thermally treated to provide copper interaction with ferroaluminium core. Particles phase composition of this powder after heat treatment at 700°C during 15 minutes is shown in figure 8.

IOP Conf. Series: Journal of Physics: Conf. Series **790** (2017) 012022 doi:10.1088/1742-6596/790/1/012022

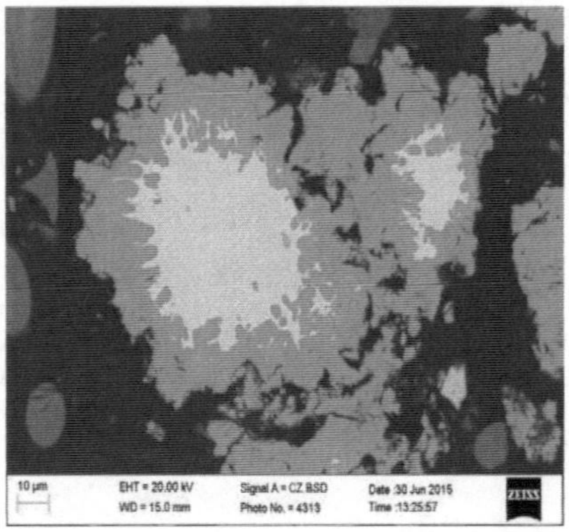

Figure 4. Iron particles plated by aluminium after solid-phase reaction, where dark gray is Fe_2Al_5 intermetallide and light gray core -Fe.

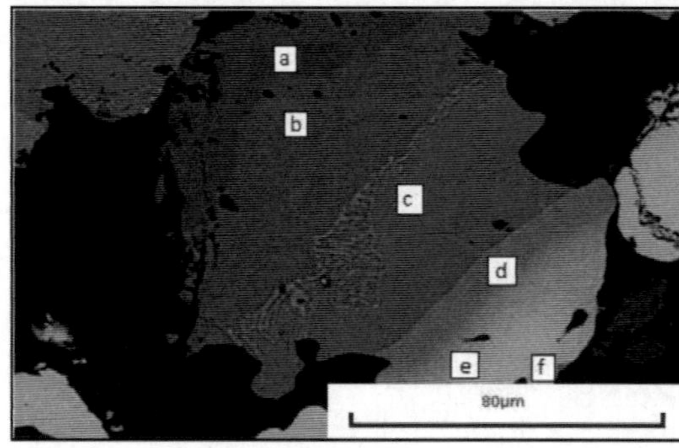

Figure 5. Iron particles plated by aluminium after 40 s heating at the temperature 850°C with following composition (at. %): a - ferroaluminide (FA) of Al 69 and the balance Fe , b - FA of Al 61 and the balance Fe , c - FA of Al 51 and the balance Fe, d - pure iron.

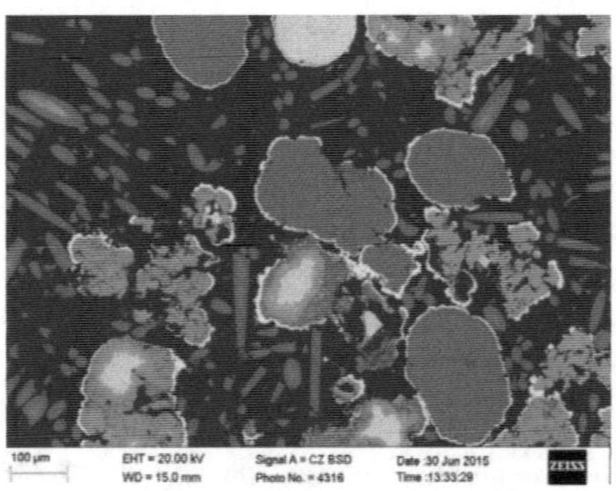

Figure 6. The Cu-Al-Fe composite powder

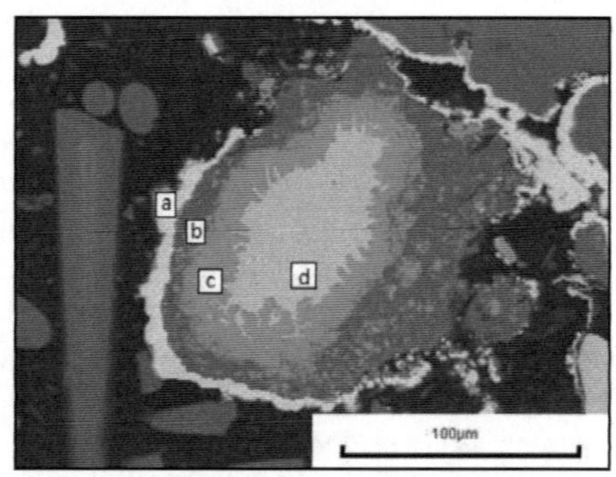

Figure 7. Phase composition (at.%) of FA powders plated by copper: a - Cu 99 and the balance Fe ; b - Al 99 and the balance Fe ; c - Al 75 and the balance Fe ; d - pure iron.

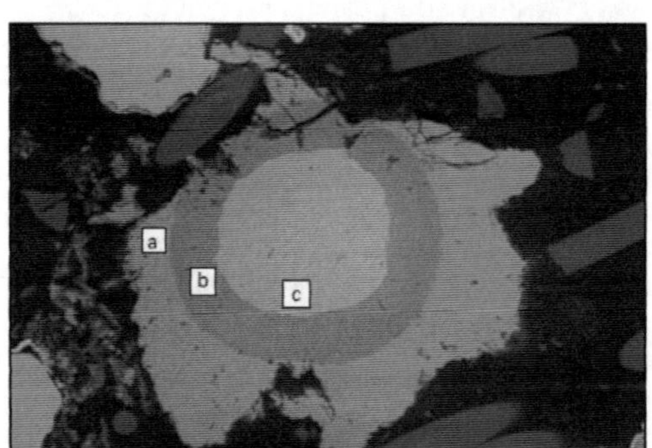

Figure 8. Phase composition (at.%) of the Cu-Al-Fe premix after final thermal treatment: a - Cu 25, Al 62 , Fe 13; b - Cu 4, Al 69 %, Fe 27; c - pure iron

4. Conclusions

High-performance vibration technique was proposed and mechanical parameters of iron powder plating by aluminium were selected. In the result of experiments the powder dispersion parameters and powder volumetric ratios were determined for efficient cladding. A number of conditions of heat treatment, namely, heating rate, top temperature and exposure time, were optimized to obtain in plating unit the Fe-Al intermetallic compounds of predetermined composition. The uniform in thickness and unbroken copper-plating of Fe-Al powders were carried out by chemical technique. Diffusion of iron and aluminium into copper shell from the Fe-Al core were revealed after heat treatment of the Fe-Al-Cu powders obtained chemically. Premixes of Fe-Al-Cu were obtained with laminated structure comprising of iron core, Fe-Al and Cu-Al intermetallide shells of variable composition.

Acknowledgements

This research was supported by the Russian Science Foundation (grant 15-13-00029).

References

[1] Meilakh A G 2009 *Steel.* **11** 90-2
[2] Romanov S M and Romanov D S 2008 *Patent RF 2336444* Anti-frictional material Romanite – UVL, method of its producing and the part of the friction unit
[3] Efimochkin I Yu, Lomov S B, Goncharov I E and Fedotov S V 2011 *Patent RF 2436857* Powder composite material
[4] Kontsevoi Yu V, Dolmatov A V, Pastukhov E A and Grigorieva T F 2015 *Russian Journal of Non-Ferrous Metals.* **1** 8-11
[5] Vatolin N A, Kontsevoi Yu V and Ckhai E V 1997 *Russian metallurgy (Metally).* **1** 95-100
[6] Kontsevoi Y V, Pastukhov E A, Ignat'ev I E, Bulanov V Y and Ignat'eva E V 2009 *Russian Journal of Non-Ferrous Metals.* **5** 507-12
[7] Goldsmith V 1965 *Hitting theory and physical properties of collided solid* (Moscow:Stroyizdat) p 448

IOP Conf. Series: Journal of Physics: Conf. Series **790** (2017) 012023 doi:10.1088/1742-6596/790/1/012023

Substantiation of Reference Method For Determining Concrete's Freeze-Thaw Resistance

S Nikolskiy, O Pertseva

[1]Peter the Great St. Petersburg Polytechnic University, St. Petersburg, Russia

E-mail: bagira@cef.spbstu.ru, olya_perceva@mail.ru

Abstract. It has been analytically proved that using concrete's rate of set ε as a measure of damage, instead of decreasing of tensile strength R, increases freeze-thaw resistance's accuracy of estimation a lot under otherwise equal conditions by the time of freeze-thaw cycling. Also it has been experimentally shown that ratio of relative decreasing R to ε in direction, perpendicular to compression, is assumed to be independent on values R and ε for a given concrete and on the ways of achieving them during mechanical or freeze-thaw cycling. Taking this into account patented methods for estimation of concrete's freeze-thaw resistance as per values R and ε received after freezing and thawing cycles of some specimens and their postliminary failure by linear compression was substantiated.

1. Introduction

Freeze-thaw resistance of concrete is an ability of water-saturated concrete specimen to maintain repeated standard thermo cycles without noticeable damage. Different types of water pressure cause concrete's freeze-thaw deterioration, such as hydraulic and osmotic pressure [1], capillary pressure [2] and other types of water influence according to existing freeze-thaw resistance theory. Decreasing strenght of construction material is associated with the processes of its water freezing (for example, rock [3]). Water gets into structure of porous bodies, separates particles and breaks coupling between them [4]. Porosity of material is determinative factor for frost resistance and durability subsequently [5,6]. So, strength of concrete could be pictured as a porosity function [7]. In order to determine the concrete mix composition, it is necessary to take into account freeze-thaw resistance. Worldwide experience offers some test methods to determine durability of concrete by freeze-thaw damage, such as Slab test [8], CDF [9], CIF-Test [10] and Cube-Test [11]. But they differ in terms of their procedures and conditions and are referenced to particular conditions.

There are two different standard types of methods of determining the freeze-thaw resistance of concrete- basic [12] and reference [13] in the Russian Federation. Considerable spread of values of strength for concrete specimens (variation coefficient ρ≈17%) in constant conditions of preparing and testing determines random spread of choosing average values $\overline{R}$ in the range of $\Delta R / R = \pm 3\rho\sqrt{n}$, where n is the volume of bath of specimens. In this case for proving the significance of relative decreasing $\overline{R}$ on 0,05…0,15 it is necessary to test more that 50 specimens and support $\Delta\overline{R} / \overline{R} = 0.068\overline{R}$, when confidence figure is equal 0,95. Therefore, main disadvantages of the basic method are high labor input and small operability by virtue of the fact that duration of the basic thermo cycle is equal 4,5 at least and $F >> 50$.

One of the existing reference methods is a Dilatometric rapid method of determining the freeze-thaw resistance of concrete [13]. This method is a prototype for the method, which has been offered by me. In this method concrete's freeze-thaw resistance is determined by the maximum relative difference of volume deformations of the tested concrete and standard specimens in accordance with tables provided in standard specification [12] taking into account concrete's type, its form and the size of specimens.

However, the results from the tables provided in state standard specification are acceptable only for Portland cement concrete and slag Portland cement concrete without surface-active additives (PEAHENS), such concretes are used extremely seldom now. Now lot of new concretes are investigated, tested and used, for example, nanomodificated concrete [14], high-strength concrete [15], concrete on the basis of fine-grained dry powder mixes [16], concrete with using recycling concrete aggregates [17] etc. In order to obtain new tables long labour-consuming experiences, which imply using, basic methods are needed [18]

2. Project's objectives

The aims of this paper are to obtain the dependence between relative decreasing in strength by rate of strain and substantiate the new method for determination of concrete frost resistance. Using of measure of damage D by the time of freeze-thaw cycling is offered for removal of disadvantages. In accordance with experimental data [19, 20] dependence $D = Ac^q$ is well approximated, where: $A = D_0$ (when $c = 1$) – value of D after first thermo cycle; c – number of thermo cycles; q – constant of material.

In accordance with this statement, evaluation of maximal measurement errors (ΔF and $\Delta F / F$) by the time of determination F is possible with help of dependence:

$$\delta R = A \cdot c^q \tag{1}$$

where $A = \delta R_1$ (when c=1) – decreasing of strength after first thermo cycle and δR – decreasing of original strength R. The first consequence from (1) covers to constant of material q:

$$q = \{\ln(\delta R_c / \delta R_1)\}/(\ln c) \tag{2}$$

After differentiation (3):

$$q\frac{\Delta F}{F} + \{\ln(F/c)\}\Delta q = \frac{\Delta R}{R} + \frac{\Delta \delta R_c}{\delta R_c} \Rightarrow \pm\frac{\Delta F}{F} - \frac{1}{q}\left\{\left|\frac{\Delta R}{R}\right| + \left|\frac{\Delta \delta R_c}{\delta R_c}\right|\right\} + \{\ln(F/c)\}\left|\frac{\Delta q}{q}\right| = 0, \tag{3}$$

where $\Delta R/R$, $\Delta\delta R/\delta R_c$ >> $\Delta R/R$ and $\Delta q/q$ are maximal relative measurement errors of determination original value R, decreasing R after c thermo cycles or values q correspondingly. By the time of using non-destructive method [21] $\Delta R/R$ =0,03 or $\Delta R/R$=0,068, when we used bathes from 25 specimens. So, $\Delta\delta R/2\Delta R$.

Therefore, $\Delta F/F$ depends on the measurement errors of strength $\Delta R/R + \Delta\delta R_c/\delta R_c$ with accumulation factor $1/q$ and measurement error of $\Delta q/q$ with accumulation factor $\ln(F/c)$. If $q = 1$ ($\Delta q = 0$), that we will get measurement error of calculation F by $\delta R/\delta R_{1c}$ from (3):

$$\pm\frac{\Delta F}{F} = \left|\frac{\Delta R}{R}\right| + \left|\frac{\Delta \delta R_c}{\delta R_c}\right|, \tag{4}$$

where $\Delta R/R \approx 0{,}03$; $\Delta\delta R_c = 2\Delta R \approx 0{,}06R$; $\Delta R_c \le [\delta R/R] = (0{,}05...0{,}15)R$.

After differentiation (2):

$$\frac{\Delta q}{q} = \frac{1}{\ln(\delta R_c / \delta R_1)} \left\{ \frac{\Delta \delta R_c}{\delta R_c} + \frac{\Delta \delta R_1}{\delta R_1} \right\}, \tag{5}$$

where $\Delta\delta R_c = \Delta\delta R_1 = 2\Delta R;\ \delta R_c = \delta R_1 c^q; \Delta\delta R_1 / \delta R_1 >> \Delta R / R$. After plugging (5) in (3):

$$\frac{\Delta F}{F} \approx \frac{1}{q} \frac{\Delta \delta R_c}{\delta R_c} \left\{ 1 + \left(\ln \frac{F}{C} \right) \left(\frac{1 + c^q}{\ln c} \right) \right\}. \tag{6}$$

Accumulation factor $\frac{1+c^q}{\ln c}$ (when $q = 1$) has a minimum when $c = 4$ and is numerically equal 3,6.

When F = 40 and $c = 4 \ln(F/c) = 2{,}3$ = 4; $\left(\ln \frac{F}{C} \right) \left(\frac{1+c^q}{\ln c} \right) \approx 8{,}3$, and $\frac{\Delta F}{F} \approx \frac{9{,}3}{q} \frac{\Delta \delta R_c}{\delta R_c}$. While $\Delta\delta R_c = \delta Rc$ (when c is too small), it is $\Delta F / F \approx 9{,}3 / q$. The result proved that contribution of measurement error of q in $\Delta F / F$ is more in 8 times than contribution of measurement error of R. It's the case, when measure of damage is a decreasing of strength. To decide this problem using of rate of strain $\varepsilon = \delta l / l$, where $l \approx 100mm$ – base length, $\delta l = l_c - l$ – the base deformation and l_c – base length after "c" thermo cycles, is offered. But for that it is necessary to determine dependence between decreasing in strength and rate of strain. So, this value is defined as $z = (\delta R / R) / \varepsilon$.

3.

To determine this dependence concrete mix was tested. Mix contained a portland cement (12%) of brand 400, sand (25%), granite crushed stone 5 … 20 mm (56%) and water (7%). 108 samples cubes with an edge of 150 mm were prepared. There points of intersection of diagonals of opposite sides were spaced far apart from each other (to 1,5 mm). Samples hardened 28 days in water at the room temperature, and then 60 days in damp sand at 18 … 26 °C. From these 108 samples, 8 samples were cycled alternately thermally and mechanically, and 100 samples were used for a frost resistance assessment by a basic method. At realization of a basic method l distance between points of each sample was measured by means of a tool microscope before tests for frost resistance or for durability. Rate of strain was counted as . Changing of distance between points at realization of a method [22] was carried out on 20±2 °C before and after cycling by means of a bracket with variable base and a measuring head of hour type (the price of division of 1 micron). Threshold loading of a water-saturated sample was determined by a way [23], registering the acoustic issue (AI) by means of the AF-15 device at cyclic loading and unloading of a sample to zero. In the first experience load of L was brought to 11 t; in the absence of acoustic issue in the course of the end of unloading value of L was increased by 5% and so until at the end of unloading there was no AE. For accepted an average of L two last cycles. The limit of long durability was found as the relation to the average area of two loaded sides. Dependence z was determined as $(L_1 - L_2)/ L_1 \varepsilon$, where L_1 is a maximal value L in the first loading and L_2 – in the second loading (fig. 1).

Results of this experiment are given in the Table 1. Dependence z is near to linear.

Table 1. Results of experimental determination dependence between R and ε after 20, 42, 84 and 105 cycles (thermocycles and mechanical cycles)

№ of specimen		**1**	**2**	**3**	**4**	**5**	**6**	**7**	**8**	**Average**
Origin value R_0, [MPa]		9,45	16,8	22,7	23,4	24,3	24,3	25,0	30,1	22,0
After 20 cycles	R_0, [MPa]	9,18	16,2	22,2	22,7	23,5	23,9	24,1	29,4	
	$\frac{\delta R}{R}\cdot 10^5$	2857	3571	2203	2991	3292	1646	3600	2326	2811
	$\varepsilon\cdot 10^3$	216	210	204	206	204	200	202	190	204
	z	13,23	17,00	10,80	14,92	16,14	8,23	17,82	12,24	13,8
After 42 cycles	R_0, [MPa]	9,03	16,15	21,9	22,3	23,4	22,9	23,9	20,4	
	$\frac{\delta R}{R}\cdot 10^5$	4482	3869	3524	4701	3704	561	4440	3222	4213
	$\varepsilon\cdot 10^3$	406	452	404	408	394	396	410	394	408
	z	11,04	8,56	8,72	11,52	9,40	14,55	10,73	8,18	10,32
After 84 cycles	R_0, [MPa]	8,10	14,7	19,1	21,2	21,0	23,0	22,6	26,7	
	$\frac{\delta R}{R}\cdot 10^5$	1429	893	1586	940	1244	1230	1253	1130	1213
	$\varepsilon\cdot 10^3$	1206	1217	1188	1196	1201	1207	1196	991	1175
	z	11,85	5,31	13,35	7,86	10,36	9,31	7,36	11,04	9,55
After 105 cycles	R_0, [MPa]	7,75	14,2	18,6	20,1	21,1	20,9	21,4	26,4	18,8
	$\frac{\delta R}{R}\cdot 10^5$	1799	1547	1806	1410	1317	1399	1440	1229	1488
	$\varepsilon\cdot 10^3$	1582	1440	1435	1534	1561	1459	1297	1196	1439
	z	11,37	10,74	12,59	9,19	8,43	9,59	11,10	10,28	10,40
$\bar{z}$ for specimen		11,40	10,76	11,04	10,73	10,21	10,30	11,05	10,34	$\bar{\bar{z}}$=10,73
R- average value oft ha long-term strength in conditions of compression, δR decreasing R after specimen cycling,								S^2	S	ρ
ε - rate of strain in the direction, perpendicular compression; $z=(\delta R/R)/\varepsilon$;								1,94	1,394	0,134
z_k-value z after 105 cycles; $\bar{z}$ - average value z for specimen; $\bar{\bar{z}}$ - average value for $\bar{z}$; S^2 –sampling variance; S-mean square deviation; ρ- variation coefficient								0,183	0,428	0,04

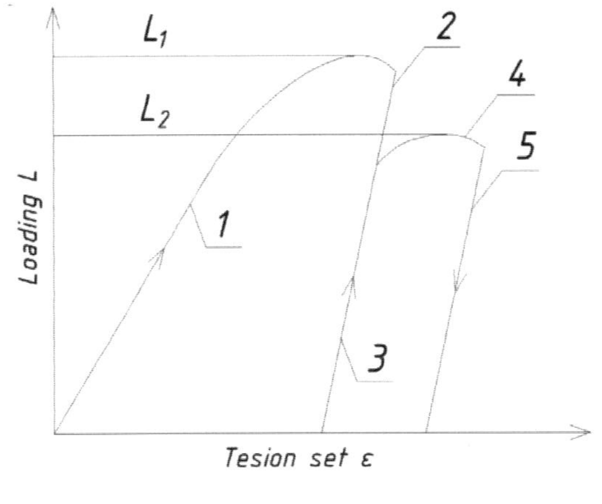

Figure 1. This is a shedule of specimen loading in conditions of monoaxial compression after thermocycles: 1 – line of first loading; 2 – line of first unloading; 3 – line of second loadind; 4 - line of second loading; 5 - line of second unloading; ε – relative longitudinal tension set; L_1 – extreme loading of the first loading; L_2 – extreme loading of the second loading.

Also results by non-destructive method were compared with results by standard method (Table 2).

Table 2. Comparison results by non-destructive method and standard method

Average value of:	Non-destructive method	Basic method
δR/R	0,1488	0,161
ε	0,01439	0,01448
z	10,4	11,1

4. Results

Results by non-destructive method differ 6,3% from results by basic method. Also this dependence is not delicate to changing some thermo cycles to mechanical cycles. So, value of z is constant and dependence of relative decreasing in strength is linear. After mathematical manipuleations it becomes clear, that using ε as a measure of damage is more accurate than using δR, in approximately in 35 times.

After taking into account theese results new method for determination of concrete frost resistance was offered by application for a patent [24].

5. References

[1] Rønning, T.F. 2001 *The Norwegian Institute of Technology Press: Trondheim*. **416**

[2] Davie, C.T., Pearce, C.J., Bićanić, N. 2005 *Proc. of the 13 ACME conference: Sheffield*. **46-49**

[3] Kurilko, A.S., Novopashin, M. D. 2005 *Journal of Mining Science*. **41** 119-122.

[4] Cherepanov, V.I., Nekrasova, E.V., Chernyh, N.A., Panchenko, Ju.F. 2013 *Construction Materials* **15** pp. 10-12. (rus)

[5] Shashank, B. 2004 *University of Tokyo Press: Tokyo*. **94**

[6] Shanshan, J., Jinxi, Z., Baoshan, H. 2013 *Construction and Building Material*. **47** 126–130.

[7] Nesvetaev, G.V., Kardumjan, G.S. 2013 *Beton i zhelezobeton* **45** 6-8 (rus)

[8] SS 137244. Concrete testing – Hardened Concrete-Frost Resistance.

[9] RILEM TC 117-FDC 2001 *RILEM Technical Committee*

[10] RILEM TC 176 2004 *RILEM Technical Committee*

[11] Bunke, N. 1991 *Schriftenreihe des Deutschen Ausschusses für Stahlbeton: Berlin,* **422**

[12] GOST 10060.1-95 (rus)

[13] GOST 10060.3-95 (rus)

[14] Ponomarev, A.N. 2007 *Construction and Building Materials* **24** 22-25. (rus)

[15] Barabanshchikov, Yu.G., Semenov, K.V. 2007 *Power technology and engineering* **36** 197-200

[16] Kalashnikov, V.I., Tarakanov, O.V., Kusnetsov, Y.S., Volodin, V.M., Belyakova, E.A. 2012 *Magazine of Civil Engineering* **34** 47-53.

[17] Gokce, A., Nagataki, C., Saeki, T., Hisada, M. 2010 *Construction and building material* **25** 2426-2431.

[18] Dikun, A.D., at al. 2005 *Construction Materials* **52** 55-56 (rus)

[19] Kuncevich, O.V. 1983 *Stroyizdat: Leningrad*. **132** (rus)

[20] Nikolskiy, S.G. Nikolskaya, T.S. 2009 *Proc. of the 3rd IC Populjarnoe betonovedenie* **35-43** (rus)

[21] Nikolskiy, S.G., at al. 2010 RU Patent 2,380,681 (rus)

[22] Nikolskiy, S.G., at al. 2009 RU Patent 2,305,281 (rus)

[23] Vorontsova, E.A. Nikolskiy, S.G., 2011 *Proc. of the XL Week of Science: St. Petersburg*, pp. **344-430** (rus)

[24] Pertseva, O.N., Nikolskiy, S.G. 2014 RU Application for a patient 2,014,116,713 from 24, May (rus)

Wall Effects induced by Ceramic in Quiescent Liquids

L Noirez[1]*, P Baroni[1]

[1] Laboratoire Léon Brillouin (CEA-CNRS), Ce-Saclay, Gif-sur-Yvette Cédex, France

E-mail: laurence.noirez@cea.fr.

Abstract: At the liquid-solid interface, the energy of the liquid is different from the bulk resulting from surface tension due to the balance between the attraction between molecules to each other (cohesion) and the attraction to the surface (wetting). While capillary effects are well known and described at the air/liquid/solid interfaces, much less is known on the effects induced in the bulk close to the wall. The present experimental study reveals that non-negligible interfacial effects can be revealed in the bulk of the liquid using the high wetting power of ceramics. Close to the wall, thermal measurements reveal a progressive temperature drop in the liquid (about 0.15°C). This zone extends up to several millimeters, creates a non-equilibrium/equilibrium interface within the liquid and is balanced at larger distances by a temperature increase. This localized effect is highlighted with strong wetting metal oxide surfaces as ceramics.

Keywords: ceramic, wetting, long range interactions, liquid/solid interface.

1. Introduction

In the bulk of the liquid, molecules are pulled in every direction by neighboring liquid molecules, resulting in a net force of zero. Close to a surface, the situation is different. The molecules at the boundary do not have other molecules on all sides of them and create some variation of internal pressure [1]. Thermal exchanges that take place in a liquid bulk close to an immersed wall or in between two walls are usually not explored, hardly modelled [2] and considered as restricted to very small length scales [3,4].Recently a submillimetre scale thermal study of liquids under flow has revealed that shearing liquids may induce a cooling effect at low shear rates in ordinary Newtonian liquids as water or low viscous polymer melts [5]. The detailed analysis of this non-equilibrium endothermic process indicates that the cooling is firstly generated at the walls and propagates in the fluid giving rise to the establishment of a thermal shear banding regime along the flow axis [5]. We report here on the situation where no external force is exerted on the liquid; i.e. the quiescent state. We examine the influence of a solid wall strongly interacting (total wetting) with the molecules of liquids as water, glycerol or alkanes on the thermal equilibrium. The liquid is observed in its quiescent bulk away from the liquid-air interface. We report on the identification of a slight temperature fall observed at the vicinity of a wetting wall followed by a slight temperature rise at larger distance from the wall. This situation is particularly observable when the liquid is in strong interaction with the wall (total wetting condition). The attraction energy exercised by the wetting surface to the liquid produces a pressure variation, in correlation with a slowing down of the molecules at the wall and correlatively creates a different thermodynamic equilibrium. The cooling zone is compensated away from the surface by a zone slightly warmer than the average temperature. These wall effects are induced in various daily used liquids and are particularly observable using zero-porosity ceramic surfaces because of their high wetting ability. These results highlight the active role of strongly wetting surfaces as ceramics and reveal long range correlations in the liquid state.

IOP Conf. Series: Journal of Physics: Conf. Series **790** (2017) 012024 doi:10.1088/1742-6596/790/1/012024

2. Experimental

This pioneering approach is carried out by using two independent methods to measure the temperature within the liquid. A thermistor probe enables a localized probe of the temperature while Infrared absorption in real-time allows a 2D-mapping of the temperature. We use recent instrumental advances that enable the detection of absolute thermal variations as small as 5.10^{-2}°C for the 2D-infrared emission and as small as 5.10^{-3} °C (relative thermal accuracy) with the thermistor. All the measurements have been carried out at room temperature without forced convection (quiescent liquid equilibrated at least 4 hours before the measurement)in a box isolated from external thermal fluctuations (conduction and convection), in the darkness and in conditions as close as possible to the equilibrium state. The whole setup is equilibrated during several hours prior the experiment. A scheme of the experimental setup is given in Fig.1a.

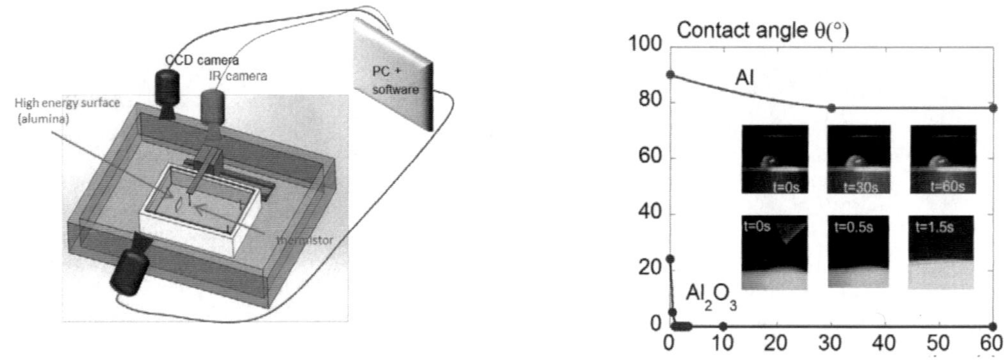

a b

Figure 1.
a) Scheme of the experimental setup: the ceramic surface is immersed in a bath filled with a liquid. The bath is itself placed in a thermally isolator container. Two optical cameras view the liquid from the side and the top of the liquid surface and an infrared CCD is positioned above the liquid surface (top view). The dimensions of the ceramic and Aluminum plates are about 15 x 20 x 1 mm^3. Three thermistors are immersed in the bath: one close to the plate and two others at fixed positions at 150mm away from the plate.
b) Measurement of the contact angle versus time for glycerol deposited at room temperature on an aluminum plate (grey circles) and on an alumina plate (blue circles). Similar stationary contact angles are found for liquid water.

Local temperature measurements: The temperatures are recorded using a 10 kΩ (room temperature) calibrated thermistor of 0.4 mm diameter. The thermal probe is fixed to a deported arm that can slowly move the probe at different locations in the liquid. Two identical thermistors placed away from the probed surface serve assimultaneous references to renormalize the temperature from small random thermal fluctuations. Each thermistor is connected to a separated Keithley multimeter (5 µA). The GPIB mode is used to enable the simultaneous recording of the temperatures. The relative reproducibility of the measurement is estimated at 5.10^{-3}°C.

Micro-thermal 2D-Imaging: The thermal mapping is based on the emissivity measurement; a medium radiates all the more that its temperature is high. The heat transfer by radiation can be expressed using the Stefan-Boltzmann law: $E= em.\sigma.A.(T^4 - T_c^4)$ where E the radiated energy, *em* the emissivity, A the radiating area, T the temperature of the sample and T_c the temperature of the surroundings. $\sigma$is the Stefan's constant. A high performance 320x240 pixels B400 FLIR Infra-Red 2D-detector equipped with a close-up lens giving a spatial resolution of 25 µm in a 8 x 6 mm^2 working area is placed at about 50 mm from the free surface of the sample filling the gap between two wetting substrates made of alumina [2]. The frame rate is 30 Hz. The resolution ellipsoid enables to probe a ± 0.15 mm depth of field. The spectral range goes from 7.5 up to 13µm and the relative temperature accuracy is of ±

0.025°C. Finally, it is reminded that the emissivity of the liquid water is 1,000. This means that water is an absolute reference (the Wien law can be directly applied without emissivity correction).

Surfaces: Two surfaces are used in this experiment: a zero porosity alumina plate of 3 mm thickness and Ra = 1.6 roughness providing a total wetting and an aluminum plate (AG3) of similar roughness providing a partially wetting (Fig.1b). We use practical Aluminum plates. Aluminum surfaces are naturally passivated by a thin oxidized aluminum film of alumina (gamma phase) [6]. The wetting degree is characterized by the contact angle method. The surfaces will be in contact with demineralized liquid water (preliminary degassed). The wetting ability is evaluated via the determination of the contact angle. It reaches zero when the wetting is total (alumina) and is around 75° in the case of the liquid water drop deposited on an aluminum substrate (partial wetting).

3. Results - discussion

Two independent techniques are used to determine the thermal profile in the bulk of the liquid at the approach of an immersed wall. This wall is either by a highly wetting surface (ceramic) or by a partial wetting surface (aluminum).We illustrate here the results obtained at room temperature on liquid water in the quiescent state. These two techniques are complementary; Infrared measurement enables an instantaneous two-dimensional mapping of the temperature distribution viewed from the free surface of the water. The thermistor enables a punctual measurement with a higher accuracy in the liquid bulk.

Fig.2 (a) displays the infrared image corresponding to the thermal emission produced through the liquid surface (confocal configuration) of a bath filled with water at room temperature in which an alumina plate has been placed. The alumina plate is vertical and totally immersed in the water. The thermal mapping reveals that the plate is surrounded by two successive static thermal waves. Close to the plate, the interaction of the liquid to the plate produces a cooling zone. This zone is followed by a second wave of higher temperature than the medium and compensating the thermal loss corresponding to the first cooling zone. This thermal distribution is stable with time and suggests that the wall induces the coexistence of two different thermodynamic states in the quiescent liquid.

The temperature of a body is resulting from the dynamics at a molecular scale. This thermal excitation produces an infrared emission due to the vibration of molecules that continuously absorb and re-emit infrared photons. The observation in the liquid of a variation of infrared emission and of the correlated temperature variation close to the surface as illustrated in Fig.2a and 2b, indicates different vibration states. The interactions between molecules are modified and the motion is affected producing close to the solid wall a slowdown of the motion responsible of the reduced infrared emission. At larger distance from the wall, the subsequent increase of the infrared emission indicates an increase of energy.

The temperature cartography revealed by infrared emission is confirmed by the thermal sensor. This variation of the infra-red emission close to the wall is corroborated by the direct measurement of the temperature using an accurate thermistor placed at different positions in the bath. The thermistor enables to probe the temperature in the liquid bulk at various depths away from the liquid surface. Fig.2b displays the typical temperature profile recorded by the thermistor as a function of its distance from the plate. The profile shows that the temperature in the liquid decreases significantly close to the ceramic wall ($\Delta T \cong 0.2°C$) and much slightly in the case of Aluminum. At larger distance from the wall ($d > 10mm$), the profile indicates a slight temperature increase.

The attraction of the wetting wall is such that the liquid exhibits close the wall, a slowing down of molecular dynamic (temperature decrease) and thus are lated lower pressure state. At distance away from the wall, the zone where the liquid exhibits a slightly higher temperature corresponds correlatively to a compressive zone. This effect that takes place within the liquid bulk can be interpreted similarly as typical pressure capillary effects usually described for air/liquid interfaces. A liquid/liquid interface is here produced by the induction of a lower pressure state close to the ceramic wall balanced by a higher pressure state away from the wall. The pressure difference between these two states creates locally an induced immiscibility in the liquid bulk. A capillary pressure separates the two states $p_c = (p_{low\ pressure} - p_{highpressure}) < 0$ following the Young-Laplace equation, the pressure difference is proportional to the induced interfacial tension.

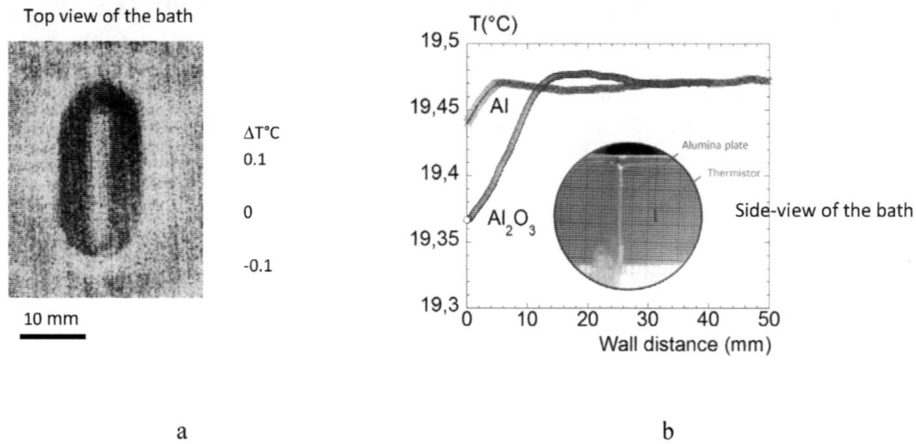

Figure 2.

a) Two-dimensional thermal cartography (top view) of the free surface of quiescent water below which a 3 mm thickness alumina plate (central vertical bar) has been immersed. The colors indicate different temperatures, the warmest corresponding to the highest temperatures while the cold ones (blue) indicate lower temperature regions. The experiment is carried out at room temperature in the quiescent state. These temperature variations are stable and persist in time during several days. Top view of the bath (recorded from the free liquid surface).

b) Temperature profile (average profile on 10 measurements of each 10min) recorded in the liquid water away from an alumina, aluminum surface respectively (at around 15mm depth) as a function of the distance from the plate. A similar profile is recorded at different depths away from the surface. Insert: photograph of the bath with the immersed plate (side-view of the bath) and of the thermistor at the approach of the plate.

Here the mechanism is complicated by the induction close to the wall of a convective motion due to the temperature gradient (Fig.3). The inductance of expansive and compressive states and of related thermal effects is also known during applied tensile strain for engineering materials in solids. Thermo-elastic effects have been extensively studied in solids. Most of them describe effects in crystalline solids relating the dynamics of the molecular vibrations. Some studies report on thermoelastic effects in glasses [7, 8]. Finally, to our knowledge, no similar effect has been reported in the liquid state probably because these effects are usually very weak and academically not foreseen. However condensed matter can sustain negative pressures because of the attractive intermolecular interactions. The wetting the surface creates a local negative pressure. Moreover recent developments in stress dynamic measurements have shown that liquids as water, alkanes or polymer melts exhibit an elastic (solid-like) behavior in restricted geometry [9]. Sub-millimeter scale stress measurements highlight an intrinsic property supporting the hypothesis of a liquid formed by a weakly bound resilient self-assembly [10, 11]. The present observation of a temperature fall still active at nearly 10 mm from the wall is the firm indication that cohesive intermolecular forces are long range correlated in the liquid bulk. Correlatively the liquid state is easily modifiable, "stretchable" at the vicinity of a high energy surface as a wetting substrate.

4. Conclusions

Though interfaces are known to crucially influence the behavior of physical, chemical, and biological systems, their contribution to the establishment of different coexisting thermodynamic states via contact is unknown or supposed negligible [3, 4]. While capillary pressure effects are well documented, much less is known concerning the possibility of modifying the thermodynamic equilibrium by wall interactions. The use of high energy surfaces as ceramic enables to highlight liquid/surface effects in the bulk that would have gone unnoticed. This experimental work evidences that the wall attraction induces the coexistence of slightly different thermodynamic states characterized by a weak, progressive and non-ambiguous temperature drop in the bulk of the liquid at the approach of the wetting surface balanced by a succeeding temperature increase away from the wall. The negative temperature gradient is around several few tenths of a Celsius degree in liquid water. The greater attraction of molecules to the wetting surface (adhesion) than to each other (cohesion) produces surface attraction that forces the cohesive nature of the liquid to behave as a slightly stretched elastic state close to the wall. Similarly as capillary effects, this lower pressure state involves the liquid cohesion and modifies the intermolecular forces at the millimeter scale. This long range effect is incompatible with a description in terms of individual molecular dynamic or via a heat transfer process but involves the cohesion forces. The cohesive forces give rise to bulk and shear elasticity. The latter have been measured in the liquid state by different authors and methods [10-14]. Only scarce theoretical approaches consider the cohesive character of the liquid state, at the approach of a glass transition [15,16], close to the wall, at several molecular layers scale [17] or macroscopically, the strength of which depends on the network size [18]. Clearly the cohesive long range intermolecular correlations should play a key role in these giant (wall) effects. The identification of a wall induced gradient of temperature points out also many questions and requires more investigation to determine the relevant parameters: what is the mechanism that governs the interactions between liquid molecules and the surface atoms? What are the parameters of the surface and the characteristics of the liquid that define these effects? What are the consequences of the temperature gradient on the flow behavior, especially in solutions? How to overcome this thermal effect at the wall? In this challenge, a multiscale description should provide new insights in understanding the true nature of the liquid state involving a description of the interaction forces at play across the wall up to the macroscopic behavior. This puzzling wall-induced effect has profound fundamental origins that remain to be identified and various possible applications as for example a selective separation in solutes or solutions [19].

References

[1] De Gennes P.G. 1985, Wetting: statics and dynamics, *Review of Modern Physics* **57** 827-861.
[2] A. D. J. Haymet and David W. Oxtobya 1981, A molecular theory for the solid–liquid interface, *The Journal of Chemical Physics* **74** 2559.
[3] A. Mari, J. Eisert 2012, Cooling by Heating, *Phys. Rev. Lett.***108** 120602.
[4] David G. Cahill, Wayne K. Ford, Kenneth E. Goodson, Gerald D. Mahan, ArunMajumdar, Humphrey J. Maris, Roberto Merlin and Simon R. Phillpot 2014, Nanoscale thermal transport, *Appl. Phys. Rev.***1** 11305.
[5] Baroni P., Bouchet P., Noirez L. 2013, Highlighting a Cooling Regime in Liquids under Submillimeter Flows*J. Phys. Chem. Lett.***4** 2026; ibid. PCT, CNRS: DI 05815-01 FR12.
[6] Bernardin J. D., I. Mudawar I., Walsh C. B. and Franses E.I, 1997, Contact angle temperature dependence for water droplets on practical aluminum surfaces, *J. of Heat Transfer*,**40**, No. 5, pp. 1017-1033.
[7] Privalko V.P., Korskanov V.V. 1999, ThermoelasticBehaviour of Amorphous Polymers Above and Through the Glass Transition Interval I. Polystyrene**,** *J. of Thermal Analy. and Calorimetry*, **55**741.
[8] Padmaja A., 1996, Pressure dependence of the thermoelastic quotient for glasses, *Inter. J. of Thermophysics*, **17** 723.
[9] Reiter G., Levent Demirel A., Peanasky J., Cai L., Granick S. 1994, *Stick to slip transition and adhesion of lubricated surfaces in moving contact,* J. Chem. Phys., **101**2606-2015.
[10] Noirez L., Baroni P. 2012, Identification of a low-frequency elastic behaviour in liquid water *J. of Physics: Condensed Matter***24**: 372101-372106.
[11] Noirez L, Baroni P, 2010, Revealing the solid-like nature of glycerol at ambient temperature, *J. of Molecular Structure,***972** 16.
[12] Derjaguin B.V., U.V. Bazaron, Lamazhapova, Tsidypov B D 1990, Shear elasticity of low-viscosity liquids at low frequencies, *Phys. Rev.* A **42** 2255.
[13] Bund A, Schwitzgebel G 1998, Viscoelastic Properties of Low-Viscosity Liquids Studied with Thickness-Shear Mode Resonators,*Anal. Chem.***70** 2584.
[14] Chushkin Y, Caronna C, Madsen A 2008,Low-frequency elastic behavior of a supercooled liquid, *EPL***83** 36001.

[15] Zaccone A, Terentjev E. 2013, Disorder-Assisted Melting and the Glass Transition in Amorphous Solids, Phys. Rev. Lett. **110** 178002.
[16] Granato A.V. 2009 Mechanical properties of simple condensed matter, *Materials Science and Engineering*A**521**: 6-11.
[17] Schoen, M., Hess, S., Diestler, D.J. 1995, Rheological properties of confined thin films, *Phys. Rev. E***52**, 2587.
[18] Volino F. 1997 Théorie visco-élastique non-extensive, *Annales de Physique*, **22** n°1 & 2.
[19] Noirez L, Baroni P., Soleaulettern 529679020215 (2015).

Numerical analysis of extensional flow through the pharyngeal duct

M Preciado-Méndez[1], M Salinas-Vázquez[2], W Vicente[2], E Brito-de la Fuente[3], G Ascanio[4]

[1]Programa de Maestría y Doctorado en Ingeniería, Universidad Nacional Autónoma de México, Ciudad Universitaria, México.
[2]Instituto de Ingeniería, Universidad Nacional Autónoma de México, Circuito Escolar
[3]Fresenius-Kabi Deutschland GmbH, Bad Homburg, Germany.
[4]Centro de Ciencias Aplicadas y Desarrollo Tecnológico, Universidad Nacional Autónoma de México, Circuito Exterior Ciudad Universitaria, México.

E-mail: mpreciadom@iingen.unam.mx

Abstract. The flow through the pharynx from the glossopalatal junction (GPJ) to the upper esophageal sphincter (UES) has been numerically investigated with a non-Newtonian fluid obeying the power-law with similar rheological indices to a contrast medium used in videofluroscopy. For that purpose, a three-dimensional model of the transport of food bolus along the pharynx has been proposed using the immersed boundaries method, which allow representing the shape of the pharynx using Cartesian grids. The pharyngeal wall has been considered to be an elastic membrane. Flow fields in terms of the axial velocity, pressure, shear rate and strain rate were obtained. Results show that the highest velocity concentrates in the central stream as the fluid enters into the pharynx. In addition, as the flow quits the pharynx, a recirculation zone appears inside the cavity, resulting in low velocity zone, which increases with the coefficient of elasticity. A strong dependence on the coefficient of elasticity was observed on the pressure fields; so that as such a coefficient increases, the pressure in the pharyngeal wall will increase. It has been also observed that the bolus head travels faster than the bolus tail, which indicates that the bolus is not only subjected to shear but also to elongation. Results from this work can be further used for a rheological characterization (shear and extension) of oral nutritional supplements for patients suffering from swallowing disorders.

1. Introduction

1.1. Role of bolus rheology when swallowing

Extensional rheological properties play an important role in processes in which the fluid is subjected to highly decelerated or accelerated flows, particularly important extensional effects are found in flows through converging or diverging geometries [1]. Many real flows, including those which are biologically relevant, are complex and can include both shear and extensional components. Extensional flows can significantly stretch macromolecules, providing orders of magnitude increases in elastic forces and extensional viscosity [2].

Rheology, defined as the science of deformation and flow of matter, is by common consent a difficult subject [3]; when it comes to the oral behavior of fluid foods analysis, rheology has been estimated as a very important factor. The flow behavior and the rate of deformation of a fluid food inside the mouth are considered as play a critical role in sensory perception and oral transportation. The rheology of complex polymeric biological fluids is vital to the correct functioning of many processes in the body.

1.2.Swallowing assubject of study

Recently, the importance of the extensional rheological properties to study the dynamics of complex fluids has become an area of important research activity, nevertheless, little information is available due to generating and sustaining uniaxial elongational flow experimentally remains a challenge [4]. The knowledge of behaviour of fluids does apply to the knowledge of biological functions, such is the case of swallowing process, which is a neurological function wherein more than fifty pair of muscles are involved. The pharynx plays a crucial role in this complex process because of its dual function since it transports both air to the larynx and the bolus to the esophagus in respiration and swallowing, respectively. Hence, the airway protection needs to be guaranteed, it is for this reason that swallowing is carried out within a second and it cannot be stopped or suspended before its completion. Furthermore, carrying out the swallowing process requires the accomplishment of a series of process, namely, closure of the nasopharynx, tongue pressure, pharyngeal peristalsis, and opening of upper esophageal sphincter, among others.

Swallowing is divided into three major stages according to the anatomic location of the bolus, namely, oral, pharyngeal and esophageal. The former one is in turn subdivided into two sub phases, preparatory and transport phase.

The oral preparatory phase occurs when food enters into the mouth aimed to be transformed in an appropriate bolus for swallowing. Subsequently, the oral transport phase begins, here, the tongue exerts a pressure against the palate, forcing the bolus to move posteriorly toward the oropharynx. This stage has a quite high importance due to the lingual activity since it helps form and maintain the bolus and also, lingual force might be considered as the main driving force for bolus flow [5]; it should be noticed that it stimulates the oropharyngeal receptors and thus, triggers the pharyngeal swallow reflex. Once this stage has been completed, it follows the pharyngeal phase, here, the airway protection and the bolus transport through the pharynx towards the esophagus are carried out, for this reason, this phase is considered as the most important and a precise synchronized work of the epiglottis to fold over the glottis and stop liquid and food from entering the trachea and the pharynx are needed. The latter phase of swallowing begins when bolus passes though the upper esophageal sphincter (UES).

2. Governing equations

A non-Newtonian shear-thinning fluid was considered in this work because that is the type of fluid used for the assessment of swallowing process as well as for treating dysphagic patients. Meng et al [6] suggested that non-Newtonian of fluids would be safer to swallow since they would allow the neuromuscular system more time to shut off air passages and reduce the risk of aspiration, for this reason it is important to simulate pharyngeal swallows of non-Newtonian boluses. A density of 1800 kg/m3 and a dynamic viscosity following the Sisko-model were used.

The Navier-Stokes equations can be written as an extension of the following compressible equations [7].

$$\frac{\partial \boldsymbol{U}}{\partial t} + \frac{\partial \boldsymbol{F}_i}{\partial x_i} = S_F \tag{1}$$

where$\boldsymbol{U}$ is a four-component vector defined by Eq. (2) and $\boldsymbol{F}_i$ are the fluxes:

$$\boldsymbol{U} = (\rho, \rho u, \rho v, \rho w) \tag{2}$$

$$F_i = \begin{pmatrix} \rho u_i \\ \rho u_i u_1 + p\delta_{i1} - 2\mu S_{i1} \\ \rho u_i u_2 + p\delta_{i2} - 2\mu S_{i2} \\ \rho u_i u_3 + p\delta_{i3} - 2\mu S_{i3} \end{pmatrix}$$

According to the equation of state for an artificial incompressible fluid the pressure is replaced in Navier-Stokes equation:

$$p = \rho c^2 \tag{3}$$

where p is the pressure and ρ is the density considered; c is the sound speed of the fluid. The values for the coefficients for the Sisko-model, Eqs.(4) and (5), were obtained from rheological measurements and are shown in Table 1.

$$\mu = \mu_\infty + k\dot{\gamma}^{(n_f-1)} \tag{4}$$

$$\mu = \mu_\infty + \begin{Bmatrix} \mu_{max}\,\dot{\gamma} < 0.1 \\ k\dot{\gamma}^{n_f-1} \quad if \quad 0.1 \le \dot{\gamma} \le 500 \\ \mu_{min}\,\dot{\gamma} > 500 \end{Bmatrix} \tag{5}$$

Table 1. Parameters for Sisko-model

Parameter	Value
k	11.0 Pa·s^n
n	0.32
μ_∞	0.42 Pa·s

S_F in Eq.(1) is the vector of source terms. For this case, the four terms in S_F are equal to:

$$S_\rho = S_{\rho u} = S_{\rho v} = S_{\rho w} = 0 \tag{6}$$

$$S_F = 0$$

Gravitational body forces were considered negligible, since bolus transport is due to the peristaltic movement [8].

2.1. Numerical resolution.

As the bolus passes through the pharynx, its cross section gets modified in an expedite form as a consequence of the peristaltic flow. Cook et al. [9] proposed aspatio-temporal shape of the pharynx, from the glossopalatal junction (GPJ) to the upper esophageal sphincter(UES), and that was the basis of this work.

To pursue with the numerical simulations, the governing equations had to be discretized in each node of the computational grid and for this reason, it was necessary to build numerically the components which allow to represent in an accurate way the physic phenomenon, namely, the pharynx geometry.

The immersion of the geometry of the pharynx, follows a process illustrated in Figure 1. The nodes that do not belong to the pharynx surface are blocked by imposing in them constant speeds or a value of zero, allowing the equations to be solved exclusively in the nodes inside the geometry, i.e. the fluid will flow only inside the pharynx. For those nodes remaining near the real body surface, within the flow area, velocity was interpolated at every time step in order to avoid stepped flow in nodes nearby the blocked areas [12]. Temporal functions (high-order polynomials) of the position of 21 reference points for 10 different times allowed obtaining the temporal evolution of the pharynx surface. From these functions, it is possible to know the approximate position of each point at every instant. Interpolations between these reference points were used to know the position of the remaining points of the mesh conforming the surface. Deriving over the time the functions corresponding to the change of position in each point on the surface of the pharynx leads to obtain the radial velocity. The temporal deformation occurring on the surface is considered to be produced only in the radial direction. Thus, the two-dimensional geometries used by Chang et al. [3] were converted into three-dimensional geometries, as shown in Fig. 2 and, which are used in this work.

The generalized coordinate system is solved by extension of the fully explicit McCormack scheme, second order in time and fourth order in space, developed by Gottlieb and Turkel [10]. The inlet and outlet boundary conditions used herein are those proposed by Poinsot and Lele [11].

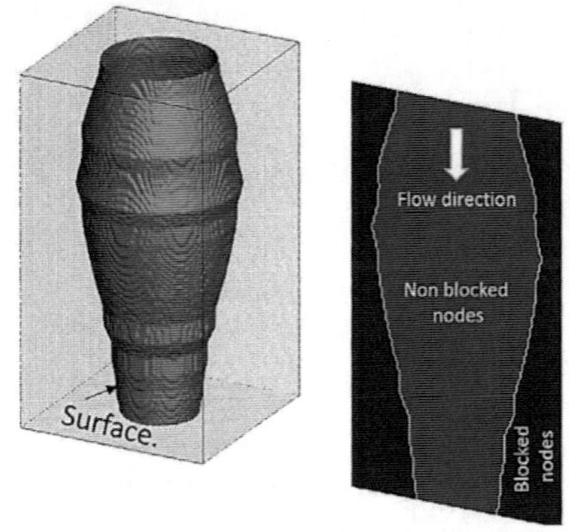

Figure 1. Representation of the immersion of the geometry of the pharynx in the computational grid

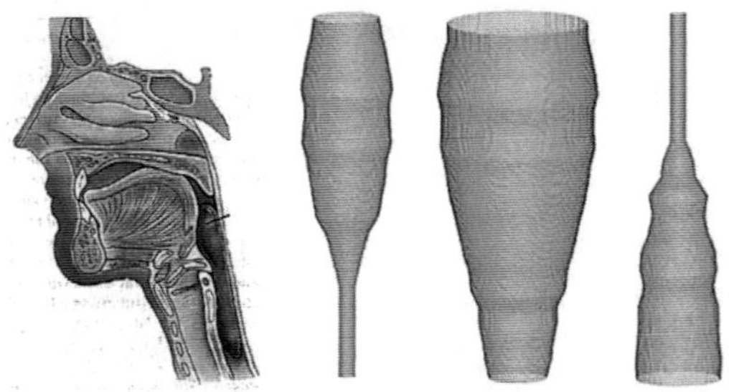

Figure 2. Illustration in 3D for the three stages simulated

3. Peristaltic flow and movable pharyngeal walls

As abovementioned, the peristaltic flow is responsible of the movement of the bolus through the pharynx. Fluid motion in elastic tubes has a high practical and theoretical relevance, however, not much results have been produced due to the complications on the calculation of simultaneous fluid-structure evolution [13].In order to consider the walls of the pharynx to be movable, an axial velocity was fixed on the surface (peristaltic wave velocity, ≈4.81 cm/s); also, a velocity normal to the wall and a wall pressure were fixed. The wall pressure was computed considering the pharyngeal wall like a non inertia membrane according to Predizzetti [13] from:

$$p_w = \tau_n - \frac{2}{Re}\left(\frac{U'_x - R'U'_r}{1+R'^2} + \frac{U_r}{R}\right) \tag{7}$$

where Re is the global Reynolds number, and U'_x is streamwise velocity and U_r is the radial wall velocity both defined by$U_x = \partial x/\partial t$: and $U_r = \partial R/\partial t$, R is the local transversal radio of the pharynx.

The hydrodynamics of the gastrointestinal duct is influenced by its mechanical characteristics. In this work, we focused on the coefficient of elasticity, K, defined by Eq. (8), hence, values of about 10^2 to 10^3Pa·m for K were taken. The membrane was considered isotropic, homogeneous, with constant wall width and a hyperelastic behavior. In Eq (7) τ_n is the wall normal stress, which is function of the principal deformation ratios and of the coefficient of elasticity K.

$$K = hE \tag{8}$$

4. Numerical method

The computational domain has as dimensions 5.0 cm × 2.8 cm × 2.8 cm. in x, y and z directions respectively are: This domain was discretized through a uniform grid distribution employing 150 × 109 × 109 nodes; the grid spacing is 0.3 × 0.25 × 0.25 mm in x, $y$and z directions, respectively. According to Hasegawa et al. [12]the total simulation time was fixed in 1.04 s and divided into three stages: From 0–0.34 s the filling stage with only the entrance (GPJ) open; from0.34–0.54 s, the intermediate stage, here both inlet and outlet are open and the emptying stage from 0.54–1.04 s where the entrance is already close and only the exit (UES) remains open.

4.1. Boundary and initial conditions

A bolus volume of 10 $cm^3$was used, being this amount one of the most commonly used in experiments and also corresponding to a mouthful; also a preexisting volume inside the pharynx of 2.5 cm^3, these value is similar to the value used by Chang et al. [3].The velocities were imposed at the inlet in order to assure the right volumetric flow in the pharynx. The oral preparatory phase is represented by a time of 0 s, where both the GPJ and the UES are closed and it is considered that all velocities are null in the entire domain.

The change of volume of the pharynx at every time instant, allowed obtaining the inlet and outlet volumetric flows. The flow rate of the inlet is obtained by the change of volume of the pharynx at the time step Δt:

$$\dot{V}_{in} = \left|\frac{\Delta Vol_{pharynx}}{\Delta t}\right| \tag{9}$$

the mean axial velocity at the inlet is given by:

$$U_{inlet} = \frac{\dot{V}_{in}}{A_{in}} \tag{10}$$

whereA_{inlet} corresponds to the perpendicular area to the inlet flow.

In the intermediate stage, t = 0.34 s to 0.54 s, it is necessary to know the volumetric flow at one of the two boundaries, and for that, a small and constant inlet and outlet volumetric flow difference was considered.

The bolus has to be propelled from the oral cavity to the pharynx, the tongue is responsible for carrying out this task. Therefore, the pressure applied by the tongue had to be considered in the simulation and itwas modeled by setting a membrane wall pressure, $P = P_{inlet}$, in the nodes close to the entrance, allowing the inlet boundary condition adapting to the pressures inside the pharynx. The membrane wall pressure at these points will be the maximum between P_{inlet} and the value obtained locally with Eq. (10).**Hiba! A hivatkozási forrás nem található.**

5. Results

Different cases were studied and the identical configuration for the temporal evolution of the pharynx was used in all studied cases. The value of the volume increased up to $V_{max} \approx 12.47$ cm^3 at $t = 0.423$ s. The value of 12.5 cm^3 was not reached (initial volume plus bolus volume), since the emptying process began at $t = 0.33$ s. Part of the bolus volume has been evacuated at $t = 0.423$ s. The coefficient of viscosity K was varied and remained the same through the whole process; the values of P_{inlet} were chosen of the maximums pressures obtained in each simulation. The values used for the most significant cases simulated are shown in

Table 2. Cases investigated

Case	**P** (Pa)	**K** $(Pa \cdot m)$
A	2.0	25
B	4.0	50
C	6.0	75
D	8.0	100

5.1. Velocity profiles

*5.1.1. Filling stage(0 s < t < 0.33 s; GPJ open and UES closed).*Regarding to the velocity, simulations showed that all the four studied cases presented a same behaviour in the first stage. Thereby, rises of the velocity were observed at the axis when the fluid (bolus) entered the pharynx and it is accelerated until reaching the pharyngeal wall which provokes also an increase of pressure.

IOP Conf. Series: Journal of Physics: Conf. Series **790** (2017) 012025 doi:10.1088/1742-6596/790/1/012025

Fig. 3 shows some profiles for the velocity in Case B at this stage, the values were close to those found by Nguyen et al. [13] who found values for the velocity of the bolus head of 37.1 ± 1.1 cm/s using the technique of videofluoroscopy. It is important to point out that the flow is not necessarily axisymmetric even though the geometries were generated from revolution surfaces as can be seen from Figure 5with an isosurface of velocity.

5.1.2. Intermediate stage (0.33 s < t < 0.54 s). It was observed in this stage that both the entrance (GPJ) and the exit (UES) are open but is the entrance that begins to close and the bolus continues its movement to the outlet. Values under 40 cm/s were observed for this case and those results are in good agreement with those reported by Hasegawa et al [12] who estimated the maximum velocity reached in the swallowing process of 50 cm/s.

*5.1.3. Emptying stage (0.54 s < t < 1.04 s).*At time t = 0.54 s the emptying phase begins and as the peristaltic contraction occurs, a recirculation zone (negative velocities) is generated and becomes larger near to the entrance where the GPJ is now closed. The zones with a highest shear rate were still found near the walls and in the recirculation zone. Recirculation zones with negative velocities were observed at the occluded zone and this was more relevant as the values coefficient of elasticity K incremented.

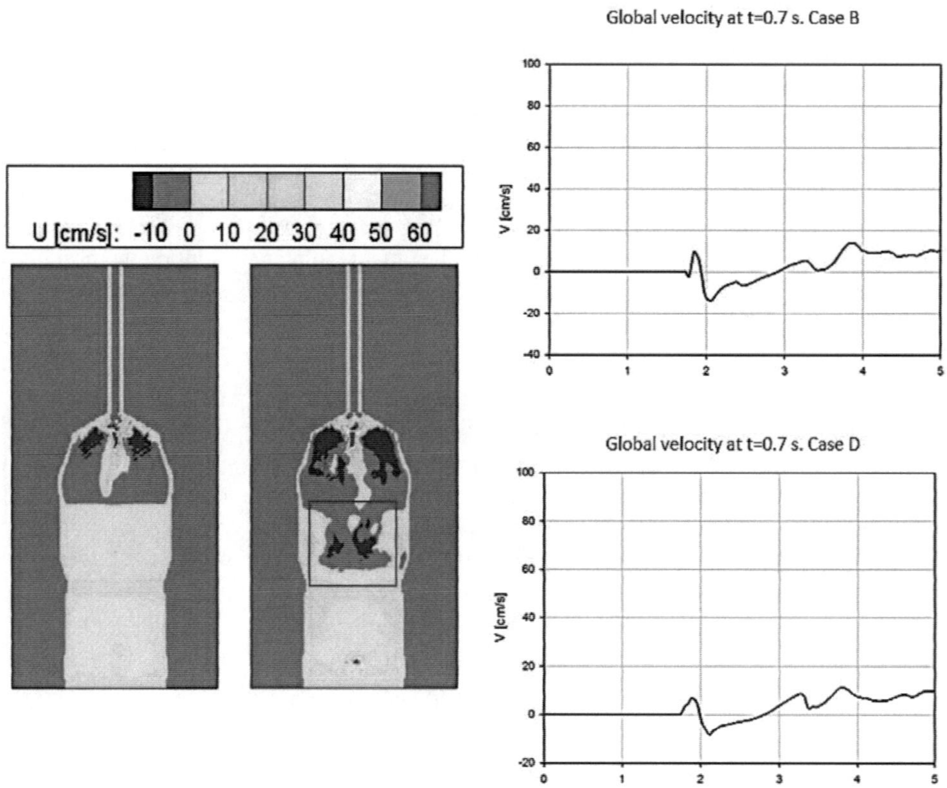

Figure 3. Velocity profiles for Case B and D (right) and global velocities at the occluded zone (t = 0.7 s)

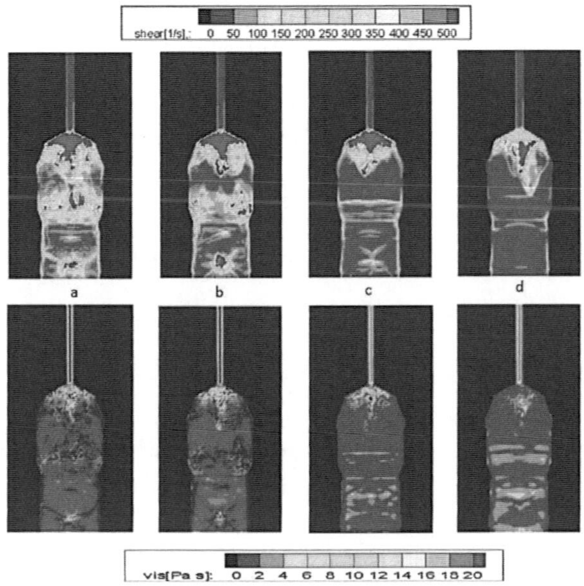

Figure 4. Shear rate (left) and viscosity (right) profiles at the occluded zone (t = 0.7 s)

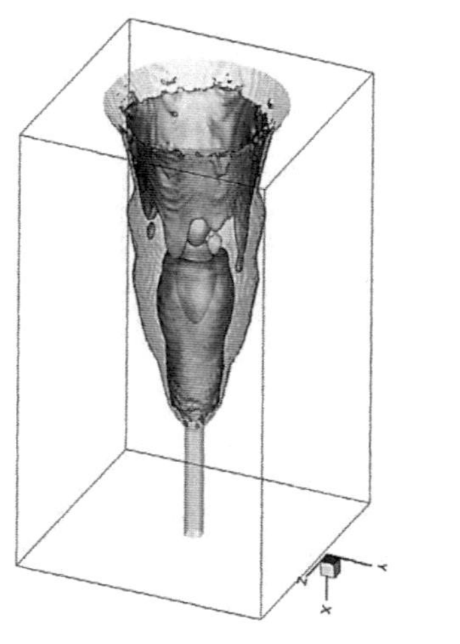

Figure 5. Isosurface of velocity at t = 0.2s

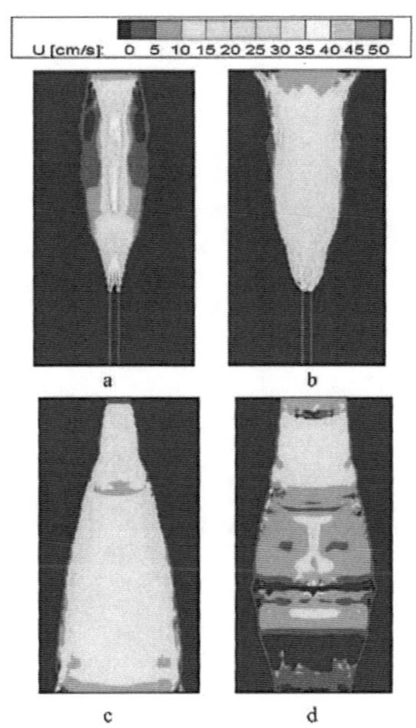

Figure 6. Case 2, velocity profiles. a) t = 0.06 s, b) t = 0.2 s. c)t = 0.34 y d) t = 0.46 s

5.2. Pressure profiles

5.2.1. Filling stage. At this stage, it can be noticed that the process is highly influenced by the pressure exerted by the tongue to initiate the movement and also by the coefficient of elasticity. As the latter increased its value, so did the pressure.

IOP Conf. Series: Journal of Physics: Conf. Series **790** (2017) 012025 doi:10.1088/1742-6596/790/1/012025

5.2.2. Intermediate stage. Figure 8shows the flow fields in terms of the pressure for Case B, the pressure in the pharynx reached values around 4 kPa which is the value estimated by Ferguson et al [14].

5.2.3. Emptying stage. The emptying process is mainly controlled by the volumetric flow, so that the elasticity level has a little influence. Under real conditions, changes in the coefficient of elasticity would create changes in the bolus clearance time.

5.3. *Extensional viscosity and extensional rate calculation.* The calculation of the nominal extensional strain rate was made by:

$$\dot{\varepsilon} = \frac{\partial u}{\partial x} \mathrm{b} \tag{11}$$

and the extensional viscosity was reckoned by the

$$\eta_{E(t,\dot{\varepsilon})} = \frac{\tau_{xx}(t,\dot{\varepsilon}) - \tau_{yy}(t,\dot{\varepsilon})}{\dot{\varepsilon}} \tag{12}$$

Central lines where plotted in order to have a general idea of the rheological behavior of the fluid used in the simulations.

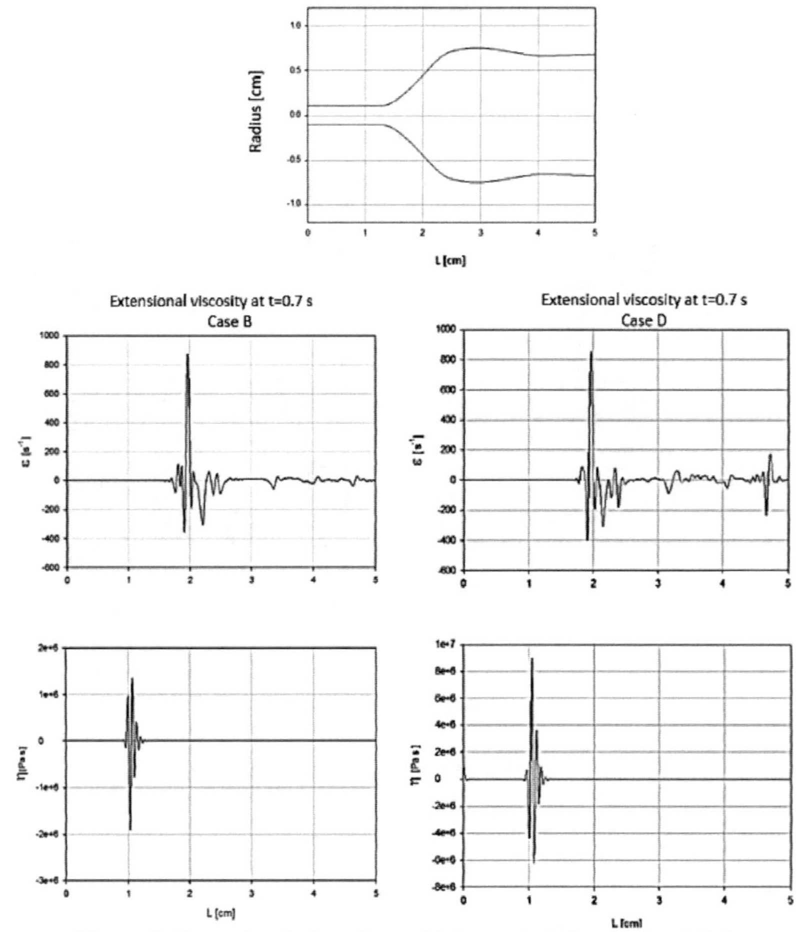

Figure 7. Extensional viscosity and t the occluded zone (t = 0.7 s)

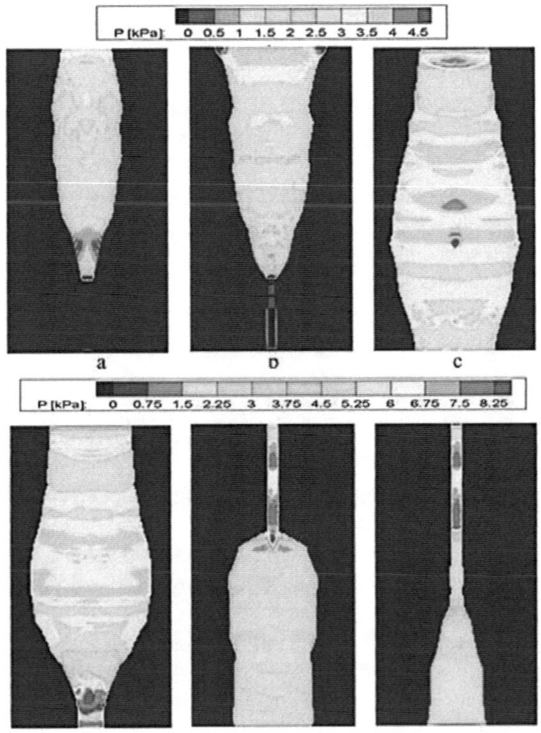

Figure 8. Pressure profiles for Case B at six different times (0.06, 0.2, 0.46, 0.54, 0.7 and 0.9 s)

Conclusion

A three-dimensional model of the transport of food bolus along the pharynx has been proposed using the immersed boundaries method, which allows representing the shape of the pharynx using Cartesian grids. The pharyngeal wall has been considered to be an elastic membrane, and its temporal shape has been taken from previous experimental works using three-dimensional interpolations. The effect of the tongue force acting on the bolus was incorporated as a pressure exerted by the pharyngeal wall close to the GPJ. The effect of the elastic coefficient on the flow behavior has been analyzed.

References

[1] P. J. Carreau , G. Ascanio, E. Brito-De La Fuente and P. A. Tanguy , "Orifice Flowmeter for Measuring Extensional Rheological Properties," *The Canadian Journal of Chemical Engineering,* vol. **80**, pp. 1189-1196, 2002.

[2] S. J. Haward, J. A. Odel, M. Berry and T. Hall, "Extensional rheology of human saliva," *Rheologica Acta,* vol. **50**, pp. 869-879, 2011.

[3] J. F. Steffe, It is important to point out that the flow is not necessarily, East Lansing: Freeman Press, 1996.

[4] P. K. Bhattacharjee, A. G. McDonnell, R. Prabhakar, L. Y. Yeo and J. Friend, "Extensional flow of low-viscosity fluids in capillary bridges formed by pulsed surface acoustic," *New Journal of Physics,* vol. **13**, 2011.

[5] A. Okada, M. Honoma, S. Nomura and Y. Yamada, "Oral behavior from food intake until termila swallow.," *Physiology and Behavior.,* vol. **90**, pp. 172-179, 2007

[6] Y. Meng, M. A. Rao and K. Datta, "Computer simulation of the pharyngeal bolus transport of non-newtonian fluids.," *Institution of Chemical Engineers Journals. Food and Bioproducts Processing.,* vol. **83**(C4), pp. 297-305, 2005

[7] M. Salinas-Vázquez, M. de la Lama, W. Vicente and E. Martínez, "Large Eddy Simulation of a flow through circular tube bundle," *Applied Mathematical Modelling,* vol. **35** (9), pp. 4393-4406, 2011.

[8] J. B. Palmer, "Bolus aggregation in the oropharynx does not depend on gravity.," *Achives of Physical Medicine and Rehabilitation.,* vol. **79**, pp. 691-696, 1998.
[9] I. Cook, W. J. Dods and R. O. Dantas et al, "Opening mechanisms of the human upper esophageal sphincter," *American Journal oh Physiology,* vol. **257**, pp. 748-759, 1989.
[10] D. Gotlieb and E. Turkel, "Dissipative two-four methods for time dependent problems," *Mathematics of computation,* vol. **30**, pp. 703-723, 1976.
[11] T. Poinsot and S. Lele, "Boundary conditions for Direct Simulations of Compressible Viscous Flows," *Journal of Computational Physics.,* vol. **101**, pp. 104-129, 1992.
[12] A. Hasegawa, A. Otoguro, H. Kumagai and F. Nakasawa, "Velocity of swallowed gel food in the pharynx by ultrasonic method," *Journal of the Japanese Society for Food Science and Technology.,* vol. **52**, pp. 441-447, 2005.
[13] H. Nguyen, J. Silny, D. Albers, Roeb E., Gartung C., G. Rau and S. Matern, "Dynamics of esophageal bolus transport in healthy subjects studied using multiple intraluminal inmedancometry," *American Journal of Physiology,* vol. **273**, pp. G958-64, 1997.
[14] D. Ferguson, A. Shuttleworth and D. Whittaker, Oral Bioscience, Edingburgh: Churchil Livingstone., 1999.
[15] A. Chorin, "A Numerical Method for Solving Incompressible Viscous Flow Problems," *Journal of Computational Physics,* vol. **135**, pp. 118-125, 1967.
[16] M. W. Chang, B. Rosendall and B. A. Finlayson, "Mathematical modeling of normal pharyngeal bolus transport: A preliminar study," *Journal of Rehabilitation Research and Development.,* vol. **35**, pp. 327-334, 1998.
[17] G. Pedrizzetti, "Fluid flow in a tube with an elastic membrane insertion," *Journal of Fluid Mechanics,* vol. **375**, pp. 39-64, 1998.
[18] W. Yang, T. Fung, K. Chian and C. Chong, "Finite element simulation of food transport through the esophageal body," *World Journal of Gastroenterology,* vol. **13**, no. 9, pp. 1352-1359, 2007.

IOP Conf. Series: Journal of Physics: Conf. Series **790** (2017) 012026 doi:10.1088/1742-6596/790/1/012026

Evaluation of thermal degradation of polymer based electronic materials by non-destructive testing

P Rafiee[1], G Khatibi[1], M Lederer[1]and M Zehetbauer[2]

[1]Vienna University of Technology, Institute of Chemical Technologies and Analytics, Vienna, Austria

[2]University of Vienna, Faculty of Physics, Vienna, Austria

E-mail: peyman.rafiee@tuwien.ac.at

Abstract. Thermal degradation of polymeric materials used in microelectronic packages was studied by means of experimental modal analysis in combination with finite element methods. The devices were subjected to vibrational loads subsequent to various stages of high temperature storage and their modal response was recorded. Statistical methods and finite element analysis were applied to quantify and evaluate the alteration of the modal response of the packages due to the degradation / delamination of the silver filled epoxy adhesive and the glass filled epoxy resin molding compound. It was shown that changes in the material properties of the molding compound due to surface oxidation is the dominant cause for alteration of the modal response of encapsulated packages exposed to high temperatures.

1. Introduction

Semiconductor packages are exposed to various environmental and loading conditions which may lead to their degradation and subsequent failure during the operational life [1]. Mechanical and interfacial adhesion properties of the epoxy resin based packaging materials like the molding compound (MC) and chip adhesive (CA) are sensitive to high temperature exposure especially in the presence of oxygen and moisture [1-3].

In this study finite element analysis (FEA) and vibrational based modal analysis were employed for detection of thermally induced damage of the epoxy based packaging material in microelectronic devices. While the experimental modal analysis provided information on the time and temperature dependent total degree of damage, FEA allowed separate calculation of the effect of thermal degradation of MC and CA on the alteration of the modal response of the packages.

2. Experimental procedure

The test devices consisted of commercial microelectronic packagesin molded and unmolded conditions provided before trimming and forming of the lead frame. The MC consisted of an epoxy resin with quartz fillers and the chip adhesive was a silver-filled electrically conductive epoxy material. Aging was performed at 250°C up to 230h at air for molded devices and 250°C up to 100h in argon atmosphere for unmolded ones. The harsh storage temperature was chosen to induce higher degrees of damage to the epoxy material[2,4].The specially designed sample holder allowed precise mounting and removal of the sample for intermittent measurements between the thermal treatments without any changes in the test conditions (Figures 1a-b). The molded chips were excited using a fixed-free set-up and a fixed-fixed set-up was used for testing the unmolded chips (Figures 1c-d). A sweep signal within the frequency range of 0.1-10 kHz was applied for excitation of the samples. The mode shapes and the frequency response were recorded by a SLDV (Polytec PSV400).

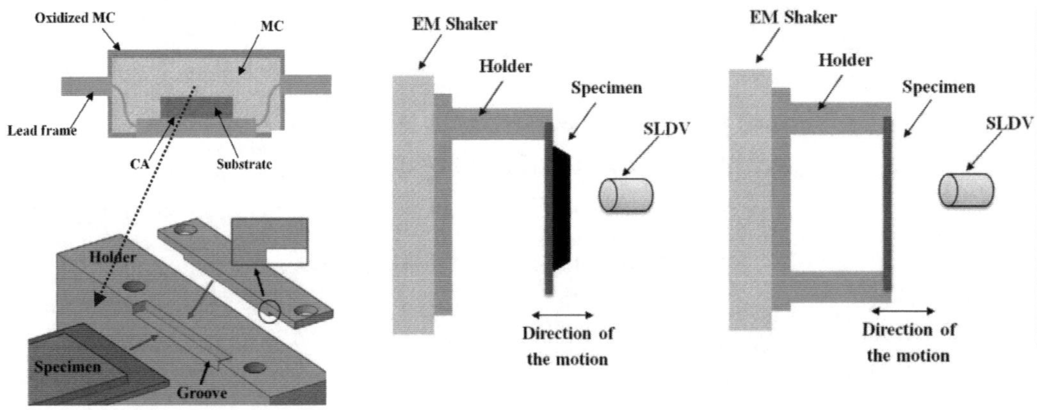

Figure 1. Schematic images of (a) molded package, (b) sampleholder for positioning of the device, (c) fixed-free set-up for molded and (d) fixed-fixed set-up for unmolded devices.

The degree of thermal damage was evaluated by intermittent failure analysis. The oxidized MC layer reached a thickness of 70 µm after 230 hours as determined from metallographic cross sections. Scanning acoustic microscopy analysis showed that minor local delamination of the CA occurred in the unmolded devices and the unmolded ones were not affected [4].Statistical evaluation of the data was conducted by ER [5] and MAC [6] methods by using equations (1) and (2) in which $g(f)$ and $r(f)$ are the resonant frequencies and ψ_X and ψ_A the mode shapes of the sample in the aged and original conditions.

$$ER = \frac{\int [g(f) - r(f)]^2 df}{\int [r(f)]^2 df} \tag{1}$$

$$MAC(A,X) = \frac{\left| \sum_{j=1}^{n} (\psi_X)_j (\psi_A)_j \right|^2}{\left\{ \left(\sum_{j=1}^{n} (\psi_X)_j (\psi_X)_j \right)^* \cdot \left(\sum_{j=1}^{n} (\psi_A)_j (\psi_A)_j \right)^* \right\}} \tag{2}$$

The accuracy of the set-up was verified by repeated mounting/testing/un-mounting of several original sample from which an average ER value of 3.9e-5 with a standard deviation of 5.3e-5 was obtained.

3. Results and discussions

3.1. Experimental modal analysis

The shift of the resonant frequencies of the molded sample as a function of aging time is presented in figure2. The minor changes in 1st and 2ndand the rapid drop of the 3rd and 4th resonance frequencies after 90 hours indicate changes in the material properties or degradation processes in polymeric material.The calculated ER values reached from 3.9e-5 for non-aged (original) samples to 0.038 for those aged up to 230 hours with an increase of about thousand fold (9.5e2).Thedependency of ER and 1-MAC values to increase in aging time is given in figure 3. The ER method appears to be more sensitive to the slight thermally induced changes of the material properties to the thermal exposureup to 100 hours.

The effect of chip adhesive degradation on the modification of the modal response is presented in figures 4a and 4b. Storage at 250°C results in thermal decomposition and mass loss of the epoxy adhesive material and on the other hand occurrence of interfacial delamination between the Si chip and the lead frame. The latter effect results in a reduction of the stiffness of the structure. The observed increase and the subsequent decrease of the ER is a result of opposite effects of the reduction of the mass and stiffness on the resonant frequency shifts (Figure 4b). Numerical analysis of this experiment showed the same trend as explained in the next section.

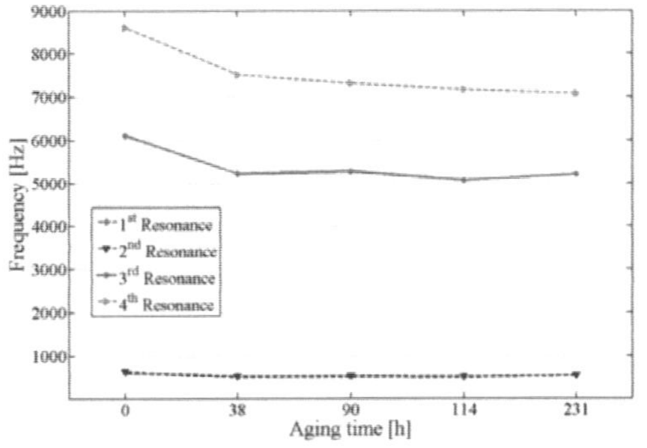

Figure 2. Variation of resonant frequencies as a function of thermal agingof molded packages.

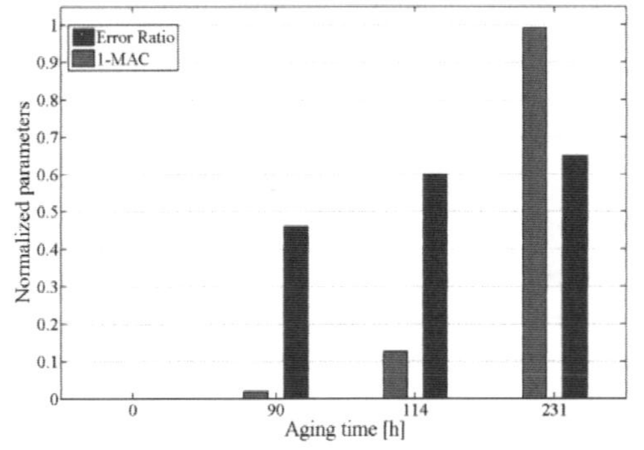

Figure 3. Variation of the modal response as quantified by MAC and ER methods.

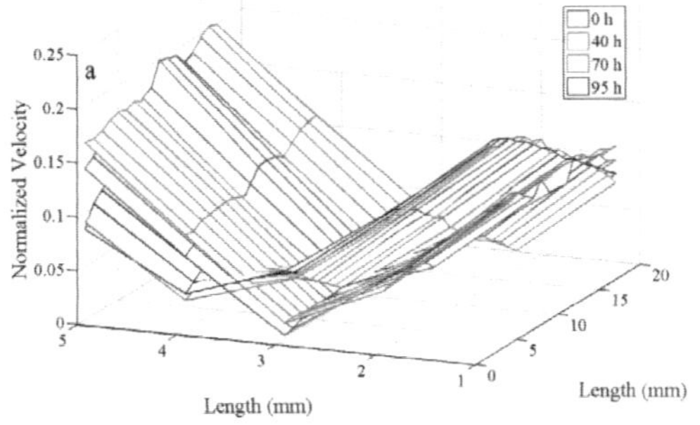

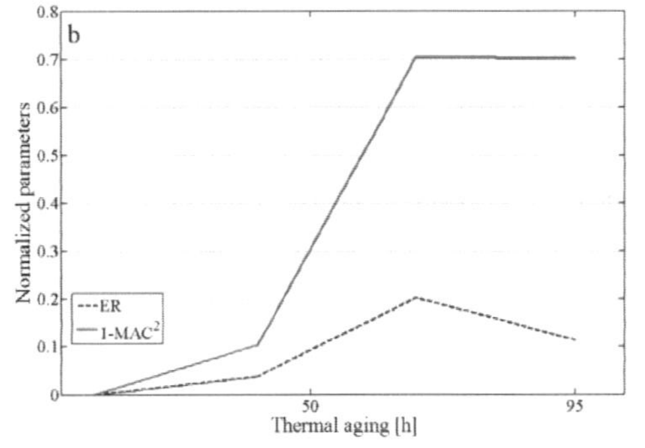

Figure 4. (a) 3D experimental plots of the 1st mode shape and (b) normalized ER and MAC values obtained forunmolded samples as a function of aging time.

3.2. Finite element analysis

The separate effect of MC and CA degradation on the alteration of the modal response was analyzed by using ANSYS software according to the simplified model presented in figure 5. The corresponding material properties are listed in table 1.Young's modulus of the oxidized layer was calculated by equation (3), where E_{OM} and E_{av} are Young's modulus of the oxidized layer and the average modulus of the MC after oxidation, L_1 and L_2 are marked as thickness of the sample and oxidized layer respectively [2].

Values of the * have been determined from [2].

$$E_{OM} = \frac{E_{av}.\frac{L_1}{2} - (\frac{L_1}{2} - L_2).E_1}{L_2} \quad (3)$$

Table 1. Material properties used in the simulations.

Layer	Density (g/cm^3)	E (GPa)	Poisson's Ratio
Lead Frame	8.92	97.6	0.33
Chip Adhesive	5.4	7.713	0.35
Si substrate	2.329	130	0.33
MC	1.74	26	0.33
Oxidized MC	1.5486	*	0.33

Decomposition and delamination of the CA was modeled by removing the corner areas of the adhesive layer.The calculated ER increasesto a value of 2.3e-03 with reduction of area up to15% and then drops to 1.8e-03 as the degree of delamination increases to about for 30%. Though the obtained ER values are rather small, but the trend of increase and subsequent decrease of the ER with reducing the chip adhesive area complies well with the experimental results shown in figure 4b.

Figure 6 shows the calculated effect of MC oxidation for a non-delaminated chip and one with 15% delamination in comparison with the experimental plot. The ER values for both models are almost overlapping showing the dominant effect of the MC oxidation on the frequency response of the devices.

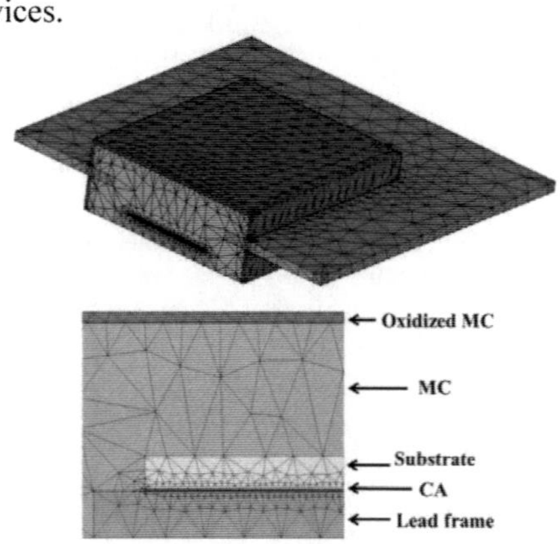

Figure 5. Cross section of the model with an enlarged section demonstrating different materials.

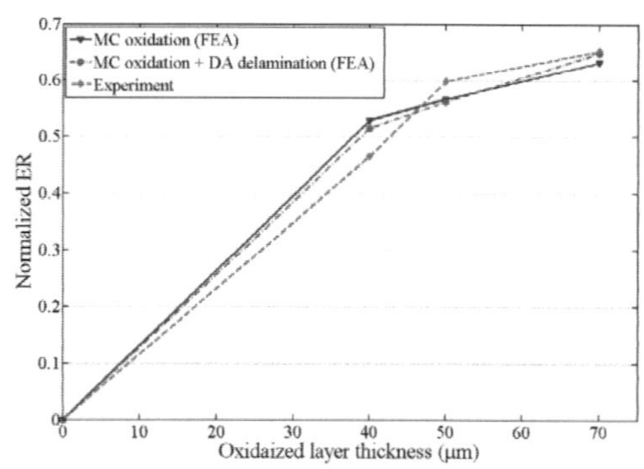

Figure 6. Comparison of experimental and calculated ER as a function of MC oxide thickness layer.

4. Conclusion

Application of experimental and numerical modal analysis for the detection of thermally induced failures and delamination effects in polymeric materials used in microelectronic packages were demonstrated. The main failure mechanisms were oxidation of the MC and delamination of the chip adhesive. Oxidation of the molding compound was determined as the most dominant parameter affecting the alteration of the modal response of the packages. The proposed method provided promising results; however, the resolution of the quantitative modal analysis depends on type, size and combination of the failure modes requiring further studies for an extended application.

Acknowledgments
We acknowledge the financial support of FFG in the frame of Comet program and the Christian Doppler Gesellschaft Austria.

References

[1] Teverovsky A, *56th Elec. Comp. and Technol. Conf.* ETCT 2006 1415

[2] de Vreugd J, Jansen K M B, Ernst L J, Bohm C and Pufal IR, *11th Int. Therm. Mech. Multi-Phys. Sim., Exp. Microelectron. Microsys.* Eurosim E 2010
[3] Wang H T and Poh Y C *33rd IEEE/CPMT Int. Electr. Manufac. Techn. Symp.* IEMT 2008
[4] Rafiee P, Khatibi G, Nelhiebel N, Pelzer R 2013 *Microelectron. Reliab.* **53** 1563
[5] Liu S and UmeI C 2003 *J. Electron. Packag.* **125** 39
[6] Ewins D J 2000 Sadhana **25** 221

Rheological properties of poly(vinyl alcohol) (PVA) derived composite membranes for fuel cells

T Remiš

New Technologies - Research Center, University of West Bohemia, Plzeň, Czech Republic

E-mail: tremis@ntc.zcu.cz

Abstract. Rheological properties of new anhydrous proton conducting membrane based on PVA, tetraethyl orthosilicate (TEOS),sulfosuccinic acid (SSA), titanium dioxide (TiO_2)was examined at various stoichiometric ratios. SSA was used as sulfonating agents to form a crosslinked structure and as proton source, whereas TEO Sand TiO_2were utilized to improve the thermal and mechanical properties of the membrane. In order to verify that all the substances were immobilized into the matrix, the membranes were analysed by means of FT-IR. The rheological, mechanical and thermal properties of the membranes were investigated using rheometer ARES G2 and thermogravimetic analyser (TGA).The analysis of mixed PVA solutions exhibited a unique behaviour of viscosity with increased crosslink density. The dynamic storage modulus G′ of dried composite membranes shows better mechanical resistance and increased tolerance to pressure applied during membrane electrode assembly (MEA).

1. Introduction

Hydrophilic PVA membranes are used in many applications, but lately they have been widely utilized as polymer electrolyte membrane (PEM) for fuel cells. In order to design a good membrane many parameters are involved, such as the properties of casting solution, preparation conditions (temperature, relative humidity and time), reactants etc. The viscosity of the polymer solution plays a key role, which not only helps to fabricate membranes of specific thickness, but also prevents the formation of pinholes and membrane delamination. Therefore, the understanding of the rheological properties of casting solutions is important for preparing appropriate membranes for fuel cells. In this study we have prepared conducting membranes based on PVA, TEOS,SSA, TiO_2 at various stoichiometric ratios and investigated the rheological properties.

2. Experimental

2.1. Material

PVA (hydrolyzed 99+%, M_w = 86000–98000), SSA (70 % in water), Titanium(IV) butoxide (97%) and were obtained from Sigma-Aldrich (USA). TEOS and hydrochloric acid (HCl, 37 wt %) were purchased from Merck Chemicals (Germany). All the chemicals were of analytical grade and used as received. Deionized water was utilized for all the membrane preparation experiments.

IOP Conf. Series: Journal of Physics: Conf. Series **790** (2017) 012027 doi:10.1088/1742-6596/790/1/012027

2.2. Membrane preparation

Composite PVA membranes were prepared by the solution casting method. At first, a 5 wt% solution of PVA in water was prepared by continuous stirring at 90 °C until the complete dissolution was achieved. The PVA solution was mixed with SSA for 24 h at room temperature. Further, TEOS mixture was prepared by mixing H_2O, HCL and TEOS in the molar ratio of 4:0.1:1, which was stirred at room temperature for 2 h [1]and added to the PVA-SSA solutin. Subsequently, TiO_2 mixture was added to the solution and the stirring was continued at room temperature for 2 h. TiO_2solution mixtures was prepared by mixing H_2O, HCL and Titanium(IV) butoxidein the molar ratio of 10:4:1. The composition of each solution is specified in Table 1. The prepared solution was poured into the automatic film applicator (TQC, Germany) equipped with a plexiglass plate and was fabricated into membranes with uniform thickness and dried at 40 °C.

2.3. Characterization

Dynamic TGAwas used to evaluate the relative thermal stability of modified PVA membranes. The TGA curves were obtained with a TGA Q500 thermogravimetric analyser (TA Instruments, USA). Samples of about 5 mg were heated at a rate of 10 °C/min in N_2 atmosphere. The weight of the samples was measured as a function of temperature.

The rheological studies of the PVA solutions were measured on a rheometer ARES G2 from TA Instruments. The polymer solutions were inserted between two paralel plate geometries (d =25 mm). For each measurement, the shear rate from 0.1 to 100 s^{-1} , for a duration 30 s was carried out at room temperature. Membranes were cut (d =25 mm) and inserted between two paralel plate geometries (d =25 mm). For these measurement angular frequency from 1 to100 $rad \cdot s^{-1}$, strain 5%, was used.

FTIR reflectance spectroscopy studies were performed on the samples using NICOLET 380-FTIR spectrometer, in a spectral range of 400 – 4,000 cm^{-1}.

Table 1. - Compositionofpolymersolution

Sample	Solutionmixture
M1	5%PVA
M2	5%PVA_0.5%SSA
M3	5%PVA_1%SSA
M4	5%PVA_2%SSA
M5	5%PVA_3%SSA
M6	5%PVA_0.5%SSA_2,5%TEOS
M7	5%PVA_0.5%SSA_2,5%TEOS_0.1%TiO_2

3. Results and discussion

3.1. FT-IR

Figure 1 shows the FT-IR spectra of the M1 and M7 membranes. It was reported that the absorption band of ester (–COO–) appeared at 1730–1735 cm^{-1}. These results indicate the spectral changes are an evidence of a crosslinking by the estherification between –OH in PVA and –COOH in SSA. The absorption band at 2369 cm^{-1} indicate the presence of sulfonic acid group, which is formed by the introduction of SSA[2].The band at approximately 1040 cm^{-1} is assigned to stretching vibrations of Si–O–Si. The absorption band, at 900 cm^{-1} can be attributed to the stretching vibration of Si–OH or Si–O– groups. A broad absorption band, situated between 3,000 and 3,600 cm^{-1}, are assigned to O–H stretching and O–H bending vibrations, respectively [3].For the spectrum of TiO_2, the peaks at 1471 and 1645 cm^{-1} show stretching vibration of Ti–O and Ti–O–Ti, respectively [5].

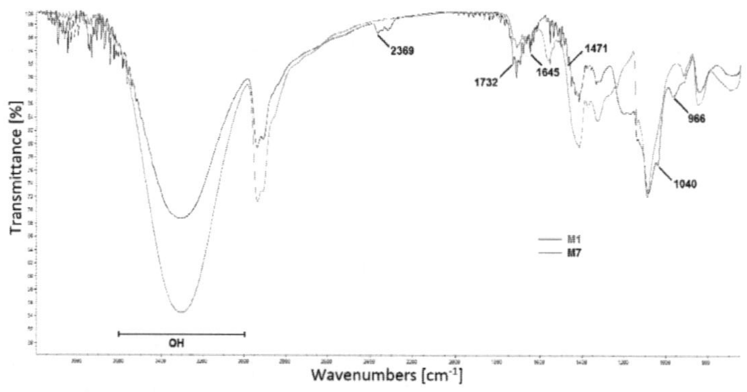

Figure 1. FT-IR spectra of composite PVA membranes

3.2. Rheology

The viscosity of the pure PVA solution changed after the incorporation of SSA (Figure 2). At low SSA content, the viscosity was initially increased due to the interaction between the acid groups of SSA and PVA. However, the viscosity was reduced slightly when the SSA loading percentage exceeded 0.5 wt%. The slight reduction in our observed viscosity can be explained by the limited dispersion behaviour of SSA in the PVA matrix. Agglomeration of particles resulted in the reduced surface area between the SSA and PVA, and caused the decrease in viscosity. The rheological properties of dried membranes were determined by means of shear experiments. Figure 3 shows the storage (G´) modulus versus frequency curves of composite PVA membranes. The storage modules are in order as expected since each additional added component means higher crosslinking extents and better mechanical properties.

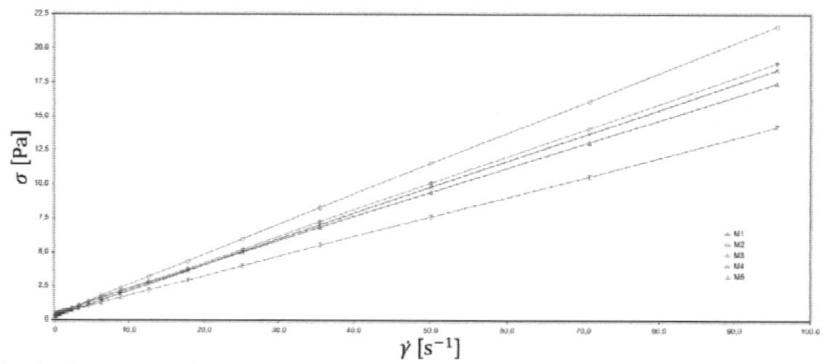

Figure 2. Rheological spectra of pure PVA solution and different SSA loadings of PVA/SSA solutions

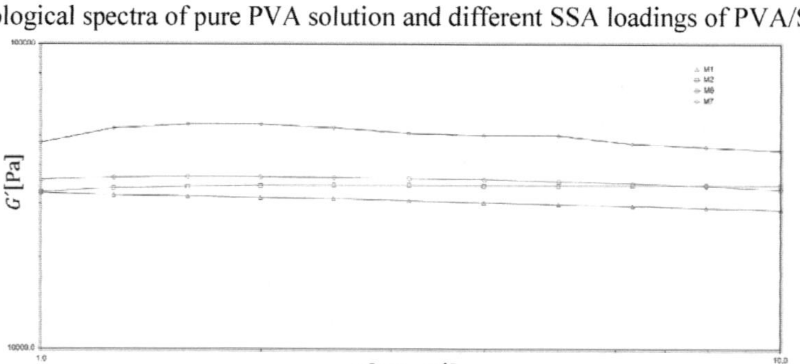

Figure 3. Frequency dependencies of the storage (G´) modulus for composite PVA membranes

3.3. TGA

Figure 4 shows thermogarmof composite PVA membranes. Thermal stability increase with each additional added component,which means higher crosslinking extentsfor the composite membrane M2 and for M6 and M7, respectively. This results correspondwith rheological properties of modified membranes and show better thermo-mechanical properties of membranes compared to pristine PVA membrane.

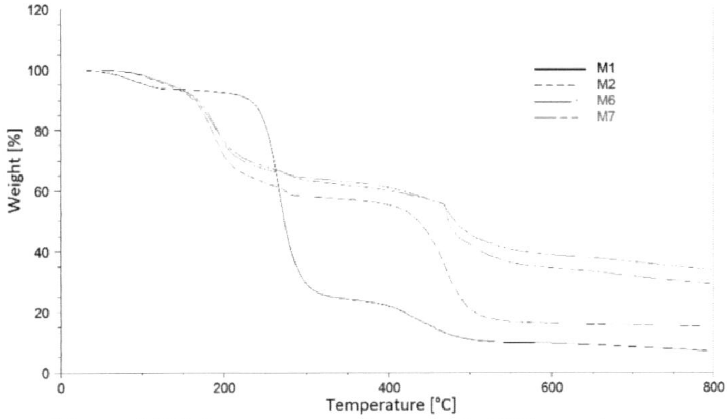

Figure 4.TG thermograms of composite PVA membranes

4. Conclusion

The results of this study can beuseful for furtherresearch of PVA based composite membranes. Rheological studies show that ideal motion of used SSA is 0.5 wt%. The incorporation of SiO_2 and TiO_2 in the PVA matrix has enhanced the mechanical and thermal properties of membranes.

Acknowledgements

The result was developed within the CENTEM project, reg. no. CZ.1.05/2.1.00/03.0088, cofunded by the ERDF as part of the Ministry of Education, Youth and Sports OP RDI programme and, in the follow-up sustainability stage, supported through CENTEM PLUS (LO1402) by financial means from the Ministry of Education, Youth and Sports under the "National Sustainability Programme I.

References

[1] Kim, Dae, et al 2007*Macrom. Res.***15**
[2] Rhim, Ji-Won, et al 2004 *J. Memb. Sci.***238** 143
[3] Katumba G, Mwakikunga B W, and Mothibinyane T R 2008*Nanosc. Res. Lett.***3** 421
[4] Tang, Shangwen, et al 2008*Carbohyd.pol.***72** 521
[5] Lou, Yuanhua, et al 2010*Pol. Comp.***31** 1184
[6] Venckatesh, Rajendran, Balachandaran K and Sivaraj R 2012*Intern. Nano Lett.***2** 1

Rheological characterization of modified foodstuffs with food grade thickening agents

I Reyes-Ocampo[1], JP Aguayo-Vallejo[1], G Ascanio[1] and MS Córdova-Aguilar[1]

Centro de Ciencias Aplicadas y Desarrollo Tecnológico, Universidad Nacional Autónoma de México, Circuito Exterior, Ciudad Universitaria, D.F., México

E-mail: marisol.cordova@ccadet.unam.mx

Abstract. This work describes a rheological characterization in terms of shear and extensional properties of whole milk, modified with food grade thickening agents (xanthan and carboxymethyl cellulose) with the purpose of being utilized in dysphagia treatment. Shear viscosity of the thickened fluids (2% wt. of xanthan and CMC) were measured in a stress-controlled rheometer and for extensional viscosity, a custom-built orifice flowmeter was used, with elongation rates from 20 to 3000 s^{-1}. Such elongation-rate values represent the entire swallowing process, including the pharyngeal and esophageal phases. The steady-state shear and extensional flow curves were compared with the flow curve of a pudding consistency $BaSO_4$ suspension (α=05), typically used as a reference fluid for the specialized commercial dysphagia products. The modified fluids presented non-Newtonian behavior in both, shear and extensional flows, and the comparison with the reference fluid show that the thickened milk prepared here, can be safely used for consumption by patients with severe dysphagia.

1. Introduction

Dysphagia is the difficulty for swallowing solids and liquids[1], which may result in malnutrition, dehydration and pneumonia[2].This alteration can appear at any age and may be related to some diseases as Parkinson and a few others[2].In many cases, modifications of rheological properties of foodstuffs and beverages are prescribed[2],taking into consideration both, the barium sulfate suspension fluids (BSS)as reference to find out the bolus viscosity that would be safe for swallowing[3-4]andalso,the National Dysphagia Diet (NDD) guidelines for thickened dietary supplements. Unfortunately, it is common that the preparation and assessment of the different thickened drinks and foods is subjective and qualitative in the majority of cases[2].

In order to produce adequate foodstuff for dysphagia patients, it is necessary to study flow in the deglutition process, for example, it has been determined that the bolus during deglutition is subjected to a significantly higher shear deformation in the pharyngeal step (shear-rates of 260-930s^{-1}) than in the esophageal phase[1,5] (~ 5s^{-1}). Furthermore, shear rheological properties of dietary food has been subjected of thorough research[4,5], while extensional response has barely received any attention, although there is definitive evidence that the bolus is also extensively stretched when passing from the mouth to the esophagus[5]. Therefore, the aim of this study is to measure and quantify not only the steady state shear flow response, but also the extensional rheological properties of two beverages thickened with xanthan gum and carboxymethyl-cellulose (CMC), all with the objective to as certain suitable concentrations for safe swallowing by severe dysphagic patients, by matching their rheological properties with those exhibited by the pudding BSS reference fluid. Once the characterization procedure is finished, parameters such as apparent viscosity, consistency coefficient (*k*-value) and flow behavior index (*n*-value) are obtained for each test fluid.

2. Materials and methods

2.1 Materials
Varibar® Pudding from E-Z-EM, Canada Inc., (Lake Success, NY) was used as reference fluid (BSS) while whole milk was used as food matrix. Food grade xanthan and CMC were used as thickening agents.Different concentrations of thickening agent (0.5 to 6 wt% for xanthan and 0.5 to 3 wt% for CMC) were used to determine the most suitable concentration required to achieve a pudding consistency.

2.2 Rheological measurements and statistical analysis
Shear viscosity tests were carried out from 0.1 to 1000 s^{-1} in a stress-controlled rheometer (MCR101, Anton Paar Physica, Austria) equipped with a plate and plate geometry (50 mm diameter, 1 mm gap). All measurements were made in triplicate at 25°C (as recommended by NDD). Flow curves of the thickened milk were matched with the pudding BSS one. Data were analyzed by a paring one way analysis of variance. Dunnett's test was used to determine whether a significant difference existed ($p < 0.05$). For the selected concentration, frequency ranges between 0.03 and 100 rad/swere assessed within the linear viscoelasticity region for each sample. Extensional viscosity was measured in a custom-built orifice flowmeter. The test was again performed at room temperature (25°C) with 10 measuring points for two intervals of strain rates 30 – 300 and 300 – 3000 s^{-1}. The parameters such as consistency coefficient (*k*-value) and flow behavior index (*n*-value) were obtained for each thickened fluid as well as the Trouton ratio as function of the effective strain rate.

3. Results and discussion

Figure 1shows the viscosity curves of the whole milk thickened with xanthan and CMC (2% wt.).For both fluids, this concentration provides the closest match with the BSS pudding although the CMC-thickened milk follows closely the BSS product trend. All samples exhibited shear-thinning behavior. Note also the significant differences exhibited by the CMC thickened milk after 100 s^{-1} and larger; the importance of such differences must be assessed for safe consumption; since these shear rates are inside the typical values of the pharyngeal steps of the swallowing process.

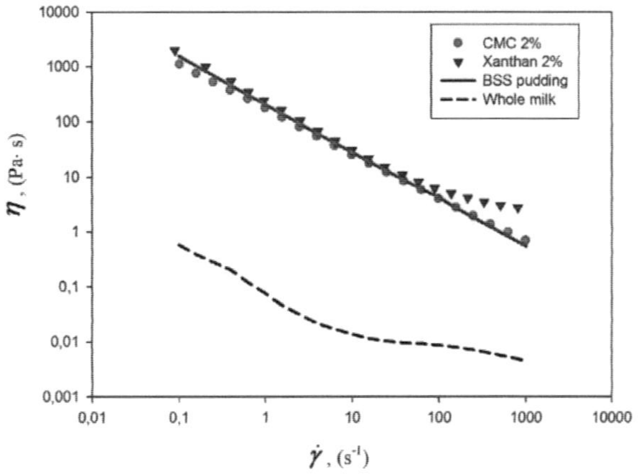

Figure 1. Steady-state viscous flow curves for whole milk, whole milk thickened with xanthan and with CMC 2 wt% and BSS pudding

In addition, the frequency test reported that both thickened milk solutions behave as a true polymer solution since storage moduli were larger than the loss ones. Extensional viscosity measurements were also performed. The results are shown inFigure2, where a decrease in viscosity with increasing elongation rate is observed for all fluids; however, whole milk thickened with xanthan shows slightly lower extensional viscosities at low elongation rates, while the CMC milk product exhibits much larger degree of softening in the higher extensional-rate zone. With such observations, it can be said that for lower extensional-rate zone, CMC thickened milk follows closer the BSS pudding trend and for higher rates, it is the xanthan product that reproduce better the reference fluid behavior. Note that even with the differences in softening trend shown by the CMC fluid, the extensional viscosity of the three fluids are of the same order of magnitude. To our knowledge these are the first reported extensional viscosities of this kind of thickened beverages.

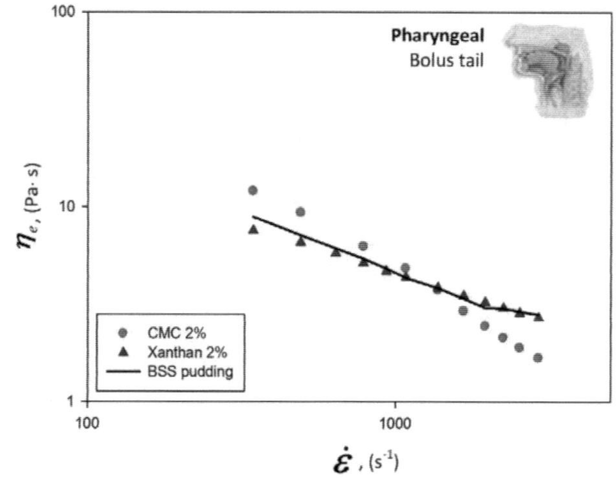

(a)

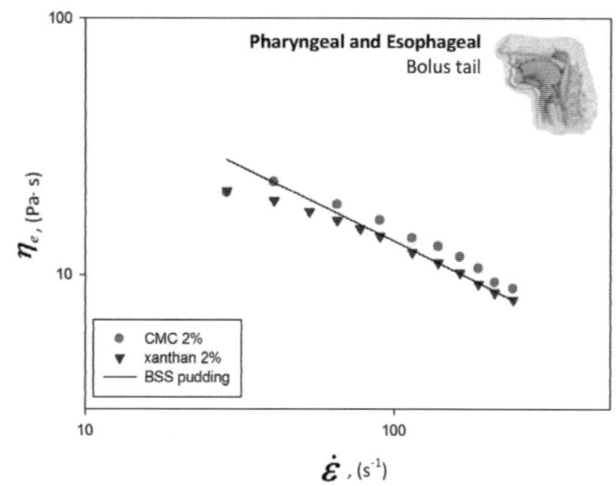

(b)

Figure 2 Extensional viscosity curves for whole milk modified with xanthan and CMC, 2 wt%: (a) at higher elongation strain rates; (b) at lower elongations strain rates

Shear and extensional data for the thickened milk solutions were fitted with the Ostwald de Waele model, giving anR^2 value greater than 0.95 for all cases (Table 1).

Figure 3 shows the Trouton ratio of the thickened milks as a function of the effective strain rate.From this figure, it can be observed that the presented viscosities are above the theoretical Newtonian limit (Tr = 3); such an observation is expected for valid measurements. For the all the tested fluids, the Trouton ration exhibits an increasing trend with increasing effective strain and again, the CMC thickened milk seems to follow closer the BSS curve. In addition, it is worth to point out that the resultspresented infigure 3show the elastic nature of the milk thickened products, which for shear-thinning fluids, an increasing trend in Trouton ratios is a definitive proof of elasticity.

Table 1. Ostwald de Waele model parameters for thickened milk solutions at pudding consistency

	Shear viscosity		Extensional viscosity			
			lower elongation rates		higher elongation rates	
Thickener	*k*	*n*	*k*	*n*	*k*	*n*
BBS	210.71	0.133	209.59	0.405	230.32	0.437
CMC 2 wt%	170.21	0.178	114.82	0.547	3499.1	0.045
Xanthan 2 wt%	238.04	0.162	107.0	0.541	145.07	0.499

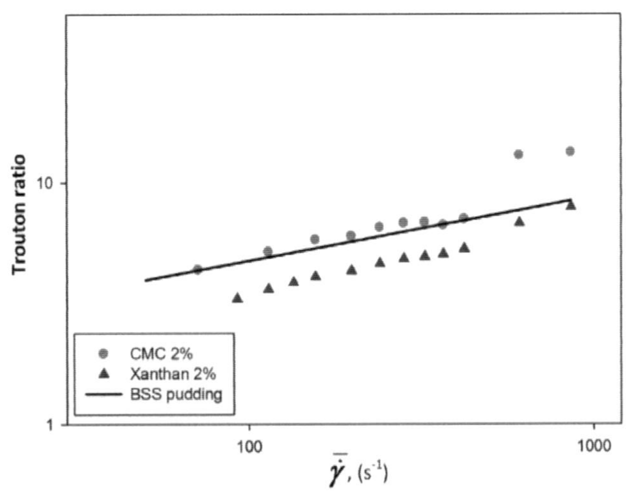

Figure 3. Trouton ratio as function of the effective strain for the BSS and pudding thickened milk solutions

4. Conclusions

A systematic technique for thickening beverages was developed to be recommended for local hospitals and nursing homes. 2%wt. of xanthan and CMC in whole milk matchedwell the steady-state flow and extensional pudding BSS curves so these concentrations may be safe for consumption by patients with a severe difficult for swallowing. CMC is the thickener agent that best fit the flow curves of BSS pudding.

References

[1] Gallegos C, Quinchia L, Ascanio G, Salinas-Vázquez M and Brito-de la Fuente E2012 Rheology and dysphagia: An overview, *Ann. T. Nord. Rheol. Soc.***20** 3-10.

[2] Brito-de la Fuente E, Ekberg O and Gallegos C 2012 Rheological aspects od swallowing and dysphagia. *Dysphagia and Treatment*ed. O Ekberg (London: Springer) pp 493-506.

[3] National Dysphagia Diet Task Force 2002. National dysphagia diet: Standardization for optimal care. American Dietetic Association. Chicago (USA).

[4] Battagel J, Johal A, Smith A andKotecha B 2002 Postural variations in oropharyngeal dimensions in subjects with sleep disordered breathing- a cephalometric study. *Eur. J. Orthodont.***24** 263-276.

[5] Nguyen H, Silny J and Albers D 1997 Dynamics of esophageal bolus transport in healthy subjects studied using multiple intraluminal impedancometry. *Am. J. Physiol. Gastrointest. Liver. Physiol.***273**,G958-964.

Hardening kinetics investigation of alkali-activated binder by small amplitude oscillatory rheometry

D Rieger*, L Kullová, M Čekalová and T Kovářík

New Technologies - Research Centre, University of West Bohemia, Pilsen, Czech Republic

E-mail: DavidPJ@ntc.zcu.cz

Abstract. In this study, the rheological behavior of geopolymeric inorganic binder was determined. This binder was synthesized by alkaline activation of mixture, comprising calcined claystone powder and milled blast furnace slag. As an alkaline activator of hardening process, the potassium silicate solution was used. For the investigation of hardening kinetics, the strain controlled small amplitude oscillatory rheometry was used with strain of 0.01%. The reproducibility and versatility of this method is demonstrated for determination of hardening process evolution. The changes of loss tangent shape were studied in this experiment and applied for determination of gelation time. All experiments were conducted at isothermal conditions in temperature range 27-70°C and parallel plate geometry. The results indicate that reaction kinetics is directly depending on temperature. The hardening kinetics was mathematically described and these calculations were compared with self-contained experiment conducted at 2°C. This experiment is described in details and the results of gelation time measurements confirmed calculated data.

1. Introduction

Inorganic polymer cements can be synthesized by alkali-activation of a variety of materials including thermally activated clays, coal fly ash and blast furnace slag to produce a hardened material with mechanical and thermal properties potentially suitable for wide range of industrial applications [1], especially for building applications, as a binder [2]. Alkali-activated pozzolan material was discovered by Purdon and Glukhovsky in the 1940s and systematically studied by them up to the 1960s [3]. Davidovits entitled this type of materials as "geopolymers" and introduced pioneering alkali-activated calcium-free systems based on calcined clays [4]. From a general point of view, geopolymers are classified as aluminosilicate binders activated by solutions of alkali metal hydroxides, carbonates and silicates. Sodium silicate solutions are widely used as geopolymer constituent controlling the dissolution processes of the solid phase via cleavage of the Si-O and Al-O bridging oxygen bonds. The dissolution rate depends on the pH, chemical composition, concentration, structural state, and distribution of silicate species in the system [5]. The following gel precipitation and solid phase formation is driven by polycondensation mechanism and this process is leading to the formation of amorphous aluminosilicate matrix [6]. The most used is kaolinite calcined to metakaoline which leads to increase of reactivity [7]. Not only calcined kaolinite, but many aluminosilicates, also containing alkaline earth oxides as CaO, can be hardened with this mechanism. Examples can be shales and other clay minerals, fly ashes, slags and theirs blends [8].

The effect of temperature in the phase of hardening was also studied by conventional furnace heating, autoclaving and by some nontraditional heating methods where the results indicate, that temperature have effect on both, the reaction rate and mechanical properties [9, 10].

The reaction kinetics was in recent years studied by conventional techniques, mainly by X-Ray diffraction, FTIR and NMR and by some non-traditional methods like micro-mechanical analysis including nanoindentation and non-direct measurements by an open porosity investigation, was used. Also rheometrical measurements are for many years used for investigation and monitoring of hardening reactions in an industry of concretes and polymers. In the concrete science, thixotropic and flow properties measurements by rotary rheometry are widely used as a parameter of workability. Rotary rheometry is sometimes used also in geopolymers for setting time estimations. But often is used high share rate and material can be disturbed. But in polymer research, oscillatory rheometry with small amplitude is helpful approach to degree of reaction estimations. Oscillatory rheometry of crosslinking polymers for tracking of sol-gel transition is widely used form many years and in recent years are pioneering researches for tracking similar transition of alkali activated geopolymer suspension solution to amorphous gel state [11, 12].

In this study, the effect of temperature on setting time of the preliminary calcined clay mineral shale, mixed together with calcium-rich slag has been investigated with using of precision rheometrical measurement in strain controlled small amplitude oscillation mode. Changes in setting times on different temperatures and estimation of reactivity on important technological temperatures are presented.

2. Materials and instrumentation

2.1. Instrumentation

The elemental analysis of the initial material was carried out by XRF using a Bruker AXS S4 Explorer spectrometer calibrated for aluminosilicate materials. The granularity of powder was measured by static light scattering technique on instrument Horiba partica LA-960 and scanning electron microscopy was conducted by microscope JEOL JCM-5000. The loss on ignition was determined by thermogravimetric measurement on TA Instruments Q500. The setting time and reaction kinetics were determined by time-resolved dynamic rheometric measurements with small amplitude on an Ares G2 rheometer from TA Instruments in plane-plate geometry.

2.2. Materials

In this study, a mixture of calcined shale and slag (SSM), supplied by Ceske lupkove zavody a.s, Nove Straseci, Czech Republic, was examined. The shale is a material rich in kaolinite mineral and pre-calcined to metakaolin state and is then mixed with fine amorphous slag for modification of setting time. The chemical composition was determined by the XRF technique and is listed in Table 1.

Table 1. Chemical composition of raw materials [mass %]

Composition	**SiO_2**	**Al_2O_3**	**CaO**	**MgO**	**TiO_2**	**Fe_2O_3**	**K_2O**	**LoI**	**Others**
SSM	46.37	28.72	15.46	3.40	2.01	0.92	0.56	1.93	0.63
Activator	17.60	-	-	-	-	-	17.13	65.27	-

Granularity of the mixture was determined by static light scattering technique with D50v of 7.046 μm and results are graphically presented in Figure 1 a). In Figure 1 b) can be seen micrograph of the mixture taken by SEM with small particles of shale and coarse particles of slag.

For alkaline activation an aqueous solution of potassium silicate with silicate module 1.61 and content of water soluble solids 34.73% was used.

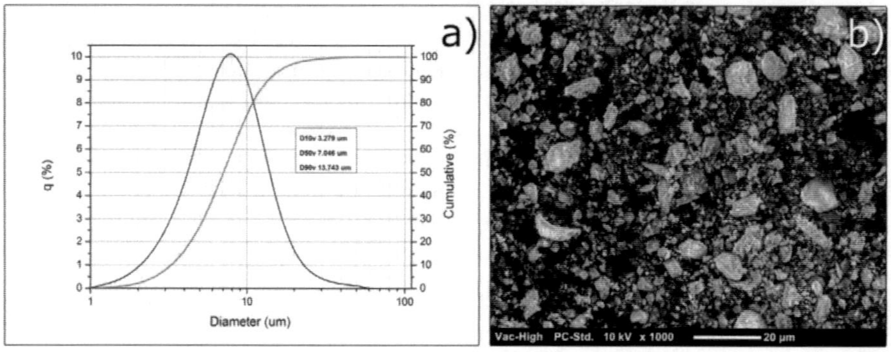

Figure 1. Particle size distribution of binder (a) and SEM image of individual particles

3. Measurements and results

3.1. Measurements of oscillatory rheometry

For determination of setting time by small amplitude oscillatory rheometry we used our procedure described earlier [12].

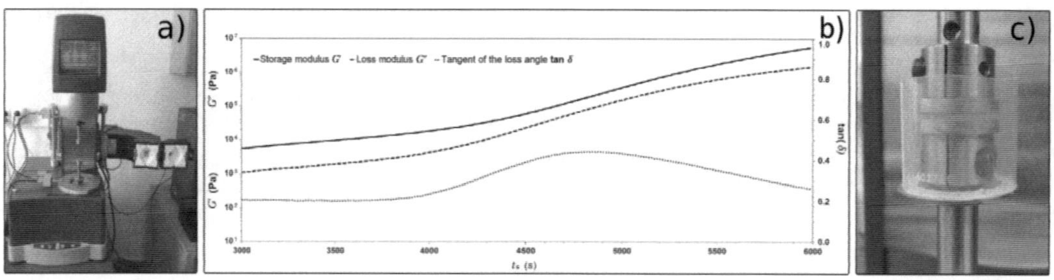

Figure 2. (a) Setup of rheometrical measurement, (b) results of paste measurement at isotherm 30°C (G', G'' – storage and loss moduli in torsion, tan (δ) – loss tangent), (c) sample with protective layer and air deflector before measurement

For all experiments standard mixing ratio of 10:8 by weight was used. This corresponds to ratio of charge balancing atoms $(K_2O+MeO)/Al_2O_3$ of 2.34 and Si/Al ratio of 1.77 in final hardened geopolymer. The mixtures were blended in laboratory vacuum mixer for 2 minutes and immediately transferred into the rheometer. The measurements were performed in plain-plate geometry so that the mixture in an amount of approximately 500 µl was placed into the middle of plain plates and the gap was set to a working distance of 0.8 mm. Then was geometry set to work distance and the edges of mixture were protected by thin layer of silicone oil to avoid evaporation of water and air deflector was used to shielding air vortex in furnace, as can be seen in Figure 2 c). After 6 minutes, isothermal heating at selected temperature in the range 27-70°C was initiated for a period of time necessary for mixture hardening. The measurements were performed at radial velocity 10 rad/sec and strain 0.01% to ensure a minimal influence on the hardening process. Subsequently, the maximum of loss tangent was traced and this point can be considered as transition of inner structure from pseudoliquid solution to inorganic polymer network. For the measurement at 2°C modified method was selected and prepared. Prepared mixtures were shock cooled between 2 metal plates precooled to temperature of 2°C and after selected time were measured. From residual time of transition was deduced the overall of transition at temperature 2°C. All measurements for better statistics were repeated 4 times.

Results of individual measurements are presented in Table 2 a) and dependencies are expressed by logarithmic plot in figure 2 b). The temperature interval of 27-40°C was used to curve fitting, because of side effects at higher temperatures. The results can be expressed by equation:

$$t = 10^{-8.15654+3592,77/T} \tag{1}$$

where t represents time of transition in seconds and T temperature in kelvins. According to this equation, the time of transition at 2 °C was calculated as 79607.85 seconds and was compared with measured value 83119.6 seconds. The measured and calculated value has reasonable difference 4.22% and so we think that this procedure can be useful for hardening predictions.

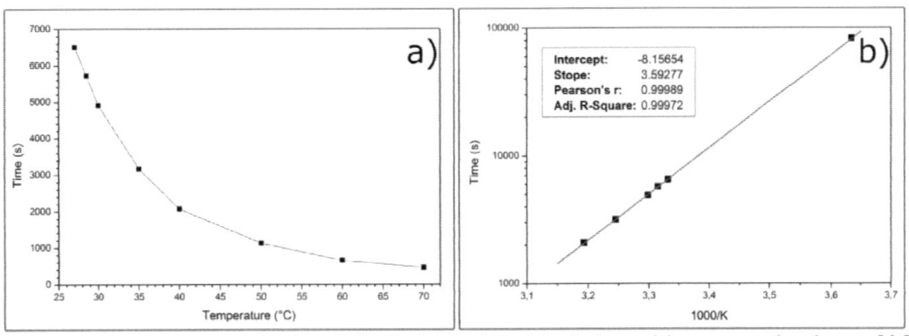

Figure 3. Average measured values of transition time (a), fitting with measured value at 2°C

4. Conclusions

In this paper, we were demonstrating oscillatory rheometry as useful method to determination of geopolymer reactivity by strain controlled measurements with small strain of 0.01%. Data collected by measurements in range 27-40°C were used to mathematical description of reactivity dependency on temperature and for measured and calculated values at temperature 2°C is difference 4.22%. From this point of view the oscillatory rheometry can be considered as reliable method for investigation of geopolymers.

Acknowledgments

The result was developed within the CENTEM project, reg. no. CZ.1.05/2.1.00/03.0088, cofunded by the ERDF as part of the Ministry of Education, Youth and Sports OP RDI programme and, in the follow-up sustainability stage, supported through CENTEM PLUS (LO1402) by financial means from the Ministry of Education, Youth and Sports under the National Sustainability Programme I.

References

[1] Pacheco-Torgal F, Castro-Gomes J and Jalali S 2008 *Const Build Mater* **22** 1305-14
[2] Duxson P, Provis JL, Lukey GC and Van Deventer JSJ 2007 *Cem Concr Res* **37** 1590-97.
[3] Purdon AO The action of alkalis on blast-furnace slag 1940 *J Soc Chem Ind* **59** 136-91.
[4] Davidovits J 1991 *J Therm Anal* **37** 1633–56
[5] Kouassi S S, Tognonvi M T, Soro J and Rossignol S 2011 *J Non-Cryst Solids* **357** 3013-21
[6] Šesták J and Foller B 2012 *J Term Anal Calorim* **109** 1-5
[7] Kovářík T, Bělský P, Novotný P, Říha J, Savková J, Medlín R, Rieger D and Holba P 2015 *Const Build Mater* **80** 98-104
[8] Kim B-J, Yi C and Kang K-I 2014 *Const Build Mater* **58** 206-13
[9] Rovnaník P 2010 *Const Build Mater* **24** 1176-83
[10] Kovářík T, Franče P, Šesták J, Rieger D, Bělský P, Kadlec J and Roubíček P 2015 *J Term Anal Calorim* **121** 517-24
[11] Steins P, Poulesquen A, Diat O and Frizon F 2012 *Langmuir* **28** 8502-10
[12] Rieger D, Kovářík T, Říha J, Medlín R, Novotný P, Bělský P, Kadlec J and Holba P 2015 *Const Build Mater* **83** 26-33

IOP Conf. Series: Journal of Physics: Conf. Series **790** (2017) 012030 doi:10.1088/1742-6596/790/1/012030

Mechanically activated fly ash as a high performance binder for civil engineering

D Rieger*, L Kullová, M Čekalová, P Novotný and M Pola

New Technologies - Research Centre, University of West Bohemia, Pilsen, Czech Republic

E-mail: DavidPJ@ntc.zcu.cz

Abstract. This study is aimed for investigation of fly ash binder with suitable properties for civil engineering needs. The fly ash from Czech brown coal power plant Prunerov II was used and mechanically activated to achieve suitable particle size for alkaline activation of hardening process. This process is driven by dissolution of aluminosilicate content of fly ash and by subsequent development of inorganic polymeric network called geopolymer. Hardening kinetics at 25 and 30 °C were measured by strain controlled small amplitude oscillatory rheometry with strain of 0.01 % and microstructure of hardened binder was evaluated by scanning electron microscopy. Strength development of hardened binder was investigated according to compressional and flexural strength for a period of 180 days. Our investigation finds out, that mechanically activated fly ash can be comparable to metakaolin geopolymers, according to setting time and mechanical parameters even at room temperature curing. Moreover, on the bases of long time strength development, achieved compressional strength of 134.5 after 180 days is comparable to performance of high grade Portland cement concretes.

1. Introduction

The ashes are one of the main by-products of human activities arising from burning of coal for production of electricity and heat in power and heating plants. Today, the worldwide production of coal burning ashes are estimated somewhere around 700 million tons per year [1]. The ash arising manly from mineral content of burned coal and principally are ashes divided to two categories, according to place of origin in combustion process. Bottom ash comprises approximately 20-25% of total coal burning ashes and takes place as a solid product in fire chamber. Remaining part of solid products are fly ashes, risen by stream of combustion products. Because of their origin, fly ash particles have very fine nature with diameters in range of 0.2 – 1000 μm [1]. In past century, the environmental and health risks associated with this fine nature of fly ash were recognized and today is fly ash mainly separated from combustion stream by electrostatic separators as a by-product.

Because of spherical shape, content of silica and aluminum oxide and mainly amorphous composition, fly ash has potential to be used as valuable industrial material. In last decades, fly ash is increasingly used in civil industry as a partial replacement of clinker in ordinary Portland cement (OPC) for their latent hydraulic properties when reacts with excess of calcium hydroxide in clinker-water mixture to precipitate of C-S-A-H gel and so on participates in final structure of hardened concrete [2]. However, despite to long term investigations, it is still only 53% of produced fly ash reused worldwide, mainly for environmental recultivations. It is because of replacement can be only partial and in many cases have blended concrete inferior mechanical performance.

Another possibility for reuse of fly ash in mass scale is process called geopolymerization [3,4]. This process consists of alkaline activation of a variety of materials including thermally activated clays

[5,6], coal fly ash and blast furnace slag to produce a solid material with mechanical and thermal properties potentially suitable for wide range of industrial applications. From a general point of view, the alkaline activation of an aluminosilicate material take place by blending with a highly concentrated alkaline solution of an alkali metal hydroxides or silicates [3,4]. In this case, the alkali solution control the disaggregation processes of the solid phase via dissolution of solid particles and cleavage of the Si-O and Al-O bonds [4]. The subsequent hardening of geopolymer binder takes place, by the polycondensation mechanism and gel precipitation [7]. These processes leads to solid phase formation and hardening of the binder. Unfortunately, in the case of fly ash, it is necessary to apply high temperatures and highly concentrated aggressive solutions for proper hardening [1,4].

In this paper, we are demonstrating possibility to use mechanical activation for improvement of fly ash reactivity in alkaline conditions. This process is composed of intensive milling and lies in high energy grinding which led to defects introduction and increase of surface area of individual grains of binder [6]. This process aid to better accessibility of reactive material and speeds up the hardening reaction of binder. The results indicate that fly ash can be converted to binder with excellent binding properties, comparable with high performance concretes. The time-resolved oscillatory rheometry was used to determination of setting time and reaction kinetics for better description of prepared binder.

2. Materials and binder preparation

In this study, the fly ash from coal power plant Prunerov II of CEZ a. s. was used. Table 1 summarizes chemical composition determined by X-ray fluorescence (XRF) spectrometer Bruker AXS S4 Explorer calibrated for aluminosilicate materials. The loss on ignition was determined by thermogravimetric measurement on TA Instruments Q500. According to result of chemical analysis, this fly ash can be classified as Class F with low amount of sulphates. The process of mechanical activation is composed of intensive milling process. The milling process was performed in laboratory vibrational mill BVM-2 for period of 150 minutes. Material after mechanical activation was investigated by static light scattering (SLS) instrument HORIBA LA-960 and consists of very fine particles, how can be seen in Figure 1 a). This result is consistent with scanning electron microscopy (SEM) observation in Figure 1 b), which was taken by microscope JEOL JCM-5000. SEM analysis reveal that particles are of irregular shape with considerable part of particles in submicrometer region. Tested binder was prepared as a mixture of mechanically activated fly ash and potassium silicate solution, with silicate module 1.61 and content of solids 35.03 %, in weight proportion of 10:8.

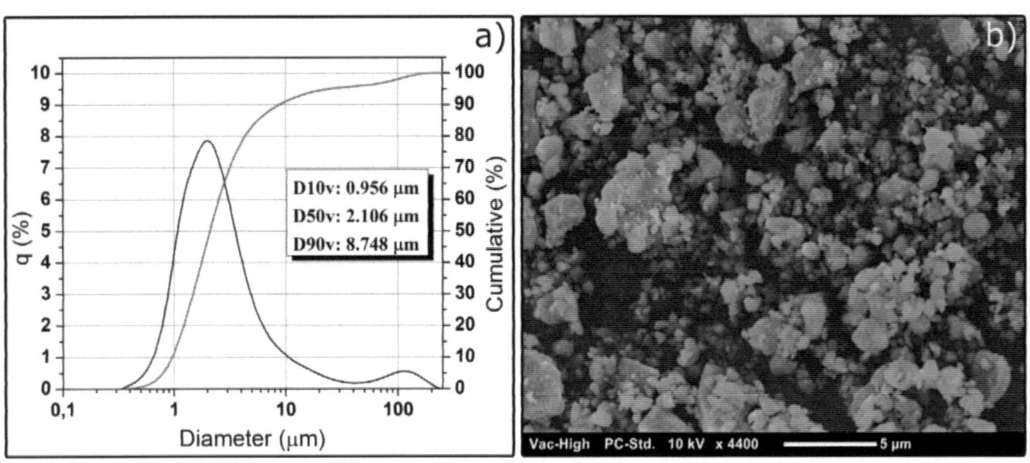

Figure 1. Fly ash after mechanical activation (a) particle size distribution, (b) SEM image of fly ash

Table 1. Chemical composition of fly ash [mass %].

SiO_2	Al_2O_3	Fe_2O_3	CaO	MgO	K_2O	Na_2O	SO_3	Others	LoI	Total
53.00	29.22	8.37	2.43	1.67	1.11	0.63	0.29	1.59	1.63	100.00

3. Measurements and results

3.1. *Determination of setting time*

The setting time and reaction kinetics were determined by time-resolved oscillatory rheometry measurements with small amplitude on a strain controlled rheometer Ares G2 from TA Instruments in plane-plate geometry according to the references [8] and according to our procedure described earlier [9]. The procedure is as follows. Mechanically activated fly ash in accurate amount of 50 g was poured into 40 g of potassium silicate solution and then were mixed in laboratory vacuum mixer for 5 minutes and immediately transferred into the rheometer. The mixture in an amount of approximately 500 μl was placed into the middle of plane plates and the gap was set to a working distance 0.8 mm. The edges of mixture were protected by a thin layer of silicone oil to avoid evaporation of water. Isothermal heating was initiated after 12 minutes for a period of time necessary for mixture hardening. The measurements were performed at radial velocity 10 rad/sec and strain 0.01% to ensure a minimal influence on the hardening process. Subsequently, the maximum of loss tangent was traced and the point of maximum value can be considered as transition of inner structure from pseudoliquid solution to inorganic polymer network. Figure 2 shows time dependences of storage (G') and loss (G") modules and of tangent delta for the alkali activated fly ash binder. Results indicate that reached setting time is comparable with metakaolin and calcined shale binders. Moreover, application of elevated temperatures was not necessary, as it is usually needed by fly ash binders.

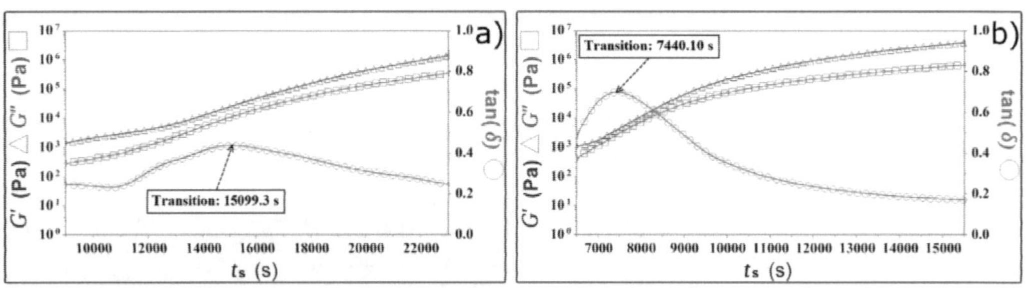

Figure 2. Results of hardening process measurements at isotherms (a) 25°C and (b) 30°C

3.2. *Mechanical properties*

For compression strength investigation, binder was prepared according to procedure described in section 2 in amount of 18 weight parts and mixed with 30 weight parts of calcined chamotte with granularity range of 0-0.5 mm. The mixing was performed in planetary mixer for a period of 10 minutes. Subsequently, the mixtures were cast in moulds with dimensions of 20×20×20 mm for compressional strength determination. Samples were cured at ambient temperature in polypropylene bags to prevent humidity loss. Mechanical properties were determined on LABTEST 6.100 SP1.

Results in figure 3 (a) indicate that mechanical strength suitable for civil engineering was reached after 3 days. After 28 days, compressional strength reached values, comparable to superior grades of OPC and geopolymer binders. But even after this period the hardening reaction is not complete and after 180 days mechanical parameters reached values comparable to so called "Ultra high performance concretes" used for demanding applications in civil engineering [2, 10].

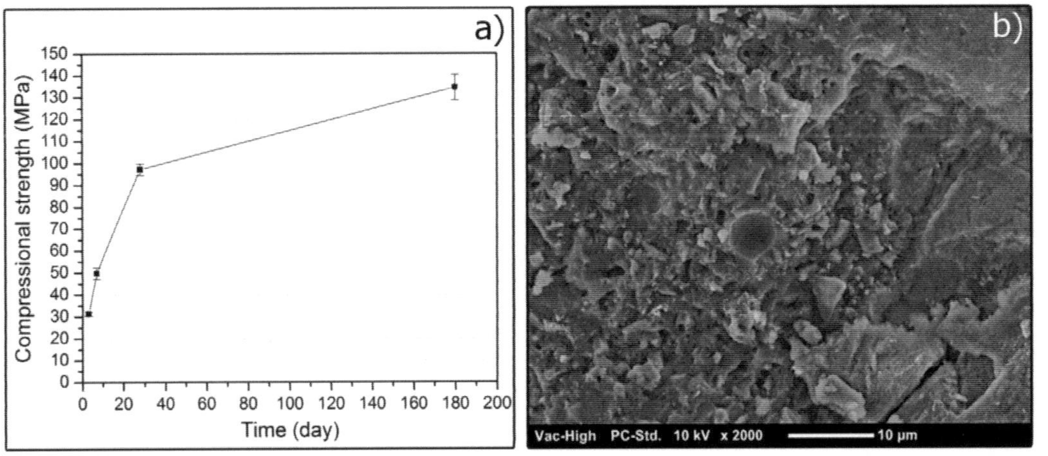

Figure 3. Evolution of compressional strength (a) and microstructure of binder after 180 days (b)

These results correspond to SEM observations. In figure 3 (b) the magnified structure of binder after 180 days and homogeneous structure of dense geopolymer binder is observable. This structure with minimum of cracks is leading to such high mechanical performance.

4. Conclusions

This study demonstrates, that fly ash can be used as suitable binder for demanding civil engineering needs and with proper treatment can outperform even high grades of cement binders. Moreover, setting time of presented binder is shorter even as metakaoline geopolymer binders. This behaviour is caused mainly by fineness of milled fly ash particles and by low porosity of particles as well. This properties leads to low demand of admixture water, rapid dissolution of fly ash particles and to fast construction of geopolymeric binding structure. How it was demonstrated, this binder have long term strength development and with 134.5 MPa in compression after 180 days can be compared with Ultra high performance grades of concrete.

Acknowledgments

The result was developed within the CENTEM project, reg. no. CZ.1.05/2.1.00/03.0088, cofunded by the ERDF as part of the Ministry of Education, Youth and Sports OP RDI programme and, in the follow-up sustainability stage, supported through CENTEM PLUS (LO1402) by financial means from the Ministry of Education, Youth and Sports under the National Sustainability Programme I.

References

[1] Tianyu X and Togay O 2015 *Ceram Int* **41** 5945-58.
[2] Yu R, Spiesz P and Brouwers H J H 2014 *Const Build Mater* **65** 140-50
[3] Pacheco-Torgal F, Castro-Gomes J and Jalali S 2008 *Const Build Mater* **22** 1305-14
[4] Davidovits J 1991 *J Therm Anal* **37** 1633–56
[5] Kovářík T, Bělský P, Novotný P, Říha J, Savková J, Medlín R, Rieger D and Holba P. 2015 *Const Build Mater* **80** 98-104
[6] Hounsi A D, Lecomte-Nana G L, Djétéli G and Blanchart P 2013 *Const Build Mater* **42** 105-13
[7] Kouassi S S, Tognonvi M T, Soro J and Rossignol S 2011 *J Non-Cryst Solids* **357** 3013-21
[8] Steins P, Poulesquen A, Diat O and Frizon F 2012 *Langmuir* **28** 8502-10
[9] Rieger D, Kovářík T, Říha J, Medlín R, Novotný P, Bělský P, Kadlec J and Holba P 2015 *Const Build Mater* **83** 26-33
[10] Dawood E T and Ramli M 2012 *Const Build Mater* **28** 193-200

IOP Conf. Series: Journal of Physics: Conf. Series **790** (2017) 012031 doi:10.1088/1742-6596/790/1/012031

Magnetorheology of dimorphic magnetorheological fluids based on iron nanorods

M Sedlacik[1] and V Pavlinek[1]

[1] Centre of Polymer Systems, University Institute, Tomas Bata University in Zlín, Zlín, Czech Republic

E-mail: msedlacik@ft.utb.cz

Abstract. The aim of this paper is to document suitability of partial substitution of magnetic carbonyl iron (CI) microspheres with iron nanorods to obtain dimorphic magnetorheological (MR) suspensions with comparable MR performance to conventional MR suspensions exclusively based on (CI) microspheres while the sedimentation stability is considerably improved. The morphology of CI and iron nanorods was analyzed via scanning electron microscopy and transmission electron microscopy, respectively, and magnetic properties via vibrating sample magnetometry. The steady shear flow and small-amplitude dynamic oscillatory shear measurements were carried out to confirm effective MR performance. The sedimentation test showed positive role of dimorphic composition of dispersed phase on the sedimentation stability.

1. Introduction

Magnetorheological (MR) suspensions are generally two-phase systems, in which micrometer-sized ferro- or ferrimagnetic multi-domain particles are suspended in a variety of carrier liquids like silicone or mineral oils [1,2]. It is worth to mention here, that MR suspensions differ from so called ferrofluids which contain nanometer-sized magnetizable particles [3]. The most important feature of MR suspensions is their ability to change viscosity in several orders of magnitude (in controllable way) during milliseconds upon the application of an external magnetic field. In the absence of magnetic fields (off-state) they behave as conventional dispersions. The application of large enough field induces a dipole moment in each suspended particle causing the particles to form columnar structures, parallel to the applied field. The formed chain-like or columnar structures restrict the motion of the fluid and, thereby, increase the elastic characteristics of the suspension. The use of MR suspensions in real systems is based on the simultaneous application of magnetic field, and shear, oscillatory or pressure driven force [4-6]. Although the mechanical loading evokes rupturing of created internal structure, the field-induced attractive forces cause "self healing" of particles alignment until the moment, when hydrodynamic forces overcome the field-induced magnetic ones and suspension starts to exhibit yielding behavior. Once the external field is removed, the columnar structures rapidly break under flow and the system immediately returns to its liquid character.

Although a large amount of various studies have been performed by now, the comprehensive qualities are still far from their wider commercialization because of large sedimentation of magnetic particles in the absence of frequent mixing due to predominant gravity forces; *i.e.*, density mismatch between dispersed particles and carrier medium. Moreover, once settled, the residual magnetic attractions and attraction forces between particles make re-dispersion difficult, which results in much lower MR effect of the system in time. Following the purpose of overcoming these limitations, several approaches such as the coating of magnetic particles with polymers [7,8], or the addition of surfactants [9], have been used. However, the addition of some non-magnetic portion can lead to reduction of field-responsive behavior of MR suspension. Recently, experimental investigation with conventional carbonyl iron (CI) spherical microparticles based MR suspension where part of the CI were replaced with magnetic wire-like nanoparticles has been performed [10,11]. Such nanoparticles play, in the so called dimorphic MR suspensions, a role of steric repulsion between the CI microparticles resulting in improved stability against their rapid sedimentation while the yield stress for dynamic applications is still maintained additionally or even higher due to the friction occurring between fibers.

In this paper, dimorphic MR suspension based on partially substituted CI spherical microparticles with iron nanorods was prepared and its MR performance as well as sedimentation stability was compared with MR suspension of the same magnetic portion loading based exclusively on CI microparticles.

2. Experimental

2.1. Materials

Carbonyl iron particles (HS grade, BASF, Germany) were selected as the main magnetic agent. The material characteristics of HS grade of CI are following: spherical shape of particles with the average size of about 2 μm, non–modified surface, and content of α–iron > 98%.

For the synthesis of iron nanorods, iron(III) chloride hexahydrate ($FeCl_3{\cdot}6H_2O$, purity ≥ 98%), sodium hydroxide (NaOH, purity ≥ 98 %), hydrazine monohydrate ($N_2H_4{\cdot}H_2O$, 64–65 %) all produced by Sigma-Aldrich (St Louis, USA), and cetyltrimethylammonium bromide (CTAB, purity % 98%) obtained from Lach–Ner (Neratovice, Czech Republic) were used. All chemical were used without further purification.

2.2. Synthesis of iron nanorods

Iron nanorods were prepared following procedure [12], with a minor modification. Briefly, 1 g of $FeCl_3{\cdot}6H_2O$ was dissolved in 20 mL of distilled water, and 1 g of NaOH, 4 mL of $N_2H_4{\cdot}H_2O$, and 0.73 g of CTAB were added into the solution. Then, the formed mixture was transferred to an autoclave which was properly sealed, heated for 10 h at 120 °C, and after that cooled down to room temperature naturally. The precipitated iron nanorods were collected on the filter, rinsed with distilled water and ethanol several times, and then dried at 80 °C under decreased pressure for 24 h.

2.3. Morphology

The morphology of CI particles was observed with scanning electron microscopy (SEM, VEGA II LMU, Tescan Ltd., Czech Republic) operated at 30 kV. Further, iron nanorods were examined using transmission electron microscopy (TEM; JEOL 1200, JEOL Ltd, Japan).

2.4. Magnetic properties

The magnetization curve of individual particles morphology was ascertained using a vibration sample magnetometer (VSM, EG&G PARC 704, Lake Shore, USA). The external magnetic field was swept from +800 to –800 $kA{\cdot}m^{-1}$ and then back to 800 $kA{\cdot}m^{-1}$. Measurements were carried out at room temperature.

2.5. Rheological properties

Two MR suspensions with the same particle concentration of 40 wt% in silicone oil (Fluid 200, Dow Corning, USA; $\eta \approx 108$ mPa·s) were prepared. The first one contained just 40 wt. % of microspherical CI particles only while the second was dimorphic MR suspension containing 35 wt. % of CI and 5 wt. % of iron nanoparticles. MR characteristics for both MR suspensions were measured using a rotational rheometer Physica MCR 502 (Anton Paar GmbH, Austria) with a Physica MRD 170/1T magneto-cell. The true magnetic flux density in the range 0 – 438 kA·m^{-1} was measured using a Hall probe (FH 51, MAGNET-PHYSIK Dr Steingroever GmbH, Germany) and the temperature was set to 25°C. The parallel-plate geometry with a diameter of 20 mm and gap of 1 mm was employed.

All the steady flow measurements in the controlled shear rate mode were performed in the shear rate range 0.1 – 300 s^{-1}. The oscillatory measurements were carried out through dynamic strain and frequency sweeps. Strain sweeps were performed in the strain range of 10^{-5} to 0.1 at a fixed angular frequency of 6.28 rad·s^{-1} in order to get the position of the linear viscoelastic region (LVR). Afterwards, the viscoelastic moduli were obtained from the frequency sweep tests (1 to 100 rad·s^{-1}) at fixed strain amplitude in the LVR. All the oscillatory measurements were performed in the controlled shear rate mode too. During each run under a magnetic field, the MR suspension was firstly sheared ($\dot{\gamma}$ = 100 s^{-1}) at zero field for 60 s to destroy previously formed structures and after the measurement the system was completely demagnetized.

2.6. Suspension stability test

Stability of MR suspensions consisted of 40 wt. % of dimorphic magnetic particles and microspherical CI particles only in silicone oil was examined by a sedimentation ratio test. The settling of the macroscopic phase boundary between the concentrated suspension and the relatively clear oil-rich phase was measured as a function of time for 24 hrs. Afterwards, the sedimentation ratio is defined as the height of particle-rich phase relative to the total suspension height.

3. Results and discussion

3.1. Characterization of prepared particles

The size and shape of microspherical CI particles (a) and iron nanorods (b) are shown in SEM and TEM images, respectively (Figure 1). These pictures clearly demonstrate the micrometer size of spherical CI particles on the one hand and nanometer size of rod-like iron particles with the geometric aspect ratio, *L*/*D*, ranging from 5 to 9 on the other hand.

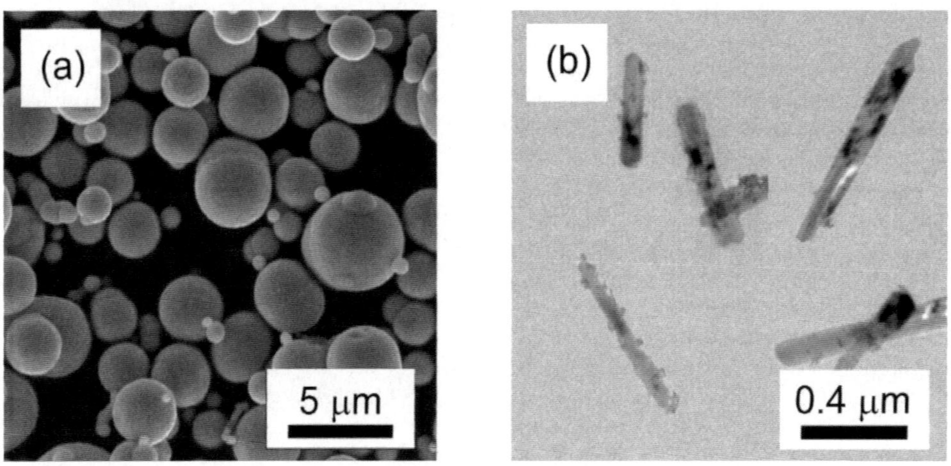

Figure 1. SEM and TEM images of (a) microspherical CI particles and (b) iron nanorods

The magnetization curve measurements of the microspherical CI particles and iron nanorods are depicted in Figure 2. The saturation magnetization of iron nanorods is expectedly lower in comparison with microspherical CI particles. However, iron nanorods in the dimorphic system more likely contribute particularly to the better MR suspension sedimentation stability and as filler between the microsphere contacts than main magnetic agent.

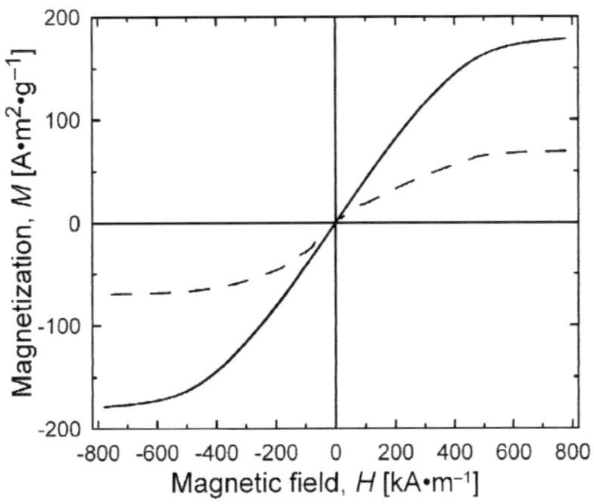

Figure 2. Magnetization curves of micropherical CI particles (solid line) and iron nanorods (dashed line)

3.2. Suspension stability

Moreover, other properties including sedimentation stability also significantly participate in the overall performance of MR suspensions and thus should be well balanced. Figure 3 illustrates the stability test for model MR suspension containing 40 wt. % of the microspherical CI particles only and dimorphic MR suspension having 5 wt. % from overall 40 wt. % substituted with iron nanorods. The dimorphic MR suspension apparently exhibits better suspension stability than that based on the microspherical CI particles only suspension, which is probably due to the increased solid friction within the suspension, *i.e.*, magnetic nanorods adsorb onto the CI microparticles through magnetostatic interactions imparting to the iron particles a brush-like effect. The addition of nanorods moreover contributes to the formation of soft sediments by avoiding short-range attractions between the large CI particles and thus allows easy redispersion of particles in the MR suspension [13].

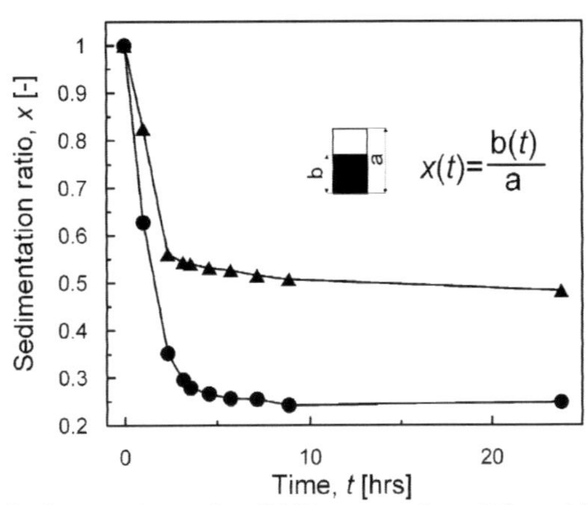

Figure 3. Sedimentation ratio of MR suspensions (40 wt. %) based on microspherical CI particles only (●) or dimorphic (▲) particles dispersed in silicone oil

3.3. MR performance

The steady shear flow of microspherical CI particles only and dimorphic particles system based MR suspensions (40 wt. %) were investigated to evaluate the influence of the magnetic dispersed phase composition on the MR performance (Figure 4). The off-state shear stress, τ_0, of dimorphic MR suspensions would increase with the presence of the magnetic nanofiber content within the system due to the residual magnetization as illustrated in reference [10], which would consequently negatively influence the MR performance expressed as $e = (\tau_M - \tau_0)/\tau_0$, where τ_M is an on-state shear stress of the suspension in external magnetic field. Nevertheless, the magnetic iron nanorods used within the dimorphic MR suspension under investigation do not exhibit too high shape anisotropy and their concentration is not excessively high at the same time to increase the off-state shear stress. Hence, in the absence of magnetic field, both types of MR suspensions exhibit nearly Newtonian behavior. In the presence of magnetic field, both systems show Bingham plastic behavior indicating that the magnetic particles were aligned into the chain-like structure sufficiently rigid to withstand certain deforming stresses without any external manifestation of flow. Typically for both MR suspensions, the shear stress increased for the entire shear rate region with the increase of external magnetic field intensity [2,3].

When magnetic field is applied, the magnetic iron nanorods are probably polarized in the contact region between two micrometer spherical CI particles, where the saturation area is expected [14]. This leads to the stronger inter-particle interactions between such CI particles. Evidently from Figure 4, both MR suspensions have approximately the same values of shear stress under the same magnetic field intensity and thus it can be concluded that although the dimorphic system contains less amount of micrometer-sized CI particles representing main component in chain-like structure formed, its MR performance is comparable to the conventional CI particles based system of the same concentration due to the improved contact area between CI particles in the chain-like structure generated under the application of external magnetic field.

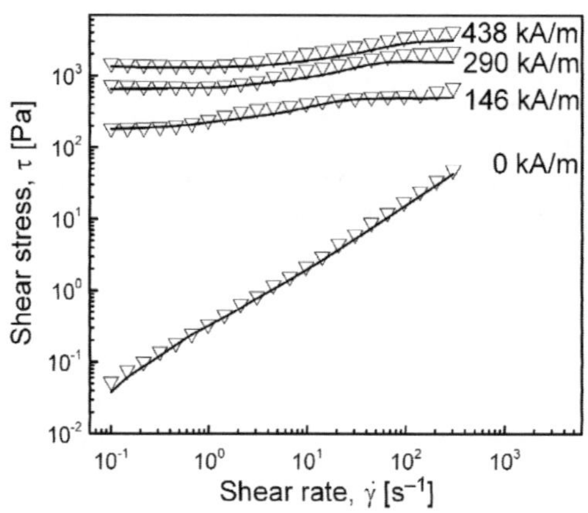

Figure 4. The shear stress, τ, vs. shear rate, $\dot{\gamma}$, dependence for 40 wt. % suspension of micro-spherical CI particles only (lines) and dimorphic system (∇) particles in silicone oil under various magnetic fields applied

The formation of internal organized chain-like structures under external magnetic field application is undoubtedly aligned with the change of viscoelastic characteristics. The storage, G', and loss, G'', moduli provide quantitative information about the magnetically induced structures in a wide range of time and frequency [7]. The dependence of G' and G'' on the angular frequency, ω, within the LVR ($\gamma = 2 \times 10^{-4}$ in our experiments) for 40 wt. % dimorphic MR suspension under magnetic field intensities of 0 and 438 kA·m^{-1} are included in Figure 5. Without the application of an external magnetic field, both moduli grow in the whole frequency range and G' slightly dominates over G'' probably due to a quite high total particle loading in the suspension. When a magnetic field is applied, however, G' becomes significantly higher due to the internal structure formation than G'' and both moduli increase rapidly in several orders of magnitude from their off-state value especially at lower angular frequencies. Moreover, G' is constant or increases slightly over wide range of driving frequencies. This illustrates a typical behavior of stiff three-dimensional network formed by magnetized particles within MR suspension which is sufficiently strong to transmit the elastic forces in such systems [15].

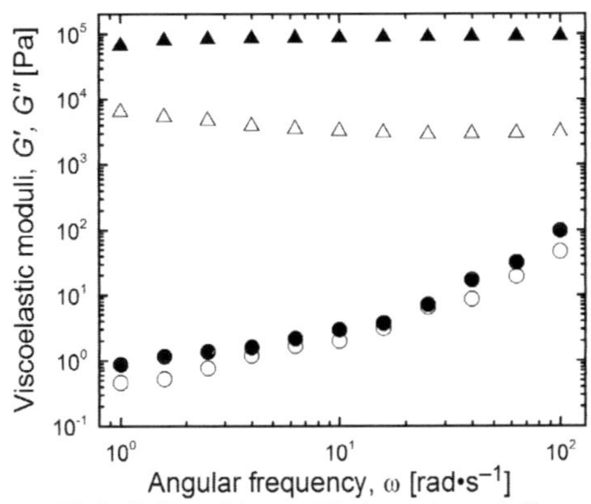

Figure 5. Storage, G', (solid) and loss, G'', (open) moduli as a function of the angular frequency, ω, in the various magnetic fields applied for 40 wt. % dimorphic MR suspension. The symbols for magnetic flux density (kA·m^{-1}): (■,□) 0; and (▼,▽) 438

4. Conclusions

A novel dimorphic MR suspension based on partially substituted micrometer-sized spherical CI particles with iron nanorods was prepared and an experimental investigation has been performed to elucidate the effect of such partial substitution on the overall MR performance as well as sedimentation stability. The dimorphic MR suspension shows typical MR characteristics including Newtonian behavior in the off-state and high values of yield stresses which are fully comparable at given external magnetic field with conventional MR suspension of the same overall concentration based on CI microparticles. In addition, the viscoelastic characteristics of dimorphic MR suspension confirmed the strong elastic behavior within the LVR due to the compact chain-like structures formed under an applied external magnetic field. Apart from promoting a comparable MR performance with microspheres-based MR suspension, the dimorphic MR suspension also exhibits better sedimentation stability due to brush-like effect of iron nanorods presented on CI microspheres.

Acknowledgements

The author M. S. would like to thank the Grant Agency of the Czech Republic (14-32114P) for financial support. This work was supported by the Ministry of Education, Youth and Sports of the Czech Republic – Program NPU I (LO1504).

References

[1] de Vicente J, Klingenberg D J and Hidalgo-Alvarez R 2011 *Soft Matter* **7** 3701
[2] Bica I, Liu Y D and Choi H J 2013 *J. Ind. Eng. Chem.* **19** 394
[3] Sedlacik M, Pavlinek V, Peer P and Filip P 2014 *Dalton Trans.* **43** 6919
[4] Kikuchi T, Oda K, Yamaguchi S and Furusho J 2010 *J. Intell. Mater. Syst. Struct.* **21** 1523
[5] Nguyen Q H and Choi S B 2010 *Smart Mater. Struct.* **19** 115024

[6] Strecker Z, Roupec J, Mazurek I and Klapka M 2015 *Int. J. Appl. Electromagn. Mech.* **47** 541
[7] Sedlacik M, Pavlinek V, Saha P, Svrcinova P and Filip P 2011 *AIP Conf. Proc.* **1375** 284
[8] Nguyen P B, Do X P, Jeon J, Choi S B, Liu Y D and Choi H J 2014 *J. Magn. Magn. Mater.* **367** 69
[9] Lopez-Lopez M T, Kuzhir P, Bossis G and Mingalyov P 2008 *Rheol. Acta* **47** 787
[10] Bombard A J F, Goncalves F R, Morillas J R and, de Vicente J 2014 *Smart Mater. Struct.* **23** 125013
[11] Ngatu G T, Wereley N M, Karli J O and Bell R C 2008 *Smart Mater. Struct.* **17** 045022
[12] Ni X M, Zheng Z, Xiao X K, Huang L and He L 2010 *Mater. Chem. Phys.* **120** 206
[13] Iglesias G R, Lopez-Lopez M T, Duran J D G, Gonzales-Caballero F and Delgado A V 2012 *J. Colloid Interface Sci.* **377** 153
[14] Ginder J M 2004 *APS News* **13** 5
[15] Tsuda K, Takeda Y, Ogura H and Otsubo Y 2007 *Colloid Surf. A-Physicochem. Eng. Asp.* **299** 262

IOP Conf. Series: Journal of Physics: Conf. Series **790** (2017) 012032 doi:10.1088/1742-6596/790/1/012032

How far are rheological parameters from amplitude sweep tests predictable using common physicochemical soil properties?

N Stoppe and R Horn

Department of Soil Science, Christian-Albrechts-University Kiel, Kiel, Germany

E-mail: n.stoppe@soils.uni-kiel.de

Abstract. A basic understanding of soil behavior on the mesoscale resp. macroscale (i.e. soil aggregates resp. bulk soil) requires knowledge of the processes at the microscale (i.e. particle scale), therefore rheological investigations of natural soils receive growing attention. In the present research homogenized and sieved (< 2 mm) samples from Marshland soils of the riparian zone of the River Elbe (North Germany) were analyzed with a modular compact rheometer MCR 300 (Anton Paar, Ostfildern, Germany) with a profiled parallel-plate measuring system. Amplitude sweep tests (AST) with controlled shear deformation were conducted to investigate the viscoelastic properties of the studied soils under oszillatory stress. The gradual depletion of microstructural stiffness during AST cannot only be characterized by the well-known rheological parameters G′, G″ and tan δ but also by the dimensionless area parameter integral z, which quantifies the elasticity of microstructure. To discover the physicochemical parameters, which influences the microstructural stiffness, statistical tests were used taking the combined effects of these parameters into account. Although the influence of the individual factors varies depending on soil texture, the physicochemical features significantly affecting soil microstructure were identified. Based on the determined statistical relationships between rheological and physicochemical parameters, pedotransfer functions (PTF) have been developed, which allow a mathematical estimation of the rheological target value integral z. Thus, stabilizing factors are: soil organic matter, concentration of Ca^{2+}, content of $CaCO_3$ and pedogenic iron oxides; whereas the concentration of Na^+ and water content represent structurally unfavorable factors.

1. Introduction

Soils are the key life resource for human beings, where food and forage crops are grown and water is cleaned by percolation. Furthermore soils provide renewable resources as well as construction grounds, and raw materials can be either extracted or captured from soils. At the same time, soils are important governing factors for local and global cycles of substances and water. Finally, soils are an essential component of ecosystems: they represent an important habitat for a giant community of plants and animals. Therefore soils are a substantial basis for biodiversity.

Nowadays degradation of soils by deformation and soil erosion are severe problems for world food supplies, hence avoidance of soil loss and deterioration as well as contamination and desertification is necessary in order to preserve beneficial soil functions.

Therefore an important consideration is the conservation of soil structure and convenient pore geometry, which guarantee exchange and supply with water and air. If the soil structure and thus pores and pathways are destroyed by external stresses (e.g. passage with heavy agricultural machines), the

IOP Conf. Series: Journal of Physics: Conf. Series **790** (2017) 012032 doi:10.1088/1742-6596/790/1/012032

infiltration of water into the soil declines. As a consequence surface runoff and soil erosion increase enormously, finally resulting in topsoil depletion. Not only the loss of fertile topsoil leads to problems, but also the formation of erosion gullies and the coverage of soils at the slope toe with colluvium. Moreover, parts of the erosion material are released into surface waters, where they deposit and accumulate. If intense sedimentation and siltation takes place in busy waterways, dredging is required. Also the large amount of pollutants and nutrients that are discharged with the sediment into surface waters is not negligible [1].

As soil structure possesses many functions, the quantification of structural stability is an important field of research and there have been ambitious projects and discussions on it for many decades [2, 3]. In classical soil mechanics the stability of soil structure is often characterized by investigating the compression and/ or the shear behavior. These analyses on the mesoscale, i.e. aggregate scale yield stability parameters e.g. precompression stress (P_c), angle of internal friction (φ) and cohesion (c) which provide information about the reaction of the regarded total soil volume towards stresses at defined boundary conditions. However, no conclusions regarding the causes of mechanical stability are possible [4]. The reasons for different strength properties especially in visually comparable soils are to be found on the microscale, i.e. particle scale and are manifested in microstructural processes and interactions. To achieve the required knowledge of the processes at the microscale, in recent years rheological studies of natural soils have received growing attention [i.a. 5, 6-8].

Since rheological analyses detect the dynamic interaction between single soil particles, entirely new insights on the combined effects of soil stabilizing factors can be gained. Due to the numerous variety of factors influencing the microstructure of soil, the following questions arise: (1) what are the relevant factors influencing microstructural stability of riparian soils? And moreover (2) can these factors be parameterized in pedotransfer functions (PTFs) and is thereby the rheological parameter *integral z* – which quantifies the microstructural stiffness – predictable?

2. Material and methods

2.1. Soil material

Marshland soils of the riparian zone of the Elbe estuary (North Germany) were used. 22 sites along the estuary were chosen to depict the gradient in salinity as well as in sodicity. At each site three soil pits were dug in the embankment foreland in dependence on the mean water levels: profile 1 above mean high tide (MHWL), where generally grassland had developed; profile 2 in the range of MHWL with partially extensive reed vegetation and profile 3 below MHWL, where vegetation is mostly lacking (Figure 1).

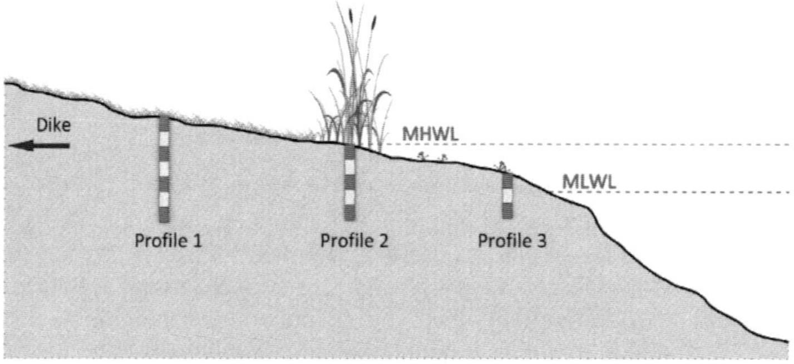

Figure 1. Schematic overview of the sampling design

Disturbed soil samples were taken from defined depths: in profile 1 in 10, 30, 50 and 70 cm; in profile 2 in 10, 30 and 50 cm; in profile 3 in 10 and 30 cm beneath ground level. The soil material was collected and transported in properly sealed plastic bags.

2.2. Standard laboratory analyses

Prior to physicochemical analyses, which were performed according to Blume, Stahr and Leinweber [9] and Schlichting, Blume and Stahr [10], samples were air dried, homogenized and sieved trough a 2 mm sieve. The screened soil material were taken to measure pH and electric conductivity (EC) in 1:2.5 (w/w) suspensions using 0.01 M $CaCl_2$ respectively deionized water. Particle size distribution was determined by sieving and sedimentation applying the pipette-method, whereby at first soil organic matter is destroyed with H_2O_2 (30%), and calcite ($CaCO_3$) is dissolved by HCl. Total soil carbon was measured by grinding a little of the sieved soil in a ball mill and afterwards about 200 mg of the material was ashed in a Ströhlein apparatus based on the principle of coulometry. $CaCO_3$ was analyzed according to the Scheibler method by gas volumetric determination of the released CO_2. The difference between inorganic and total carbon corresponds to the organic C. Exchangeable cations were extracted by 0.1 M $BaCl_2$. Concentrations of Ca^{2+}, Mg^{2+}, K^+ and Na^+ were measured using an atomic absorption spectrometer.

The total amounts of pedogenic iron oxides (Fe_d), which develop during aging of soils, were determined by extraction with 0.3 M Na-citrate solution, 1 M $NaHCO_3$ solution and Na-dithionite according to the method described by Mehra and Jackson [11]. The iron concentration was determined by atomic absorption spectrometry as well.

2.3. Rheometry

Amplitude sweep tests (AST, oscillatory mode) with controlled shear deformation were conducted with a modular compact rheometer MCR 300 (Anton Paar, Ostfildern, Germany) with a profiled parallel-plate measuring system at a plate distance of 4 mm. The configurations of the tests were carried out in accordance with Markgraf [12] (Table 1).

Table 1. Configuration parameters amplitude sweep test with controlled shear deformation

parameter	
plate distance (h)	h = 4 mm
shear deformation (γ)	γ = 0.0001 to 100%
frequency (f)	f = 0.5 Hz ($\omega = \pi$ 1/s)
temperature (T)	T = 20 °C (regulated by Peltier unit)
measuring points	30 pts.
test duration	appr. 15 min.

Air dried, homogenized and sieved (< 2 mm) samples from all sites and depths were repacked into small cylinders (45 cm^3) to a standard bulk density (ρ_t = 1.4 Mg m^{-3}) and capillary saturated with deionized water to achieve quasi-saturated conditions (0 kPa). Subsequently a portion of the prepared cylinders were drained and equilibrated to defined matric potentials (-6 or -15 kPa). Thus rheological analyses for every sampled depth were performed at three different matric potentials.

A small amount of soil (4 cm^3) was taken from each cylinder with a spatula and put on the fixed measuring plate of the rheometer for the actual measurement. After lowering the upper plate to the defined measuring gap (4 mm), the normal force was checked to maintain an undisturbed quasi-elastic soil structure (i.e. $F_N \leq 12$ N). Additionally a waiting time of 40 sec prior each test was implemented. Operation of the rheometer, monitoring of the ASTs and generation of the rheological parameters (G′, G″, tan δ, integral z) were carried out by the software Rheoplus/32 V3.21. Per depth and dewatering state five replicates were performed. Before and after each test the water content (w/w) was determined.

2.3.1. Loss factor tan δ and integral z.

According to Markgraf [12] soil can be defined as viscoelastic substance. Hence amplitude sweep tests are an appropriate instrument to investigate the structural character of soils. During ASTs soil samples experience a gradual depletion of microstructural stiffness (Figure 2). This degradation of internal structure can be characterized by the well-known rheological parameters storage modulus (G′), loss modulus (G″) and the linear viscoelastic range (LVE). Moreover the microstructure of soils can be specified by the loss factor tan δ, which equals the ratio of G″ to G′, i.e. the imaginary part to the real part of the deformation energy. If tan δ < 1, G′ prevails over G″, and the soil exhibit a quasi-elastic behaviour: applied deformation leads to full or at least partial microstructural regeneration. When tan δ = 1 the flow point ("cross-over") is reached (G′ = G″). If the flow point is exceeded a viscous character predominates, tan δ > 1 (i.e. G″ > G′) and irreversible microstructural collapse occurs [13].

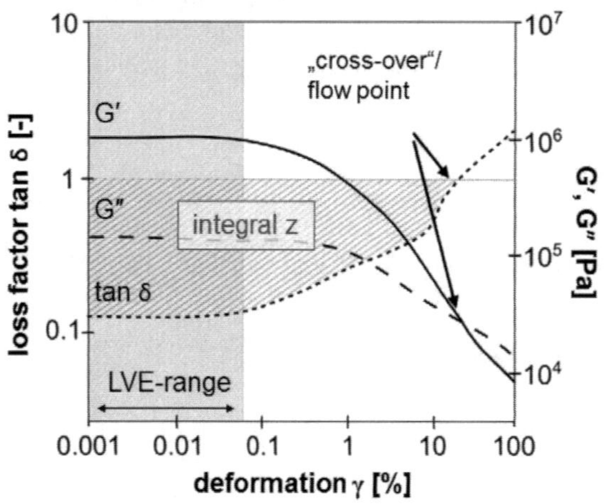

Figure 2. Diagram of an amplitude sweep test as a function of deformation γ. Loss factor tan δ serves as analogue expression of quasi-elastic (tan δ < 1) or viscous (tan δ > 1) behaviour. For further comparison the integral z can be calculated for the interval from γ = 0.001% to "cross-over" as difference of the integrals of the 1-function and the tan δ function.

For further comparison of the quasi-elastic range (tan δ < 1) in absolute terms, the dimensionless area parameter integral z can be used. The greater the value of integral z, the more elastic resp. rigid the soil is [7]. Thus integral z quantifies the structural strength in consideration of loss of elasticity due to increasing shear deformation over time. The integral of tan δ(γ) with tan δ = 1 as defined limit on the ordinate was calculated from equation (1).

$$\int_{0.001}^{\text{"cross-over"}} (1 - \tan\delta)\,d\gamma \qquad (1)$$

2.4. Statistics

The statistical evaluation was accomplished with the free software environment R (Version 3.1.1) [14] based on mixed models [15, 16] by means of analyses of covariance and variance (ANCOVA resp. ANOVA).

After testing all eligible chemical parameters on multicollinearity the explanatory variables shown in Table 2 have been selected as initial input parameters to run ANCOVA. Subsequently performed ANOVA permits the identification of the significant predictor variables.

Table 2. Selected input parameters

parameter	*acronym*	*unit*
electric conductivity	EC	mS cm^{-1}
organic matter	OM	g 100 g^{-1}
calcite content	$CaCO_3$	g 100 g^{-1}
pedogenic iron oxides	Fe_d	g kg^{-1}
calcium concentration	Ca^{2+}	$cmol_c$ kg^{-1}
sodium concentration	Na^+	$cmol_c$ kg^{-1}
water content	Θ_g	w/w
depth beneath ground level	depth	cm

3. Results

The rheological response of soils under oscillatory conditions is influenced strongly by particle size distribution and water content (Figure 3). The texture effect is depicted by the median of grain size for all top soils (10 and 30 cm depth) in Figure 3: under quasi-saturated conditions (0 kPa) the integral z values become greater the greater the median of grain size gets (figure 3a). Because all discovered textures, which could differ concerning further chemical parameters, were taken into account, the coefficient of determination (R^2) indicates just a weak association between integral z and matric potential depending on texture. However, it can be stated that in fine-grained soils the microstructural stiffness is smaller than in sandy soils when saturated with water.

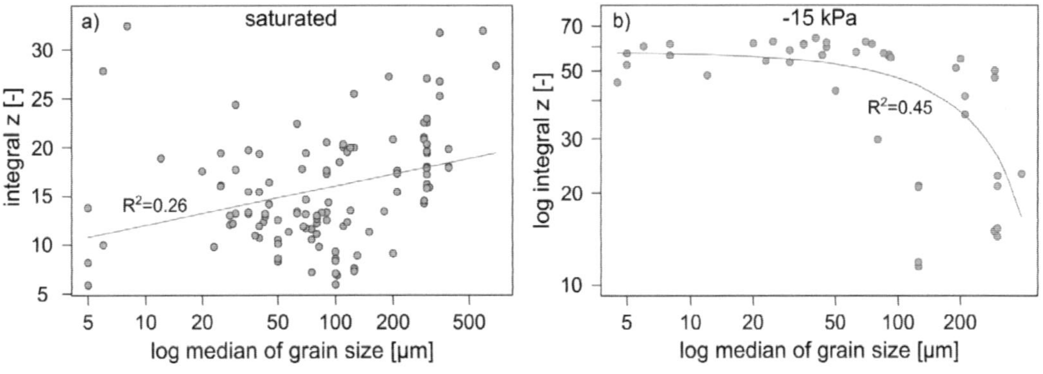

Figure 3. Texture effect depicted by the median of grain size for all top soils (10 and 30 cm); a) quasi-saturated conditions and b) matric potential of -15 kPa

At a matric potential of -15 kPa first of all greater integral z values were determined and in addition the integral z values become greater the smaller the median of grain size is (figure 3b). Consequently the desiccation has a stabilizing effect on the soil microstructure. This effect becomes more pronounced, the more fine-grained the soil is.

Due to the considerable impact of particle size distribution, the textures were grouped according to the German soil texture classification (Figure 4).

Based on the defined texture groups the statistical analyses (ANCOVA, ANOVA) were conducted and the physicochemical factors shown in Table 3 have been identified to significantly affect microstructural stability.

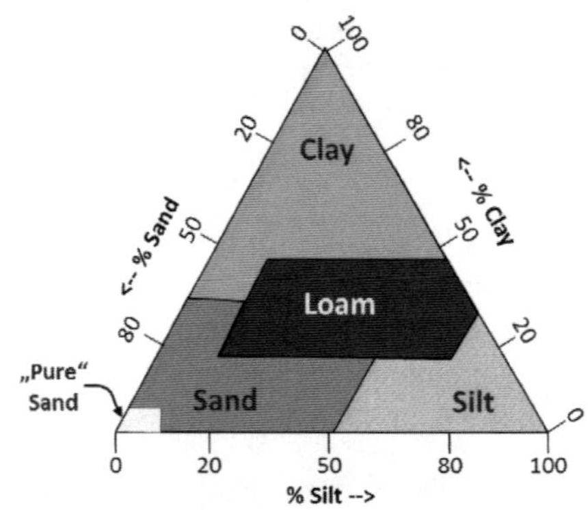

Figure 4. German soil texture classification (after [17])

Table 3. Physicochemical factors significantly influencing soil microstructure.

parameter	*acronym*	*unit*
organic matter	OM	g 100 g^{-1}
calcite content	$CaCO_3$	g 100 g^{-1}
pedogenic iron oxides	Fe_d	g kg^{-1}
calcium concentration	Ca^{2+}	$cmol_c$ kg^{-1}
sodium concentration	Na^+	$cmol_c$ kg^{-1}
water content	Θ_g	w/w
depth beneath ground level	depth	cm

Furthermore, the determined statistical relationships between the rheological parameter integral z and parameters influencing microstructural stiffness facilitated the development of pedotransfer functions, which allow a mathematical estimation of the microscale stability parameter integral z for the defined texture groups. Since the effect of the individual parameters vary depending on texture, those parameters, which have a minor impact on the soil microstructure, remain unconsidered, without deterioration of model quality resp. accuracy of PTFs. Hence, not every parameter is required for each defined texture group to predict the integral z value. Table 4 summarizes the developed PTFs for each texture group.

Table 4. Pedotransfer functions to predict microstructural stiffness through mathematical estimation

texture group	*pedotransfer function*	R^2
sand	integral z = 40.2 + 3.2 OM + 2.1 Ca^{2+} + 2.7 Fe_d - 1.2 Θ_g - 0.1 depth	0,64
loam	integral z = 67.9 + 4.6 OM + 0.2 $CaCO_3$ - 1.4 Θ_g	0,56
silt	integral z = 41.3 + 6.8 OM + 0.5 $CaCO_3$ + 2.3 Fe_d - 1.2 Θ_g	0,50
clay	integral z = 46.7 + 5.0 OM + 2.3 $CaCO_3$ - 3.6 Na^+ - 0.8 Θ_g - 0.2 depth	0,62

It turns out that when the combined effects of the physicochemical parameters are considered, the stabilizing factors are: soil organic matter, concentration of Ca^{2+}, content of $CaCO_3$ and pedogenic iron oxides (Fe_d); whereas the concentration of Na^+, water content and the depth beneath ground level represent structurally unfavorable factors.

4. Discussion

4.1. Texture effect

Under oscillatory stress a characteristic degradation of internal structure depending on particle size distribution takes place. The textural effect, especially the importance of particle shape and mineral surface properties, was already emphasized by Markgraf and Horn [7] and Baumgarten [18] and is reflected in the results presented here. Coarse-grained textures exhibit a fast reduction of elasticity with increasing deformation at first. But as a consequence of relative motions (rotation, interlocking, reorganization) coarser particles are able to attain a more stable resp. an energetically more favorable arrangement, whereby the complete collapse of microstructure is delayed.

The more fine-grained the soil is, i.e. the higher the silt and/ or clay content, the less often interlocking between particles occurs and elasticity decreases gradually with increasing shear deformation [18]. While rounded and angular particles show rolling shear behavior [19], an increase in platy particles leads to sliding shear behavior [20], assuming a completely homogenized system. Sliding shear behavior and formation of smooth slickensides results in a reduced shear strength [21] caused by depletion of internal structure.

However, because of its smaller particle size, chemical surface properties prevail in clayey soils additionally to the induced shear behavior. Therefore clayey soils show an elastic character over a wide deformation range, until elasticity suddenly continuously decreases. Particularly the rheological properties may vary substantially because of diverse clay mineral associations, which react differently to shearing [22].

Furthermore, excess pore water pressure may arise under mechanical stress, which reduces effective stress at almost constant total stress. Convex menisci promote the displacement of particles relative to one another and therefore deformation of microstructures during shear strain. Depletion and/ or redistribution of positive pore water pressures are dependent on hydraulic conductivity, tortuosity of pore space and hydraulic gradient, thus reallocation of excess pore water pressures is closely connected with particle size distribution and (micro) aggregation of soils: short flow paths and/ or high conductivity favor the reduction of positive pore water pressure. Concave and thus contractive menisci are established when a new equilibrium is regained [23].

However, rheological characteristics and depletion of soil microstructure are not exclusively regulated by soil texture. They are further influenced by varying soil chemical features and especially varying water contents by which the rheological properties are modified.

4.2. Physicochemical parameters

The obtained findings corroborate that the microstructure of soils is the product of a variety of chemical, physical and biological interactions. Taking the combined effect of physicochemical features into account it appears that soil organic matter, concentration of Ca^{2+}, calcite content ($CaCO_3$) and pedogenic iron oxides (Fe_d) represent stabilizing factors, whereas the concentration of Na^+, water content and the depth beneath ground level stand for structurally disadvantageous factors.

The positive effect of the beneficial factors is mainly attributable to their gluing and cementing effect. Soil organic matter influences soil structure on different scales and by several mechanisms due to its complexity [24]. Extracellular polymeric substances (EPS), metabolites and other breakdown product produced by microbial decomposition act as binding agent, stick soil particles together and enhance soil (micro) aggregation [24, 25]. Ca^{2+} provokes the formation of a stable card house structure [26] and various authors reported on the stabilizing effect of Ca^{2+} [e.g. 27, 28]. Thereby the formation of Ca^{2+} bridges represents a long-term effect [29], which is further intensified through the addition of organic substances [30]. In this regard, it appears plausible that $CaCO_3$ is also included with a positive sign in the transfer function, as it serves as a Ca^{2+} source. Cementation of mineral particles also occurs with an increasing content of iron oxides [31]. Strong interactions between iron oxides and organic substances benefit the formation of organo-mineral complexes, which on the one hand stabilize (micro) aggregates and on the other hand soil organic matter [32]. Markgraf and Horn [33] found the formation of pseudosand in association with the cementation by iron oxides. Moreover they reported a significant reduction in microstructural stiffness after extracting pedogenic iron oxides from the soils under study.

The pedotransfer function for clayey soils includes sodium with a negative sign, constitute the well documented dispersive effect of Na^+ [e.g. 34], which is attributed to its monovalency and its big hydrated radius. Although it is generally known that in natural soils the soil moisture is insufficient to completely hydrate exchangeable cations [31], Na^+-clay-connections are inherently unstable, as Na^+ mainly forms ionic bonds [35]. Finally, the water content negatively affects the elasticity of soils: increasing water content weakens the microstructure of soils due to lower menisci forces.

4.3. Menisci forces

The soil microstructure is weakened by increasing water saturation. Furthermore, the more fine-grained the soil material is, the more sensitive it reacts towards shearing under saturated conditions. Under saturated conditions additional water layers are absorbed, which enlarge the distance between individual particles and act as lubricant. The more water layers are attached to the mineral surface, the greater the mobility of single particles [36]. If chemically active particles are available, existing electrostatic interactions are reduced to that effect and finally capillary suction disappears when complete saturation is reached [37].

On the contrary, increasing desiccation stabilizes the microstructure of soil by water menisci forces, whereby fine-grained soils show a stronger reaction towards dewatering. In unsaturated soils, i.e. at negative pore water pressure, water menisci between soil particles provoke a contractive force such that the matric potential acts as a binding force between particles and stable connections are formed at rest [38]. Therefore any change in matric potential also causes a change in pore water pressure. By successive drainage soil particles are pulled together [39] and apparent cohesion increases [40], accordingly the resistance of soils against external stresses is enhanced [41]. The observed stabilizing effect of menisci forces on the microscale was also found by other authors [18, 42, 43].

5. Conclusion

The microstructural stiffness of the investigated riparian Marshland soils was characterized by amplitude sweep tests with controlled shear deformation. The results show a clear dependence on soil texture and water content. Depletion of internal structure takes place in a characteristic manner influenced by particle size distribution. Coarse-grained textures show a fast reduction of elasticity with increasing deformation at first, but a delay of microstructural collapse is possible due to interlocking and/ or rotation of single particles. The higher the silt and/ or clay content the less often interlocking between particles occurs. Hence, elasticity decreases gradually with increasing deformation. Desiccation stabilizes the microstructure of soils through amplified menisci forces, whereas increasing water saturation weakens the microstructure. Moreover the presented results confirm that the microstructure of soil is the product of a variety of processes and that the consideration of the combined effects of physicochemical features is appropriate and allowed the development of the plausible pedotransfer functions.

References

[1] Wissenschaftlicher Beirat Bodenschutz 2002 *Ohne Boden - bodenlos* (Berlin: BMU)
[2] Bronick C J and Lal R 2005 *Geoderma* **124** 3-22
[3] Warkentin B R 2008 *Adv Agron* **97** 239-72
[4] Hartge K H and Horn R 2014 *Einführung in die Bodenphysik* (Stuttgart: Schweizerbart´sche Verlagsbuchhandlung)
[5] Torrance J K 1999 *Appl Clay Sci* **14** 199-223
[6] Ghezzehei T A and Or D 2001 *Soil Sci Soc Am J* **65** 624-37
[7] Markgraf W and Horn R 2009 *Progress in Management Engineering,* ed L P Gragg and J M Cassell (New York: Nova Science Publishers, Inc.) pp 237-79

[8] Holthusen D 2010 *Fertilization induced changes in soil stability at the microscale revealed by rheometry* (Christian-Albrechts-Universität zu Kielt: Institut für Pflanzenernährung und Bodenkunde, Dissertation)
[9] Blume H-P, Stahr K and Leinweber P 2011 *Bodenkundliches Praktikum. Eine Einführung in pedologisches Arbeiten für Ökologen, insbesondere Land- und Forstwirte, und für Geowissenschaftler* (Heidelberg: Spektrum Akademischer Verlag)
[10] Schlichting E, Blume H-P and Stahr K 1995 *Bodenkundliches Praktikum. Eine Einführung in pedologisches Arbeiten für Ökologen, insbesondere Land- und Forstwirte und für Geowissenschaftler* (Berlin, Wien u.a.: Blackwell Wissenschafts-Verlag)
[11] Mehra O P and Jackson M L 1960 *Clay Clay Miner* **7** 317-27
[12] Markgraf W 2006 *Microstructural Changes in Soils. Rheological Investigations In Soil Mechanics* (Christian-Albrechts-Universität zu Kielt: Institut für Pflanzenernährung und Bodenkunde, Dissertation)
[13] Mezger T G 2010 *Das Rheologie Handbuch: Für Anwender von Rotations- und Oszillations-Rheometern* (Hannover: Vincentz Network GmbH & Co. KG)
[14] R Developement Core Team R: A language and environment for statistical computing http://www.R-project.org./
[15] Laird N M and Ware J H 1982 *Biometrics* 963-74
[16] Verbeke G and Molenberghs G 2008 *Linear Mixed Models for Longitudinal Data* (New York: Springer Verlag)
[17] Ad-Hoc-Arbeitsgruppe Boden 2005 *Bodenkundliche Kartieranleitung* (Stuttgart: E. Schweizerbart´sche Verlagsbuchhandlung)
[18] Baumgarten W 2013 *Soil microstructural stability as influenced by physicochemical parameters and its environmental relevance on multiple scales* (Christian-Albrechts-Universität zu Kielt: Institut für Pflanzenernährung und Bodenkunde, Habilitationsschrift)
[19] Cho G C, Dodds J and Santamarina J C 2006 *Journal of Geotechnical and Geoenvironmental Engineering* **132** 591-602
[20] Smith D W and Reitsma M G 2002 *Environmental Geomechanics,* ed L Vulliet, *et al.* (Lausanne, Schweiz: EPFL/ Centre Midi) pp 27-44
[21] Li Y R, Aydin A, Xu Q and Chen J 2012 *Ksce J Civ Eng* **16** 1152-9
[22] Dolinar B and Trauner L 2007 *Eng Geol* **92** 88-96
[23] Peth S, Rostek J, Zink A, Mordhorst A and Horn R 2010 *Soil Tillage Res* **106** 317-28
[24] Tisdall J M and Oades J M 1982 *J Soil Sci* **33** 141-63
[25] Oades J M 1984 *Plant Soil* **76** 319-37
[26] Lagaly G 1993 *Tonminerale und Tone. Struktur, Eigenschaften, Anwendungen und Einsatz in Industrie und Umwelt,* ed K Jasmund and G Lagaly (Darmstadt: Steinkopff Verlag)
[27] Wuddivira M N and Camps-Roach G 2007 *Eur J Soil Sci* **58** 722-7
[28] Paradelo R, van Oort F and Chenu C 2013 *Geoderma* **200** 40-4
[29] Six J, Bossuyt H, Degryze S and Denef K 2004 *Soil Till Res* **79** 7-31
[30] Baldock J A, Aoyama M, Oades J M, Susanto and Grant C D 1994 *Aust. J. Soil Res.* **32** 571-94
[31] Schwertmann U and Niederbudde E-A 1993 *Tonminerale und Tone. Struktur, Eigenschaften, Anwendungen und Einsatz in Industrie und Umwelt,* ed K Jasmund and G Lagaly (Darmstadt: Steinkopff Verlag) pp 212-65
[32] Pronk G J, Heister K and Kogel-Knabner I 2011 *Soil Sci Soc Am J* **75** 2158-68
[33] Markgraf W and Horn R 2007 *Soil Sci Soc Am J* **71** 851-9
[34] Sumner M E and Naidu R 1998 *Sodic soils: distribution, properties, management and environmental consequences* (New York: Oxford University Press Inc.)
[35] Rengasamy P and Olsson K A 1991 *Aust. J. Soil Res.* **29** 935-52
[36] Santamarina J C and Shin H 2009 *Meso-Scale Shear Physics in Earthquake and Landslide Mechanics,* ed Y H Hatzor, *et al.* (Boca Raton, London u.a.: CRC Press)
[37] Al-Shayea N A 2001 *Eng Geol* **62** 319-42

[38] Osipov V I 2014 *Water Resour+* **41** 801-18
[39] Horn R 1990 *Soil Till Res* **17** 265-89
[40] Santamarina J C 2001 *Proceedings of the Symposium of Soil Behavior and Soft Ground Construction, in honor of Charles C. Ladd*: MIT) pp 1-32
[41] Kézdi Á 1974 *Handbook of Soil Mechanics, Vol. 1: Soil Phyics* (Amsterdam: Elsevier)
[42] Czibulya Z, Szegi T, Michéli E and Tombácz E 2014 *Int J Agric Sci Technol* **2** 22
[43] Holthusen D, Peth S, Horn R and Kuhn T 2012 *J Plant Nutr Soil Sc* **175** 535-47

IOP Conf. Series: Journal of Physics: Conf. Series **790** (2017) 012033 doi:10.1088/1742-6596/790/1/012033

Solvents of pus-medicines with physical-chemical aggressive action

A Urakov[1,2], N Urakova[1], A Reshetnikov[1], M Kopylov[1], L Chernova[1]

[1]Department of General and Clinical Pharmacology, Izhevsk State Medical Academy, Izhevsk, Russia
[2]Department of Thermal Deformation Processes, Institute of Mechanics, Ural branch of RAS, Izhevsk, Russia

E-mail: urakoval@live.ru

Abstract. In laboratory and clinical conditions was studied rheology of pus and sulfuric tubes after their interaction with aqueous solutions of drugs from different pharmacological groups. It is shown that solutions of almost all medicines can influence or not influence on their rheology, because local action is determined not by the name, dose or route of administration of medicines. It is established that only physical-chemical properties of fluids and physical-chemical factors of their interaction with dense pus can give them the ability to dissolve or thickening pus. We found that deliberate change physical-chemical properties of medicines solutions from various pharmacological groups, namely, raising the temperature to +42°C, increasing the alkalinity above pH 8.1 and aeration as for example by introducing carbon dioxide under pressure of 0.2 ATM, or by introducing hydrogen peroxide in 0.5 - 3%, turning them into solvents of pus, ear wax and sulfuric tubes. Discovered that solutions of drugs with such physical-chemical activity may turn thick pus and solid sulfur tube in a homogeneous liquid after a few minutes after injecting them into these biological mass.

1. Introduction

It is no secret that any infection is dangerous the appearance of pus, it becomes a source of germs and the cause of the destruction of healthy tissue in the body of a diseased person. The germs multiply very rapidly, so the suppurative focus is growing very rapidly. In this regard, for effective treatment of purulent diseases need as soon as possible to save the patient from pus. While it is best to remove all the pus out, then it is advisable to wash with an antiseptic solution, disinfectant or cauterizing agents. Quick completion of this task in the clinic is possible only at a superficial or outer location of the purulent focuses. However, when the location of the purulent focus in the body cavities in the depth of the soft and hard tissues cannot quickly remove the pus out, because the pus is sticky and viscous. In such situations, the pus cannot be removed using surgical tools, because it is dangerous to the health of the patient, and with medication, because special solvents pus not developed [18].

Previously, we reported that the nature of the local action of drugs on tissues of our body is determined primarily by physical-chemical factors of the interaction [9,13,15]. It was shown that the conformity of the physical-chemical properties of drugs and physical-chemical properties of tissues prevent local irritating action of the drugs and vice versa [20]. The larger the difference between medicines and tissues from each other on the physical-chemical properties, the stronger change their aggregate condition in the local interaction [21]. In this case, we were able to identify physical-chemical factors, that determine the state of aggregation and the rheology of thick pus [1, 10, 17].

What physical and chemical properties of drugs are most important for turning the drugs in solvents of pus ?

To date, there is no doubt that one of the most important factors for physical and chemical dissolution thick pus is the local hyperthermia [1, 5, 17]. Our results show that in real life there are no conditions under which the temperature of all body parts would be equal to 36.6° C. The reality is that the temperature of various body parts of different patients in hospital can vary ranging from +24 to +42° C [20]. It is shown that solutions of medicines are introduced in a body at room temperature (as a rule, at a temperature of +24 - +26°C) [10, 12, 13,]. However, some medicines may be of lower or higher temperature, as some medicines are deliberately cooled, because they are stored in the refrigerator at a temperature of +4 - +8°C, and some medicines are heated due to ignorance, as there is no temperature control even in the presence of heat sources (solar rays, infrared radiation from incandescent lamps and medical devices) [20]. At the same time Arrhenius and Van't Hoff proved that 10°C temperature change can change the chemical reactions flow rate from 2 to 4 times [7].

It is shown, that very important factor for the rheology of blood and pus is alkalinity [1, 16, 18]. Very important factor for physical move of the biological mass is gravity [3, 4, 6, 19]. The reason is that the human body is constantly changing its position in space, but almost all of its body tissues move insignificantly relatively to each other. Therefore, we assume that a man and his body are relatively mobile, but pus and cerumen are relatively stationary inside his body and these products may remain at the same place for a long time for two reasons. First, due to the high friction, and secondly, due to the high viscosity.

In addition, it is assumed that other important factors for medical dissolution of pus can be duration of local action of drugs and its gas saturation [2, 4, 8, 10, 11, 12].

2. Materials and methods

We studied pus rheology in 10 patients with tuberculous empyema, in 5 patients with purulent peritonitis and in 10 patients with hearing loss. Pus rheology was studied through visual observations in vivo and in vitro before and after the introduction solutions of medicines in pus with regard to the volume of interacting masses, the duration of their interaction, gravity, specific gravity (density), local temperature, pH and osmotic activities. Pus was taken for analysis before and after 15 minutes after the start of its interaction with solutions of medicines began.

We have identified the most important physical and chemical characteristics about 200 quality medicines produced by pharmaceutical companies in the form "Solution for injection". The medical drugs which were examined belonged to various pharmacological groups. The medicine quality was studied by taking into account the manufacturers and serial number of medicine. To analyze medicine quality we compared quality characteristics values which had been specified by the manufacturer in the medicine passport with the observed ones. Additionally, we identified osmotic activity with cryoscope using vapor-pressure OSMOMAT-030 RS brand osmometer of ANSELMA Industries (Austria) company. Behind them we controlled acidy (alkalinity) of solutions and purulent masses defined with strips of the universal indicator paper company Lachema.

When evaluating the quality of drugs solutions we were taking into account their osmotic activity and the degree of gas saturation. The visualization of gas bubbles in aerated solutions was held in visible spectrum of radiation through visual observation on the eyes and ultrasound method using an ultrasound device "ALOKA SSD-ALPHA 10" using convection sensor with a frequency of 3 to 7 MHz [3, 19].

An additional the dynamics of temperature in different parts of the body in 25 patients was studied using advanced infrared radiation with the help of thermal imager TH91XX (NEC, USA). All patients were being treated in Izhevsk Hospitals within the period from March 2008 to October 2015. Further data processing was performed with Thermography Explorer and Image Processor software programs. Ambient temperature in the examination room was 24 ± 25°C, the temperature range in the thermal

IOP Conf. Series: Journal of Physics: Conf. Series **790** (2017) 012033 doi:10.1088/1742-6596/790/1/012033

camera was set as 25 - 36°C. At the examined area we determined the mean temperature, standard deviation and distribution of isotherms at the thresholds of 34, 32, 30, 29 and 25°C.

Statistical analysis of the results was performed using the BIOSTAT software program using standard methods.

3. Results

Purulent mass, extracted from the pleural cavity of patients with pleural empyema, had the consistency of opaque, shapeless, thick, viscous and sticky mass of yellow-grey. Then the purulent mass is then maintained this state after placing them in transparent laboratory glass vials at room temperature. The test tubes during the study were table stands motionless in a vertical position with the bottom down. Purulent mass flowed down the walls of the tubes to the bottom and was in the bottom of the tubes still.

Sulfuric tubes, extracted from patients ears with hearing loss, had the form of a solid and opaque flagellum gray-brown in color with rounded ends. Then sulfuric tubes maintained this state at room temperature after placing them in Petri dishes.

The results showed that the careful infusion to purulent masses and sulfur tubes of known aqueous solutions at comparable volumes at a temperature of +24 or +37° C did not lead to the dissolution of pus and cerumen in the last 15 minutes of observation. Moreover, the studied biological mass always remains unchanged at the bottom of the dish, and injected fluid does not mix with them and remained transparent. Moreover, between biological mass and the fluids remained clear border of separation of environments and biological mass always sank when lowering them inside the well-known liquids and never surfaced to the top. In particular, during 15 minutes of interaction at a temperature of +24° C boiled water, tap water, water for injection, a solution of 0.9% sodium chloride solution 5% glucose solution 20% glucose solution 40% glucose solution 10% sodium chloride, 10% solution sulfatsila-sodium, a solution of 20% sulfatsila-sodium, a solution of 0.02% furatsilina solution of 0.5% chlorhexidine, 70% solution of ethyl alcohol, and solutions of 4% sodium bicarbonate was not mixed with pus and with sulfuric jams. Therefore, the rheological properties and other physico-chemical characteristics of pus and sulfur tubes, and their shape and state of aggregation remained unchanged.

A slight decrease in the viscosity of thick purulent masses after 15 minutes of interaction with fluids was observed only when the increase temperature to +42°C.

Then we have defined the indicators for the acidity of water and solutions of antiseptics and plasma substitutes. The obtained results showed, that all of them, except solution of 4% sodium bicarbonate and solutions of 10% and 20% sodium sulfatsil, have a pH below 7.0, and so are acidic. In particular, indicators of acidity of tap water, boiled water, water for injection, solution of furatsiline 1:500, solution of 3% hydrogen peroxide, solutions of 0.9% and 10% of sodium chloride are the pH range of 5.0 - 6.0, and the performance of the acidity of solution of 0.25% novokaine and solutions 5%, 20% and 40% glucose are in the pH range of 2.7 to 4.0 (Table 1).

In the next series of experiments in vitro at a temperature of +24°C we studied the change of rheology of thick pus throught 15 minutes after administration in pus water or water solutions of medicines after preliminary artificial amplification acidification to pH 2.0 or alkalinity to pH 12,0 by introducing them accordingly hydrochloric acid or sodium hydroxide. Our results showed that only alkalization decreases viscosity purulent masses.

So, after 15 minutes after infusion to thick purulent mass of an equal volume one of the following "hyperacid" liquids (they all have a pH 2.0) - water for injection or water solution of one of the following means: solution of 0.9% or 10% sodium chloride, solution of 10% or 20% sulfatsil sodium, solution of 0,02% furatsilin or 0.5% chlorhexidine, purulent masses remain practically unchanged: thick, sticky and viscous. At the same time, 15 minutes after infusion to thick purulent mass of an equal volume one of these liquids with a pH value of 12.0 thick and sticky pus turns into a liquid with a good fluidity.

Table 1. Values of acidity (alkalinity) of some antiseptics and plasma substitutes

No.	Liquid medication	Manufacture, Series No.	Values of acidity (alkalinity) (pH)
1	Tap water		5,95 ± 0,11*
2	Drinking water from the kettle		6,05 ± 0,07*
3	Water for injection in vials of 2 ml	SE " Lvivdialect" Series No. 1151205	5,02 ± 0,08*
4	Sol. 0.9% Sodium chloride for injection 200 ml	JSC «Galichpharm» Series No. 300604	5,34 ± 0,04*
5	Sol. 10% Sodium chloride for external use 200 ml	Pharmacy No. 131 (Izhevsk)	6,05 ± 0,09*
6	Sol. Furatsillin 1:5000 500 ml	Pharmacy No. 131 (Izhevsk)	5,70 ± 0,10*
7	Hydrogen peroxide 3% 40 ml	LLC «Ватхэм-pharmacy» Series No. 841101	5,45 ± 0,25*
8	Sol. 5% Glucose for injection 500 ml	JSC «Biosynthesis» Series No. 48102000	4,05 ± 0,09*
9	Sol. 20% Glucose for injection 500 ml	«Novosibirskpharm» Series No 30703	3,50 ± 0,15*
10	Sol. 40% Glucose for injection 500 ml	JSC «Dalhimpharm» Series No. 210101	2,70 ± 0,07*
11	Sol. 0,25% Novokaine for injection 500 ml	Pharmacy No. 131 (Izhevsk)	4,05 ± 0,03*
12	Sol. 10% Sodium sulfatsil 10 ml	FSUE «Moscow endocrine factory» Series No. PN 001084/01	7,90 ± 0,10*
13	Sol. 20% Sodium sulfatsil 10 ml	FSUE «Moscow endocrine factory» Series No. PN 001084/05	8.30 ± 0.15*
14	Sol. 4% Sodium bicarbonate 200 ml	Pharmacy No. 131 (Izhevsk)	8,20 ± 0,09*

Note: * - $P < 0,05$, $n = 5$.

In parallel, we studied the features of dissolution of thick and sticky pus and sulfuric tubes under the influence of water and water solutions of medicines in the conditions of increased their turbulence, which was reached by heating and mechanical wobbled from side to side model cavity (test tube), filled commensurate amounts of pus and one of the investigated solutions. It is shown that the continuous rocking from side to side tubes with interacting environments, namely - with pus and with liquids with high alkalinity (water for injection, solutions of 0.9% and 10% sodium chloride, solution of 10% and 20% sulfatsil sodium, solution of 0,02% furatsilin at pH 12.0), causes continuous reciprocal progressively offset fluid, which is the top layer, with respect to the stationary mass of thick pus and sulfuric tubes, which is the lower layer, and accelerates the process of liquefying of pus and cerumens on the dividing line between interacting environments.

However, it appeared that artificially supported for 30 min increased turbulence all of the above fluids don't lead to absolutely complete dissolution of pus and sulfuric tube.

On the basis of obtained results it was concluded that the above fluid does not possess the ability to effectively dissolve thick pus and cerumen impactions. At the same time, such physical and chemical factors of their local interaction, as hyperthermia, high alkalinity and high turbulence may underlie the nonspecific ability of drugs to dissolve the thick solid pus and cerumen impactions.

We checked the validity of these assumptions. For this was studied the dynamics of aggregation state thick pus at a temperature of 24°C after infusion solutions of 4% and 10% sodium bicarbonate (pH 8.1 and pH 8.3, respectively) or solutions of 2.4% and 24% aminophylline (pH 9.0 and pH to 12.0, respectively). Our results showed that solutions of sodium bicarbonate and solution of aminophylline good dissolve thick pus. The results showed, that 15 minutes after the introduction of these solutions into test tubes with thick pus thick pus were completely dissolved and transformed into a homogeneous liquid.

After that, we investigated the role of high turbulence and temperature of the drugs solutions to dissolve thick. To this end, the dynamics was studied the state of aggregation of thick pus after the transfer to it of an equal volume solution of 4% sodium bicarbonate at a temperature of +24, +37 or +42°C under continuous rocking of the tubes. The results showed that complete dissolution of pus occurred at a temperature of +24, +37 and +42° C, respectively after 15, 12.5 and 12 minutes. Therefore, high turbulence and temperature solutions and turn them into solvents of thick pus.

Consequently, moderate and high alkalinity, high turbulent and thermal activity of drugs solutions gives them a reasonable physical-chemical aggressiveness, therefore, is able to turn them into solvents of thick pus.

Then, we were define the value of specific gravity, osmotic and acid activity of purulent masses, obtained in patients with purulent peritonitis, purulent pleurisy, purulent conjunctivitis, purulent rhinitis and purulent abscess. It was found that the pus has a value of specific gravity in the range of 1.030 - 1.040 g/cm^3, has osmotic activity within 280 - 300 mOsmol/l of water and has an acid activity in the range of pH 5.8 - 6.2.

After this was determined the specific weight of liquids is known, which the traditional use for sanitation of purulent cavities. The results showed that they all have a specific gravity less than 1.30 g/cm^3. In other words, the specific weight of the sanitizing solutions was less than the specific weight of purulent masses. These data allowed us to explain why the pus is always drowning in these solutions, and solutions are always over flowing with pus and do not penetrate inside it.

The extensive research has shown that the solutions of sodium bicarbonate in concentrations above 4%, have specific weight, exceeding 1.040 g/cm^3 (meaning that they are more «heavy» than the «heaviest» pus), pH 8.1- 8.2 (that is, are alkaline) and osmotic activity above 450 mOsmol/l of water (i.e. are weak hypertonic solutions).

These data suggest that solutions of sodium bicarbonate at a concentration of above 4% will be able to sink into the festering mass under the action of gravity. It was also assumed that such "heavy" solutions can be implemented on top of the thick purulent masses and it's possible that they can dissolve the thick pus, because they are alkaline buffers. In order to increase the ability of solution of 4% sodium bicarbonate to dissolve thick pus, we decided to heat the solution and give it the activity is high turbulence.

It is clear that the maximum permissible hyperthermia can be given a solution to its simple heating up to +42°C, but the maximum high turbulence needed to accelerate the process of liquefaction and dispersion of pus, cannot be achieved by «manual» jiggle the capacity of interacting environments from side to side. Therefore to give a solution of the maximum possible turbulence decided to increase the concentration and the pressure of gases, in particular, due to carbon dioxide similarly carbonated mineral water and due to the hydrogen peroxide. We hypothesized that the high saturation by gas of the solution 4% sodium bicarbonate will have a powerful aggressive action on pus, because warm, heavy, alkaline and high osmotic action of solution is to be able to intensively penetrate into a pus and high saturation solution by gas and rapid formation of bubbles of carbon dioxide within a pus is to be able to blasting it from the inside.

To test this hypothesis, we initially injected carbon dioxide under excessive pressure of 0.2 ATM in a solution of 4% sodium bicarbonate and then heated carbonated solution to a temperature of +42°C. After that entered a warm carbonated solution into a test tube, in which there was pus. It turned out that this solution immediately entered a turbulent interaction with pus and turned it into a festering foam. The foam was expelled outward from the tube almost in its entirety. The process of foam formation and its ejection from the tube out was very intense and resembled a Geyser.

Then we found that high turbulence and high saturation by gas of warm solution of 4% sodium bicarbonate gives him ability to effectively and safely disposed of pus out of purulent fistula when pancreatic necrosis [2, 4, 10, 16].

Similar results were obtained with sulfuric jams. It turned out that the injection of aerated and warm solution of 4% sodium bicarbonate inside the sulfuric tube also quickly destroys the tube. Sulfuric tube begins to increase in size due to the formation within it of a plurality of gas bubbles. In fact, sulfuric tube begins to foam. However, it breaks into several small pieces, which soon also begin to foam, so very quickly all sulfuric tube is falling apart and turns into a foam dirty-brown color [].

Consequently, the high dense, high alkalinity and high osmotic activity, that is inherent in a solution of 4% sodium bicarbonate, ensures a high ability to dilute thick pus. In addition, as shown by our results, ability to dissolve thick pus of this solution can be enhanced by high temperature and high turbulence.

Following this, it was decided to study the rheology solution of 0.9% sodium chloride and peculiarities of its interaction with pus in peritonitis after introduced the gas carbon dioxide excessive pressure of 0.2 ATM and heated solution to a temperature of +42°C. The results showed the unexpected benefit of sanitizing activity aerated solution of 0.9% sodium chloride solution. The fact that this solution formed gas bubbles, provide a visualization of the vector flow of fluid in the peritoneal cavity using ultrasound. In addition, it is shown that the visualization of ultrasound move process of gas bubbles in a moving solution of 0.9% sodium chloride in the abdominal cavity allows you to monitor and change the flow of a fluid due to changes in the location of the patient's torso in space together with a cavity. The point is that changing the location of the torso and abdomen in space allows you to change the direction of fluid in the abdominal cavity, necessary to wash better chosen site. The ultrasound provides visualization move the liquid on the change of movement of gas bubbles [3, 19].

Therefore, the formation of gas bubbles in solutions of sanitizing agents can enhance their ability to dissolve (or destroy) the pus, as the formation of gas bubbles in liquids creates conditions cold "boiling".

The data obtained by us testify that we are in half-step beyond the creation of a completely new group of drugs - solvents of pus. In this regard, we continued the research. The intuition prompted that to complete the formation of the basic physicochemical properties of new drugs is not enough antiseptic with a more reliable and a proven ability cold boiling. The analysis of known funds have allowed us to choose for these purposes, the hydrogen peroxide.

Then a solution of 4% sodium bicarbonate we added hydrogen peroxide in a concentration of 3%, and then heated the solution to a temperature of +42°C. Studies have shown that this solution possesses all the properties necessary for medicinal means - solvents of pus. The results showed that the infusion in a test tube with thick pus equal number of solution of 4% sodium bicarbonate and 3% hydrogen peroxide at a temperature of 42°C leads to the fact that all the pus is dissolved and after 5 minutes, turns into a turbid liquid. Thus the solution additionally has antiseptic and deodorizing activity.

Thus, it was found that medicines having the following physicochemical characteristics: high temperature, high alkaline, high turbulence and high ability to lead liquid in a state of cold boiling, may improve rheology of liquid, viscous and dense biological tissues. Leader of improvement of rheological properties of thick pus is a solution of sodium bicarbonate and hydrogen peroxide, which additionally contains carbon dioxide under excess pressure of 0.2 ATM like food, fizzy drinks.

The mechanism action of drugs of this group on the pus boils down to the fact that under the influence of alkaline drugs activity hydrolysis of proteins and saponification of fats in the festering mass. It dilutes the colloidal intercellular environment in the pus and facilitates the penetration of drugs inside the pus. The medicines liquefy the pus by increasing its temperature. In other words, the hot solutions heated pus, and therefore it melts (melt). And, finally, the drugs have the potential to cause interstitial cold boiling, that is, to form gas bubbles within the purulent masses. This is the reason that drugs make the pus into the foam. In this case, the gas bubbles can occur due to the fact that the gas can be inside solvents of pus under high pressure, or oxygen gas can emerge from the hydrogen peroxide by the enzyme catalase, which is always available in the purulent masses.

Today we can announce that the data has allowed us to develop several new hygienic preparations for the dissolution of thick purulent masses in purulent diseases, such as empyema, peritonitis, rhinitis, sinusitis, conjunctivitis, osteomyelitis, otitis and hearing loss with occlusion of the auditory canal of a sulfuric stopper [2, 3, 4, 5, 6, 8, 14].

Here are the recipes of several solvents of pus.

1. **Means for liquefaction the thick and sticky pus** (RUS Patent No. 2360685):
   - Hydrogen peroxide is 2,7 – 3,3 %
   - Sodium hydrogen carbonate 5,0 – 10,0 %
   - Water for injections The rest

   This means is designed to dissolve very thick and sticky pus and to remove it like a geyser from cavities and tracts.
2. **Hiper fizzy and hyperosmotic antiseptic** (RUS Patent No. 2331441):
   - Hydrogen peroxide is 2,7 – 3,0 %
   - Sodium chloride 0,9 - 10.0 %
   - Carbon dioxide to create excess pressure of 0.2 ATM. at $+8^0$ C
   - Water for injections The rest

   This tool is used to dissolve thick purulent masses and removing them like a geyser from the cavities and sinus tracts.
3. **Powerful solution for epibulbar instillations** (RUS Patent No. 2452478):
   - Hydrogen peroxide 0,55 – 1,0 %
   - Sodium bicarbonate of 1.0 – 1.5 %
   - Lidocaine hydrochloride 0,5 – 1,0 %
   - Water for injections The rest

   The tool is intended for rehabilitation of the organ of vision and skin purulent conjunctivitis, the gluing of the eyelids and the lacrimal accumulation of stones.
4. **Remedy for fistula sanitation in infected pancreonecrosis** (RUS Patent No. 2455010):
   - Sodium chloride 0,9 %
   - Sodium gidrofosfat 0,142 %
   - Sodium dihydrogen phosphate 0,120 %
   - Water for injections The rest

   This tool is intended for repeated and prolonged rehabilitation fistulas in necrotizing pancreatitis.
5. **Means for removal of cerumen** (RUS Patent No. 2468776):
   - Hydrogen peroxide 0.3 - 0.5 %
   - Sodium bicarbonate 1.7 - 2.3 %
   - Water for injections The rest

   This means is designed to dissolve and remove the ear tube of their external auditory canal during hearing loss.

It should be noted that the strongest solvent and the destroyer of pus can be safely applied only for the rehabilitation of open purulent wounds, because the drug is very quickly embedded in pus and rapidly turns it into foam. Pus begins to foam right before your eyes and the volume of foam increases several times. So the strongest solvent of pus cannot be inserted into the deaf purulent cavity.

4. Conclusion

It is shown that the physical-chemical characteristics of the solutions of drugs can have a significant influence on the rheological properties of pus under local interaction with him. Found that aqueous solutions with a high temperature, alkaline, relatively heavy, hyperosmotic activity and hiper aeration capable of dissolving thick purulent masse.

There is every reason to believe that we have developed a new group of medicines, which was called "Solvents of pus". Drugs - solvents of pus differ from all other preparations that are heated to +42° C aqueous solutions 0.5-10% sodium bicarbonate and 0.5-3,0% of hydrogen peroxide, which is potentially ready to cold boiling. The most powerful drug in this group is the carbonated solvent pus, which contains carbon dioxide at a gauge pressure of 0.2 ATM.

References

[1] Dementiev V B et al. 2009 *Chemical Physics and Mesoscopy* ***11 (2)*** 229 – 234
[2] Chereshnev V A et al. 2008 Hyper-gazation and hyper-osmotic antiseptic. Invention RU Patent 2331441 C2
[3] Chereshnev V A et al. 2008 Method of peritoneal dialysis from fizzy solution. Invention RU Patent 2336833 C2
[4] Kozlov A E et al. 2012 Agent For Fistula Sanitation In Infected Pancrenecrosis. Invention RU Patent 2455010 C1
[5] Strelkov N S et al. 2002 The method of treatment of long-term psychological wounds. Invention RU Patent 2187287 C2
[6] Strelkov N S et al. 2007 Method For Treating Pleural Empyema Cases. Invention RU Patent 2308894 C2
[7] Urakov A L 1989 *Science and Life* ***9*** 38-42
[8] Urakov A L et al. 2009 Tool to thin thick and sticky pus. Invention RU Patent 2360685 C2
[9] Urakov A L et al. 2010 Amethod Of Rapid Assessment Of Drugs' Washover Activity. Invention RU Patent 2389013 C2
[10] Urakov A L, Urakova N A, Yushkov B G, Zabokrickiy N A, Hausknecht M Y 2011. *Bullitin of Ural Academy Medical Sciences* ***1 (33)*** 84 – 87
[11] Urakov A L, Urakova N A, Novikov V E, Boiarintseva A V, Gaisina L F 2011 *Recept* ***6 (80)*** 26 – 31
[12] Urakov A L, Novikov V E, Yushkov B G, Urakova N A, Gaysina L F 2011. *Ural Medical Journal* ***6*** 142 – 145.
[13] Urakov A L, Urakova N A, Kozlova T S 2011 *Bullitin of Ural Academy Medical Sciences* ***1 (33)*** 105 – 108
[14] Urakov A L et al. 2012 Method And Means For Removal Of Sulphur Plug. Invention RU Patent 2468776 C2.
[15] Urakov A L, Urakova N A 2012 After injection bruising, infiltrates, necrosis and abscesses can cause medicines because of the lack of control of their physical-chemical aggressiveness. *Modern Problems of Science and Education* ***5*** 5 - 7; URL: www.science-education.ru/105-6812 (Date of access: 14.10.2015)
[16] Urakov A L, Urakova N A 2013 Original hygiene for prevention of postoperative adhesions, efficient liquefaction of thick purulent masses, sulfuric tubes and tear stones. *Modern Problems of Science and Education* ***1*** URL: http://www.science-education.ru/107-7607 (Date of access: 04.12.2013)
[17] Urakov A, Urakova N, Chernova L 2013 *Journal of Materials Science and Engineering* B ***3***

(*11*) 714 – 720
[18] Urakov A, Urakova N, Kasatkin A, Chernova L 2014. *Journal of Chemistry and Chemical Engineering* ***8 (01)*** 61 – 65
[19] Urakov A L, Kasatkin A A, Urakova T V 2014. *Ultrasound and Functional Diagnostics* ***4*** 61 – 65
[20] Urakov A L 2015 *Építőanyag – Journal of Silicate Based and Composite Materials* ***67 (1)*** 2 – 6. http://dx.doi.org/10.14382/epitoanyag-jsbcm.2015.1
[21] Urakov A, Urakova N 2015 *Journal of Physics: Conference Series* ***602*** 012043. http://dx.doi.org/10.1088/1742-6596/602/1/012043

Temperature and blood rheology in fingertips as signs of adaptation to acute hypoxia

A Urakov[1,2], N Urakova[1], A Kasatkin[1], V Dementyev[2]

[1]Department of General and Clinical Pharmacology, Izhevsk State Medical Academy, Izhevsk, Russia
[2]Department of Thermal Deformation Processes, Institute of Mechanics, Ural branch of RAS, Izhevsk, Russia

E-mail: urakoval@live.ru

Abstract. It is found that the absolute values of temperature and color infrared image of the fingers and palms in healthy volunteers and in patients with hemorrhagic shock are in the same range, so they don't represent precisely their condition. It turned out that what really matters is the dynamics of temperature and color infrared image of the palms after cuff occlusion test. In healthy volunteers and in patients with high resistance to shock, there is a rise in temperature and change in color from blue to red in the infrared image of the fingers and palms for 1 - 1.5 minutes after elimination of ischemia, but in patients with low resistance to shock there is a decrease in temperature and expansion of blue color in the palm surface in the infrared image. On the other hand, to assess the resistance to hypoxia in a fetus inside a mother's womb it is proposed to determine the duration of the period of the fetus stationary state during apnea period in pregnant women by using ultrasonography or during period of uterus tonic contractions during childbirth. We found that if fetus has high resistance to hypoxia, the duration of stationary state during the apnea or uterus contraction is greater than 20 seconds, and after exhaustion of reserves for adaptation to hypoxia it approaches zero. It is shown that growing hypoxia causes decrease in the local temperature of the central area of the skull and vice versa.

1. Introduction

It is very difficult to diagnose the beginning of hemorrhagic shock and recovery from it in each patient based on generally accepted ideas about the danger of massive blood loss and the body response to it. Nevertheless, the concept of hemorrhagic shock can be significantly changed today. In this regard, we propose the following amendments:

1. Clinical manifestations of effects of massive blood loss and their dynamics are individual for each patient.
2. The same amount of blood loss results is different response in different patients depending on many internal and external factors.
3. Hemorrhagic shock is a condition that people inherited from warm-blooded animals that developed it in the process of evolution as a defense (adaptive) mechanism aimed at preserving life after blood loss.
4. Massive blood loss and hemorrhagic shock are not the direct causes of biological death in warm-blooded animals and humans.

5. Hypoxic brain injury is the immediate cause of death in all patients regardless the cause of circulatory arrest.
6. Maintaining sufficient reserves for adaptation to hypoxia is a measure for favorable recovery from blood loss and protective role of hemorrhagic shock at any stage of its development.
7. The results obtained from diagnostic test for the patient's reserves for adaptation to hypoxia, can greatly improve the forecast accuracy concerning blood loss danger, hemorrhagic shock diagnosis and successful recovery.

To test predictive validity of finger temperature for survival in adult patients with hemorrhagic shock, we conducted a research to study the dynamics of fingers and hands temperature by means of infrared thermography. At the same time we studied the dynamics of the local temperature of fingers pads in healthy adults during voluntary apnea and cuff occlusion test. Along with this, we studied the dynamics of the rib cage and fins activity in aquarium fish with acute hypoxia and monitored the dynamics of the motor activity of the chest and limbs in adults and fetuses in the mother's womb. We also monitored the local temperature of fetal head during labor using infrared thermography.

As a result, we have developed new functional tests that have been tested in a pilot clinical trial [1-7]. We showed the importance of infrared technology for the detection of ischemia, hypoxia and hypoxic damage to the cerebral cortex in a number of scientific reports and academic articles. In this research we studied the relationship between the duration of ischemia and hypoxia of organs and tissues and their local temperature [8].We were able to find out that the use of infrared thermography and ultrasound improves the diagnostic accuracy. Besides, well-timed application of new functional testing provides information about the patients' resistance to hypoxia and hemorrhagic shock, improves forecast accuracy, facilitates evaluation of the degree of hypoxic damage to the cerebral cortex, and enables the detection of vital signs during clinical death with the possibility of surviving and the treatment effectiveness assessment [9, 10].

2. Materials and methods

We studied blood rheology in 24 adults healthy volunteers,5 blood donors, in 35 patients diagnosed with traumatic hemorrhagic shock. Blood rheology was studied through visual observations in vivo (inside peripheral vessels) by ultrasonic and infrared thermography before and after voluntary apnea, cuff occlusion test and blood donation 400 ml. In the period from 2009 to 2015 we studied the dynamics of local temperature of body surface using infrared thermal images in healthy adult volunteers, pregnant women, women in labor, babies during delivery, newborns immediately after birth, in adult blood donors after 400 ml of blood loss and in adult patients after massive blood loss.

Dynamics of local body temperature in adult volunteers was studied prior to, during and after voluntary apnea lasting from 15 to 90 seconds in 10 healthy volunteers (Group 1); after 2 minutes of provocable ischemia in the right hand (shortened cuff occlusion test) in 14 healthy volunteers (group 2); in 5 blood donors after 400 ml of blood loss (group 3); and in 35 patients diagnosed with traumatic hemorrhagic shock who were treated at the anesthesiology and intensive care department (group 4). Hemorrhagic shock was classified with the use of ATLS system. In order to induce ischemia, a cuff was placed on the arm and inflated to a pressure 30 mmHg above systolic blood pressure; this pressure was held for 2 minutes. Occlusion was performed within 3 hours after the patient's admission to intensive care. The infrared thermal images of the palm and fingers surface were recorded prior to, during and after the occlusion test with a time interval of 30 seconds.

At the same time we studied the dynamics of local temperature in hands in 10 healthy volunteers prior to and 30 minutes after cold test application. The cold test implied that the hand was immersed in melting snow water for 2 minutes. In the first series of observations, study was conducted in sober volunteers. In the second series of observations, study was conducted in the same volunteers, but in 30 minutes after they took 40° ethyl alcohol solution at a dose of 0.4 g/kg body weight (Russian vodka "Stolychnaya") on an empty stomach.

Another series of observations were carried out in pregnant women in their 20-24 and 30-34 weeks of pregnancy in women's clinics. We studied the dynamics of the local temperature in pregnant women fingers during voluntary apnea, which lasted from 15 to 30 seconds. At the same time, we also examined the motor activity and radiation properties of fingertips of the fetuses in mothers' wombs. The dynamics of the women fingers temperature was studied using a thermal imaging camera, while the dynamics of fetal movement and his fingers radiation properties was studied using ultrasound investigation. In the study, 100 pregnant women had normal pregnancy (control group) and the other 25 pregnant women (study group) had signs of placental insufficiency of 1B degree. Later, 35 of 125 women were admitted to maternity hospitals in their 37 - 41 weeks of pregnancy for term labor. The dynamics of the local temperature of the fetal head skin during labor and the newborns body temperature after birth was studied by means of thermal imaging device.

Ultrasound examination of pregnant women and fetuses was performed by means of ALOKA SSD - ALPHA 10, Medison SonoAce-600-C, which is the expert class medical device, and standard convex sensors with the frequency of 3–7 MHz [1,3,6]. After completing the standard ultrasound examination, and taking voluntary informed consent from a pregnant woman, apnea test was applied using M.Yu. Gauskneht method for assessing the fetus resistance to fetal hypoxia [1].

Local skin temperature changes were documented with photographs. Infrared monitoring of hands temperature was performed by using ThermoTracer TH9100XX (NEC, USA) thermal imager. Ambient temperature of the examination room was 24 - 25°C, the temperature window of the thermal camera was set to the range of 25 to 36°C. The obtained data were processed using the Thermography Explorer and Image Processor software. A region of interest was defined over the entire hand using the outline of the hand to separate the hand temperature from the background. From this region of interest, we determined the mean temperature, standard deviation and distribution of isotherms at the thresholds of 34, 32, 30, 29 and 25°C. In addition, spot temperatures of the finger tips were measured. Mean finger temperature was calculated by averaging the finger tips temperature of the 2nd to the 5th finger. The thumb was excluded from temperature measurements as it usually carried a pulsometer.

At the same time, we studied the dynamics of fins color, motor activity and lifetime in aquarium fish, including 150 blue tetras and 150 guppies with acute hypoxia. In addition, their defecation time, water color change, and water clarity were recorded. Acute hypoxia was achieved by placing the fish in water inside a sealed container at a certain temperature without access of air. Before the beginning of the research, its protocol was approved by the Ethics Committee of Izhevsk State Medical Academy, based on the principles set in the Declaration of Helsinki of the World Health Organization. All volunteers signed an informed consent for voluntarily participation in the study.

3. Results

As no similar studies had been carried out before, we had high hopes that the monitoring of the motor activity and the local temperature of the peripheral parts of the body would arm us with new information that would allow to develop methods for evaluating the patients' resistance to the effects of acute massive blood loss, and increase the accuracy of diagnosis and forecasts. To check these assumptions, we conducted initial experiments with aquarium guppies and blue tetras. In order to identify the patterns we simulated acute and potentially fatal hypoxia resulting in aquarium fish death. The research was carried out at different temperatures. In several series of experiments, we investigated the dynamics of fish motor activity continuing until its biological death. We also measured their life duration after cutting off air supply to water in which they were held. At the same time we studied the dynamics of fetal movement inside pregnant women wombs during voluntary apnea. Studies were conducted by means of ultrasound examination in pregnant women in their 20-24 and 30-34 weeks of pregnancy.

The results revealed the perfect analogy between the dynamics of motor activity in fish and fetuses under the conditions of acute hypoxia; they also made it possible to find out diagnostic symptoms of the exhaustion of their reserves for adaptation to hypoxia. In particular, it appears that when cutting off air supply to water with fish and to the respiratory system of pregnant women during their voluntary apnea, chaotic motor activity of fish and fetuses in the mothers' wombs is followed by a period of complete and continuous immobility. Our results showed that if fish and fetuses have sufficient reserves for adaptation to hypoxia, their bodies remain still for a longer period of time than in fish and fetuses with insufficient reserves for adaptation to hypoxia. Besides, we found out that when the reserves for adaptation to acute hypoxia are exhausted, both fish and fetuses suddenly increase their motor activity, which is manifested as follows. Fish spread out its pectoral fins, and its oscillating fins movement becomes more frequent and increased to the maximum, its mouth is widely open, and there is also quickening and increase in ventilating movement of gill arches. Fetuses spread their arms away from the body, unclench their fists, straighten their fingers, and demonstrate sudden chest rise.

In addition, in experiments with fish, we found out that as soon as the reserves for adaptation to hypoxia were depleted, the fins color changed. To explain this phenomenon, we assumed that the color of the fish fins during hypoxia changes because of the changes in its blood color. The matter is that in normal conditions the blood which is delivered to the fins is oxygenated, but under the conditions of lack of oxygen and after the reserves for adaptation to hypoxia are exhausted the blood, which is delivered to fins, lacks oxygen. At the same time, the blood enriched with oxygen is bright red and the blood which lacks oxygen is dark cherry red.The dependence of fish fin color on oxygen level in its blood was demonstrated by experimental results which were obtained while introduction of hydrogen peroxide in water with fish. It showed that the introduction of hydrogen peroxide in water with fish after they have exhausted its reserves for adaptation to hypoxia restores its fins color, prevents heavy motor activity and prolongs fish survival under the conditions of total lack of oxygen in water [67].In addition, we carried out laboratory experiments studying dynamics of color and ultrasound echogenicity of preserved donor blood. It was shown that introduction of hydrogen peroxide solution into donor venous blood enriched it with oxygen and, in fact, turned it into arterial blood. The blood lost its dark cherry color and became bright red. At the same time the ultrasound echogenicity of the blood was increased.

Based on these data, we assumed that a similar change in the color of blood and hands soft tissues might be observed in fetuses with intrauterine hypoxia. In consequence, fetuses' skin color and ultrasound echogenicity of their fingers soft tissue might be changed. Since today there are no safe methods for studying the dynamics of color of fetus fingers skin inside his mother's womb, we decided to investigate fetuses' fingers using ultrasound examination. For this purpose the ultrasound examination was done using conventional transabdominal techniques. Ultrasound echogenicity of fingers was studied during voluntary apnea in pregnant women in the second half of pregnancy and during pushing contractions in women in the last stage of labor. Our results showed that in normal conditions fetuses are highly resistant to intrauterine hypoxia and can easily withstand a period of voluntary apnea in pregnant women and pushing contractions in labor for 20 seconds. At the same time their fingertips preserve stable ultrasound echogenicity. In particular, in pregnancy the fetuses don't move during apnea, their fingers are clenched into fists, and their fingertips preserve stable ultrasound echogenicity.

However, in case of low fetal resistance to intrauterine hypoxia which might be caused by placental insufficiency, fetuses don't remain calm during voluntary apnea period in pregnant women or the pushing contractions period in labor. In particular, in 5 - 10 seconds after the beginning of apnea period in pregnant women or pushing contractions in women in labor, fetuses started to make breathing movements with their chest and hands; they unclenched their fists and straightened their fingers. At the same time fetuses demonstrated a reduction in ultrasound echogenicity in their fingers pads. Moreover, in several seconds after terminating apnea in pregnant women and pushing contractions in labor, ultrasound echogenicity in fetal fingers pads became normal.

These results allowed us to assume that during hypoxia, some fetuses with low resistance to hypoxia might demonstrate the increase in the reduced haemoglobin in blood up to very high levels. In its turn, it might cause acrocyanosis in fetus' fingers and become the reason of the heat emission reduction in his fingertips. Moreover, the temperature drop in fingertips might be not just an effect from hypoxia, but also a way to adapt to it. The matter is that the survival of biological objects under acute hypoxia might be increased by means of hypothermia. The protective effect of cold in hypoxia was proved by experimental results in such cold-blooded animals as fish. It appeared that the life duration in fish after cutting off the air supply to water was more than 2 times longer when the temperature of water was 15°C than when this temperature was 25°C.

Following this, we assumed that local hypothermia may naturally occur in a human fetus after he is expelled out of the mother's womb. When the baby comes out head first, it is his head that might be the first part of the body to be cooled by the surrounding air. To check this assumption, we studied the dynamics of the local temperature of fetal heads after their passing through the birth canals. The results showed that initially the fetal head had the same temperature as his mother's body, regardless of his resistance to hypoxia. But later the temperature of the fetal head surface started to decrease. The longer the fetal head was exposed to the surrounding air at room temperature, the lower the temperature was.At the same time we found out that in normal conditions the head of live fetus was displayed by thermal imager mainly in the yellow, orange and red colors, and an area of local hyperthermia might be observed on the top of the fetal head. The temperature in this area might be 0.5 - 4.0°C more than the temperature of the areas close to it. It is shown that the local hyperthermia area is located above the central suture or frontal fontanel of the skull. If located above the central suture of the skull, the hyperthermia area has oblong shape. In case of complications of pregnancy and labor, and, in particular, when the fetus' resistance to intrauterine hypoxia is extremely low, the head of a live fetus is displayed on the thermal imager screen in the yellow, orange, and red colors as well, but in some fetal heads a local hypothermia area can be observed. In particular, the newborns with severe hypoxic damage, who were born in meconium-stained amniotic fluid, had areas of local hypothermia over central suture or frontal fontanel of the skull. At the same time the area of local hypothermia over central suture had an elongated.

At the same time we studied the dynamics of the local temperature and ultrasound echogenicity in fingers' pads in healthy adult volunteers and pregnant women using special functional tests, and in adult patients of both sexes with conditions including acute massive blood loss. The results showed that the dynamics of radiation properties of fingers' pads reflected the body response (adaptive capacity) to acute ischemia, acute hypoxia and massive blood loss, so it can be used to diagnose the patients' condition. The results also reveal that infrared thermography, which is done by means of a thermal imaging camera, provides a non-invasive monitoring of the fingers' pads local temperature and can be considered as an independent method for diagnosing the condition of trauma patients' with massive blood loss and acute hypoxia.

The results showed that the infrared and ultrasonic monitoring of the radiation properties of fingers' pads in adults provided urgent information about their body response to the apnea and the compression of the blood vessels in the arm. We found out that in 12 - 15 seconds after the cessation of breathing or blood supply to the hands, the local temperature in fingers' pads begins to decrease. After that, if the period of ischemia is prolonged by more than 2 times, ultrasound echogenicity in fingers' pads begins to decrease. In addition, the decrease in local temperature is directly proportional to the duration of ischemia period, and the area of decreased temperature extends proximally.

Consequently, infrared thermography, which is done by using a thermal imaging camera, is more sensitive and much more informative compared to the ultrasound examination of the fingers' pads echogenicity. Since infrared thermography done by using a thermal imaging camera is safer than the ultrasound examination, further studies of the radiation properties of fingers were conducted only by using a thermal imaging camera. We carried out a series of observations of the dynamics of local temperature in the fingers of healthy adult male volunteers during acute ischemia in their arms caused by compression of blood vessels by applying a tourniquet to their upper arms. The results show that after compression of the vessels, local temperature in the fingertips drops very quickly in all fingers except the thumb. The process of local decrease in temperature in fingers starts with the decrease in temperature in their pads. In particular, a decrease in local temperature by more than 0.1°C in the finger pads was observed within 12 seconds after applying the tourniquet in different men.

It is shown that extending the ischemia period results in a decrease in temperature, at the same time the area ofhypothermia increases in each of the 4 fingers, spreading from their tips towards the palm. Hence, the area of local hypothermia is always observed only in the fingertips at the beginning of ischemia period. It is noteworthy that first the decrease in temperature is observed in the pad of the longest finger. Since in most people the middle finger is the longest one, generally it is the middle finger pad that cools down first during ischemia. After that, if the ischemia period is prolonged, the area of local hypothermia can be observed in other fingers' pads. The area of local hypothermia, which is initially localized in the fingertips, first spreads to the distal phalanges, and then (if the ischemia period continues) to the middle phalanges in 4 fingers. The thermal imaging camera displays the finger phalanges in which the temperature was decreased in blue, while phalanges where no temperature decrease was observed are displayed in yellow and orange colors.

It is important that after removing the tourniquet, the fingers' areas with decreased temperature demonstrate temperature rise. The temperature in fingers returns to its initial values after 30 - 60 seconds. Later we could observe a steady increase in local temperature in all fingers and in the palm to the values greater than the initial ones. Moreover, an area of local hyperthermia was observed in each finger in the place of the local hypothermia area. We found out that in healthy men, the period of local temperature rise in fingers and palm, observed in 2 minutes after applying the tourniquet to the upper arm, lasted about 5 minutes after its removal. The same dynamics of local temperature in fingers was observed when applying the tourniquet to the upper arm in pregnant women. Later, a series of observations of the local temperature dynamics in fingers was carried out in healthy adult male volunteers and pregnant women during voluntary breath holding. The results showed that apnea, as well as compression of blood vessels resulted in common dynamics of temperature decrease in fingers.

In particular, a decrease of more than 0.1°C in local temperature of fingerpads was observed in a group of healthy pregnant women in 13 seconds after beginning voluntary apnea, and in the group of healthy male volunteers in 15 seconds after it. Concurrently, the images of 4 fingers' pads (except the thumb pad) in the thermal imaging camera changed their colors from yellow and orange to blue. The other areas of the fingers, as well as the palm and forearm in both men and women remained colored.

After the restoration of breathing, the local temperature in fingers increased in all men and women involved in the study. The temperature in the area of local hypothermia in each finger returned to initial values in 15 - 30 seconds after the restoration of breathing. In the subsequent period of time, the temperature in all the fingers and palm increased to the values greater than the initial ones. The period of local hyperthermia in fingers lasted for about 3 minutes in healthy men and women, after that the temperature in fingers and palms became normal.

These results confirmed our assumption that, all other things being equal, the occurrence of local hypothermia in the distal phalanges of the fingers might be a symptom of not only hypoxia and ischemia, but also of lack of blood in the body, caused by exsanguination, and in particular, by massive blood loss. Now we had to find an explanation for the reason for local temperature decrease in fingers during hypoxia. First we have found the answer to the question of why the compression of blood vessels in the upper arm causes temperature decrease in the fingertips. The matter is that the exposed hands are surrounded by air at a temperature of 25°C. Therefore, a part of the hands heat is continually going into the air. In these circumstances, fingers and palms can only stay warm if they are continuously warmed up. Here, warm blood is a natural heat transport fluid. That is why compression of the blood vessels in the upper arms stops the delivery of warm blood to the hands, and in these conditions they begin to cool down. Along with this, in accordance to the laws of physics, temperature decrease starts from the peripheral areas of the hands.

It was more difficult to answer the question of why the dynamics of temperature decrease in fingers observed as a result of cessation of breathing, was the same as in case of compression of the blood vessels in hands. To answer this question, we referred to the common idea that the cessation of breathing reduces the oxygen level in blood and increases the carbon dioxide concentration. This, in turn, excites specific chemoreceptors in the central nervous system. Excitation of chemoreceptors can reflexively trigger a general protective and adaptive body response aimed at preventing the brain hypoxia. Reflex spasm of blood vessels in the peripheral parts of the body might be a part of complex adaptive changes, as the body redistributes arterial blood to the benefit of heart and brain. Moreover, reducing blood delivery to hands and other peripheral parts of the body might play a protective role, especially for the brain. Exposed hands continuously radiate the heat to the outside, so reducing the blood delivery to them at an air temperature of 25°C results in their cooling. Here, the temperature decrease begins with the fingers' pads. Consequently, the temperature decrease in fingers observed during apnea is a result of a reflex spasm of blood vessels. Thus, local cooling of fingers' pads may be a diagnostic symptom of the organism response to hypoxia, namely, the ability of the body to cause a reflex spasm of blood vessels in the peripheral parts of the body.

In order to check this assumption, we decided to find out whether the reflex spasm of blood vessels plays a key role in the occurrence of local hypothermia in the fingertips. To simulate the reflex spasm of blood vessels in the fingers we used cold test. In order to reduce the reflexive effect of cold on the muscle tone of blood vessels, we used a food product with a general anesthetic effect, namely, "Stolichnaya" vodka (an ethanol solution with 40 % of ethanol). The dynamics of the local temperature in fingers after the cold test was observed in 10 healthy adult male volunteers in normal conditions and in 30 minutes after the intake of ethyl alcohol, at a dose causing mild alcoholic intoxication. The results showed that in normal conditions (in healthy sober men) cold test (immersing hands in cold water at a temperature of +3 - +4 °C for 2 minutes) causes cold spasm of blood vessels, strong painfulness and prolonged local hypothermia in the distal finger phalanges. However, cold test does not cause cold spasm of blood vessels, painfulness and prolonged hypothermia in fingers in the same men under alcoholic intoxication (specifically, in 30 minutes after intake of ethyl alcohol at a dose of 0.4 g/kg on an empty stomach).

After that we studied the dynamics of local temperature in the hands during voluntary apnea and cuff occlusion test in the same volunteers under alcoholic intoxication. The results were really amazing! We found out that alcoholic intoxication postpones, slows down and reduces the local hypothermia in the fingers' pads during apnea and in hands ischemia. Consequently, a spasm of blood vessels in the hands might occur reflexively as part of the body adaptation to hypoxia. Clinically, it can be observed as a decrease in local temperature in the fingers, if they are exposed and surrounded by air at a room temperature. Therefore, monitoring the dynamics of local temperature in the fingers, in the conditions when ambient temperature is below the body temperature, can be used for non-invasive diagnostics of the body response to hypoxia.

At this stage of the research, the study resulted in the following findings. In normal conditions (in healthy and sober volunteers) a sudden stop of lungs ventilation with breathing gas depletes peripheral parts of the body by means of reflex spasm of blood vessels. Avascularization of the peripheral parts of the body reduces the delivery of warm blood and oxygen, which are sent to vital organs instead. Therefore, reflex spasm of blood vessels in the peripheral parts of the body can be considered as a part of the universal protective and adaptive response of the body aimed at preserving arterial blood supply to the brain under the conditions of sudden decrease in oxygen reserves. This response may be observed as local cooling of the fingers. Therefore, local cooling in the fingertips observed during apnea can be considered as a diagnostic symptom of the organism response to acute hypoxia. We assumed further that the protective mechanism that triggered a spasm of blood vessels in the peripheral parts of the body would be universal. The body activates it not only during apnea and ischemia, but also in any other acute critical conditions that might cause hypoxic brain damage. Therefore, acute massive blood loss might trigger this protective response in healthy and sober people.

To check this assumption, we studied the dynamics of the local temperature in the hands of blood donors while collecting venous blood and during 70 minutes after it. The observations were performed during regular donation of 400 mL venous blood in 5 healthy male donors by using a thermal imaging camera. Donors were in the supine position during the entire period of collecting the blood. Blood collection took 10 – 15 minutes. After donating, the donors were kept in the supine position for 15 minutes more. After that, donors stood up and walked or stayed in the sitting or standing position. In other words, their body was held in an upright position.

Our observations showed that the dynamics of the local temperature in the fingertips during the blood collection from the donors was similar to the dynamics of the temperature decrease in fingertips during apnea and cuff occlusion test in healthy and sober volunteers. However, after the blood loss, the fingers' pads of the donors remained cold for a long period of time, while the fingers' pads of the adult volunteers after apnea and ischemia became warm very quickly. In particular, we found that the temperature in all fingers' pads of the right hand in all donors was decreased by more than 0.1°C after taking approximately 100 ml of blood from the vein (the blood volume which is usually taken in 2-4 minutes after the introduction of the injection needle into the cubital vein of the left hand). Then, in all donors, the temperature in each distal phalanx of the right hand was progressively decreasing together with the increase in the volume of blood drawn from the vein of the left hand. At the same time, the area of local hypothermia was increasing in each fingertip. This increased cooling of fingertips and an increase in local hypothermia areas was observed until 400 ml of blood was drawn from veins.

Later, after the bleeding had stopped, the local temperature, shape, location and size of the local hypothermia area in the fingertips of the right hand remained essentially unchanged for over 60-minute observation period. We observed the dynamics of the local temperature and the size of local hypothermia area in the donors' fingers up to the moment when 400 ml of blood was drawn and the bleeding stopped. We analyzed the results and found out that in the final stage of blood collection, local hypothermia in the fingertips increased by 1.0 – 1.5°C, and the size of the hypothermia area in the distal phalanges increased by more than 3 times. However, we continued observing the temperature in the donors' fingers. Within 60 minutes after the loss of 400 ml of venous blood by each donor, we studied the dynamics of local temperature in the fingers of their right hand during and after the application of *shortened cuff occlusion test* (occlusion 2 minutes). The results showed that in 1 hour after the 400 ml blood loss, cuff occlusion test resulted in the same dynamics of local temperature in the fingers, as in normal conditions (i.e. prior to the blood loss) in healthy volunteers.

In particular, we found that in 11 - 12 seconds after we started applying the shortened cuff occlusion test, further drop of more than 0.1 °C was observed in the decreased temperature in the fingers' pads in each donor. At the same time the local hypothermia area in the distal phalanges of the fingers extended. After the cuff occlusion test the dynamics of the temperature in fingers' pads was diverse: immediately after the cuff occlusion test the temperature rose, and within 1 minute became greater than its initial value (before the cuff occlusion test), then it remained at this level for about 3 - 5 minutes, and then finally returned to its initial value. Thus, an acute blood loss of more than 100 ml triggers an organism's protective and adaptive response aimed at preserving arterial blood supply to the brain in sober and healthy adult subjects. The fingers are among the first body parts to respond to this signal that triggers a reflex spasm of blood vessels, reduces warm arterial blood delivery and basically causes ischemia and hypothermia of fingers. At a room temperature, reduced delivery of warm blood to the exposed fingers is observed as common dynamics of the decrease in the local temperature in fingers' pads and distal phalanges. At the same time a local hypothermia area is formed in each finger pad, and after the blood loss it is preserved for more than 1 hour. After observing the donors we found that acute blood loss changed the dynamics of the heat radiation in the fingers' pads; and infrared monitoring of local temperature performed by using thermal imaging camera provided this information in real time. But that was not our most important discovery. What is more important is that the use of thermal imaging camera and shortened cuff occlusion test allows us to detect the body's additional reserves for adaptation to acute hypoxia! Using a thermal imaging camera made it easy to reveal them and assess the body's resistance to hypoxia!

It turned out to be quite simple in reality: after acute but harmless blood loss the availability of unused reserves for adaptation to acute hypoxia is manifested by the following changes in the local temperature in the hands: at the end of the shortened cuff occlusion test we can observe the further drop of the temperature in the cooled fingers' pads and the increase in the local hypothermia area in the distal phalanges. On the contrary, after the cessation of ischemia, the temperature in the fingertips increases, and the size of the local hypothermia areas in the fingers' distal phalanges decreases.

After that, we had to detect the lack of reserves for adaptation to hypoxia in trauma patients with acute massive blood loss.

To complete this task, we observed the dynamics of the local temperature in the hands in 35 patients admitted to the hospital with traumatic hemorrhagic shock (average patients' age was 46 ± 11 years). According to ATLS, 10 patients' condition was classified as II class haemorrhage, 22 patients' condition was classified as III class haemorrhage, and the condition of 3patients was classified as IV class haemorrhage. All patients were provided with medical treatment in accordance with the recommendations. Along with this, surgical haemostasis was achieved in all patients. The results of infrared monitoring of the dynamics of the local temperature in fingers showed that at the admission, the temperature of distal phalanges of the fingers was 3 - 6 °C less than the temperature of proximal phalanges and the palm in all 35 patients. Local temperature of the fingertips was in the range from 25°C to 29°C. In this case, the absolute values of the local temperature of the patients' finger pads had no prognostic value, because such temperature values were occasionally observed in healthy people without blood loss as well. However, the values of the local temperature in the fingers' pads observed in these same patients when using shortened cuff occlusion test were quite different.

The results of the observations showed that in 8 patients out of 10 with II class haemorrhage and in 1 patient out of 25 with III class haemorrhage the use of shortened cuff occlusion test caused diverse changes in the fingertips local temperature, which were typical for healthy subjects and blood donors (Figure 1). Therefore they were included in the group of patients who did not completely use their reserves for adaptation to acute hypoxia. That is why, all other things being equal, we predicted a higher probability of their recovery.

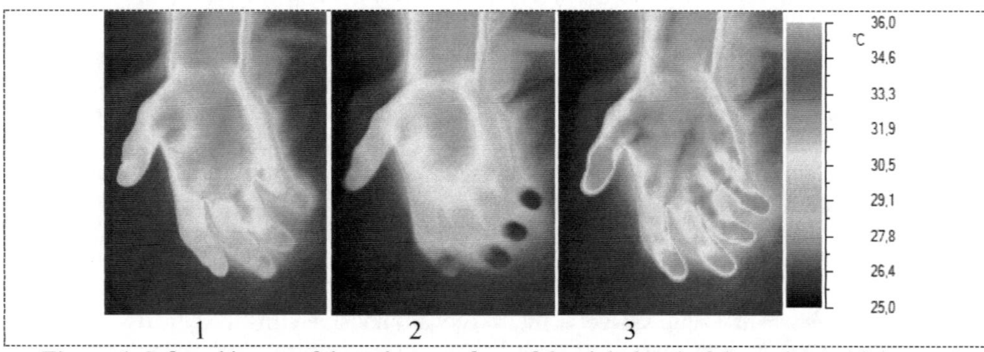

Figure 1. Infrared image of the palmar surface of the right hand of the patient with hemorrhagic shock (L. Male, 28 years) at admission to the hospital (surviving patient):
1 - at the time of hospitalization before the shortened cuff occlusion test,
2 – immediately after the 2-minute shortened cuff occlusion test,
3 – 60 seconds after 2-minute shortened cuff occlusion test

In other patients the use of shortened cuff occlusion test did not cause diverse changes in the local temperature in the fingers' pads. Based on these data, we concluded that the body of these patients completely depleted its reserves for adaptation to acute hypoxia. That is why, all other things being equal, we predicted a low probability of their recovery. The results showed that after receiving treatment in the hospital, all 9 patients in whom the use of shortened cuff occlusion test resulted in diverse changes in local temperature in fingertips, survived. On the other hand, from 26 patients who had no evidence of adaptive response in their fingers when using shortened cuff occlusion test, only 22 patients survived. Thus, four patients died within 48 - 72 hours after admission to the hospital.

4. Conclusion

By using thermal imaging camera we found out that acute blood loss, as well as the voluntary apnea and cuff occlusion test, result in local hypothermia in the exposed fingers' pads. At the same time, the increase in the volume of lost blood, as well as an increase in the duration of apnea and hand vessels occlusion, increases hypothermia and local hypothermia area in fingers, which is consistently expanding first to the entire distal phalanx, and then to the middle phalanx of each finger. Therefore, growing local hypothermia and expansion of its area in fingers, detected by a thermal imaging camera, are suggested to use as a diagnostic symptom for increased hypoxia, all other conditions being equal.

It was found that local hypothermia was observed in blood donors' finger pads after a 400 ml blood loss, and it was preserved for more than 1 hour later. The use of shortened cuff occlusion test after collecting blood from donors first facilitates hypothermia, and then results in postocclusive hyperthermia in their fingers. The similar diverse changes in the local temperature in fingers are observed in healthy people after breathing restoration in 0.5 - 2 minutes after apnea, and after recovery of blood supply to hands when using shortened cuff occlusion test. At the same time, we have shown that in patients with acute massive IV class haemorrhage (30 - 40% of blood loss) the use of shortened cuff occlusion test does not change the dynamics of progressively growing and expanding local cooling in fingers. Therefore, diverse changes in local temperature in exposed fingers detected by a thermal imaging camera after the use of shortened cuff occlusion test (i.e. decrease in the local temperature at the beginning and its increase later) are suggested to be considered as a diagnostic symptom of availability of reserves for adaptation to hypoxia. This diverse change in local temperature in finger pads in a patient after acute massive blood loss allows us to conclude that he has sufficient reserves for adaptation to acute hypoxia and predict high probability of his survival.

References

[1] Urakov A L et al. 2011. Method for assessment of foetus resistance to hypoxia by MY Hausnekht. Invention RU Patent 2432118.

[2] Kasatkin AA et al. 2013. Method of determining microcirculatory injuries in case of shock and efficiency of anti-shock treatment. Invention RU Patent 2480183.

[3] Urakov A L et al. 2014. NA Urakova's method for antenatal assessment of foetal adaptation to repeated hypoxia. Invention RU Patent 2529377.

[4] Urakov A L et al.2014. Method for assessing body compensatory response to acute hypoxia. Invention RU Patent 2531924.

[5] Urakov A L et al. 2011. Method of determining stage of hypoxic injury and probability of reanimation by AL Urakov. Invention RU Patent 2422090.

[6] Urakov A L et al.2014. Method for assessing foetus resistance to obstetric hypoxia Invention RU Patent 2511084.

[7] Radzinskij VE et al. 2014. Method of protecting fetus from hypoxic damage in labour Invention RU Patent 2503414.

[8] Urakov A L 2015The change of physical-chemical factors of the local interaction with the human body as the basis for the creation of materials with new properties. *Epitőanyag – Journal of Silicate Based and Composite Materials* ***67 (1)***2–6.

[9] Urakov A, Urakova N. 2015Rheology and physical-chemical characteristics of the solutions of the medicines. Journal of Physics: Conference Series ***602*** 012043. http://dx.doi.org/10.1088/1742-6596/602/1/012043

[10] Urakov A L et al.2014 Infrared thermographic investigation of fingers and palms during and after application of cuff occlusion test in patients with hemorrhagic shock. *Thermology international*. ***24 (1)*** 5-10.

Measurement of whole blood of different mammalian species in the oscillating shear field: influence of erythrocyte aggregation

U Windberger[1], Ch Pöschl[1], S Peters[1], J Huber[2], R van den Hoven[3]

[1]Department Biomedical Research, Medical University of Vienna, Austria
[2]University Clinic for Ruminants, Kremesberg, Veterinary University Vienna, Austria
[3]Clinical Unit of Equine Internal Medicine, Veterinary University Vienna, Austria

Email: ursula.windberger@meduniwien.ac.at

Abstract. This is the first systematic analysis of mammalian blood of species with a high (horse), medium (man), and low (sheep) erythrocyte (RBC) aggregability by small amplitude oscillation technique. Amplitude and frequency sweep tests (linear viscoelastic mode) were performed with blood from healthy adult volunteers, horses, and sheep in CSS-mode. Blood samples were hematocrit (HCT) adjusted (40%, 50%, 60%) and tested at 7°C, 22°C, and 37°C. Generally, storage modulus (G′) increased with HCT and decreased with temperature in each species, but the gradient of this increase was species-specific. The lower dependency of G′ on the equine HCT value could be a benefit during physical performance when high numbers of RBCs are released from the spleen. In sheep, an HCT-threshold had to be overcome before the desired quasi-static condition of the blood sample could be achieved, suggesting that the contact between RBCs, and between RBCs and plasma molecules must be very low. The frequencies for tests under linear viscoelastic condition were in a narrow range around the physiologic heart rate of the species. In horse, time-dependent influences concurred at frequencies lower than 3 rad.s^{-1}probably due to sedimentation of RBC aggregates. In conclusion, blood is a fragile suspension that shows its best stability around the resting heart rate of the species.

1. Introduction

The rheology of blood is influenced by the quantitative and qualitative properties of blood cells, by the constitution of blood plasma, and by the attracting forces between blood cells and blood plasma. Compared to particle systems or dispersions, the viscosity of whole blood (WBV) is low, even for high cellular fraction such as 40- 45%, which is the physiological hematocrit (HCT) of man. This is the result of specific red blood cell (RBC) properties.

RBCs have a high surface-to-volume ratio that allows viscous deformation in narrow vessels (1). Exposure to shear forces causes a deformation from the relaxed biconcave towards folded shapes (2-4).The RBCs return to their biconcave resting shape after disappearance of the shearing force, and even exhibit a shape memory effect (5). The RBC membrane is not static, but an active ATP-driven structure. Not only the typical shape, but also dynamic fluctuations of the cell membrane at physiological body temperature are increased in relation to metabolic and thermal energy (6). Furthermore, RBC membranes use a tank thread motion to absorb hydrodynamic stress - a process that facilitates laminar blood flow and suspension stability within the blood vessel. Tank thread frequency starts at a certain shear rate threshold (7, 8), which is dependent on the viscosity of the suspending medium (9, 10). Another dynamic feature of RBC membranes is that they enable the cells to sustain a rolling motion in shear flow, which avoids energetically costly deformation (8). It should be kept in mind that the volume concentration of RBCs changes dynamically throughout the body with the lowest HCT values being present in the microvasculature. As a result, the viscous resistance of blood flowing through narrow capillaries is only 20% higher than that of pure plasma (11).

RBCs form aggregates, called as Rouleaux, at low volume flow (12-15).These aggregates break up when shear rates increase at higher flows. There is remarkable diversity in RBC aggregability among the mammalian species (16). For instance, RBC aggregation is too low to be measured easily in cow, sheep, goat, mouse, and rat blood. In contrast, in horse and other equidae, the physiological RBC aggregation is as high as it would be in inflammatory disease in man (17, 18). RBC aggregation affects venous vascular resistance (19), and clinical cases associated with high RBC aggregation indicate specific diseases (20). The impact of RBC aggregability for in-vivo blood flow is discussed controversially. Although there is evidence that high RBC aggregation blunts the parabolic blood velocity profile (21), moderate RBC aggregation rather seem to promote blood flow (22, 23), at least in comparison to non-aggregating RBC suspensions. It is generally agreed that in-vivo viscosity rises when intravascular shear forces are low enough to allow RBC aggregation.By considering blood flow as Poiseuille flow (24), the lowest wall shear rates are calculated for the postcapillary venules. RBC aggregates have been seen with high-speed video microscopy in the venules after passing the postcapillary region (25). The physiological relevance of RBC aggregation lies in the phase separation phenomenon of composite fluids in narrow tubes reaching a critical diameter. Due to the axial migration of RBCs in the tube flow (26), a marginal cell-free layer (CFL) is formed, whereby the magnitude of the width of this layer (CFLW) is linked to RBC aggregability. This CFL reduces the endothelial shear stress, and subsequently, an intrinsic response mediated by vasoactive factors starts to adjust the vascular diameter (27, 28). This in turn modulates the distribution of RBCs in subsequent vessels (29-31).

Rheological properties of blood are usually measured by viscosityas a function of shear rate. Methods to test properties of singular RBCs were recently summarized (32). Blood was characterized as a shear thinning viscoelastic fluid, showing different degree of thixotropy in relation to RBC aggregation. Typically, at low shear rates, RBC aggregates and clusters, while at high shear rates singularly suspended RBCs contribute to plasma enhancement (33).It must be noted that low shear viscosity reflects the specific texture of blood at low shear which is given by the sum of all attracting forces in the respective sample, rather than showing RBC aggregation alone.

In contrast to rotational flow, small amplitude oscillating flow (SAOS) can be used to study fluids under quasi-static conditions. An application of this method to human blood has been described recently (34). The rheological material characterization by SAOS tests the association of components within a sample and includes the forces between the filler materials plus the forces between filler and matrix. Applying this principle to blood, a specific "texture" of blood will result from the forces between RBCs, and from the quality of the contact between RBCs and plasma molecules.

In the present study we tested human, equine and ovine blood at small amplitude oscillatory shear flow. As expected from the differences in RBC aggregability among these species, we observed that shear moduli varied between these three species, being highest in the horse at physiological HCT. In fact, it was difficult to measure ovine shear moduli, which likely is the result of the minute RBC aggregability of this species. It is worth noting that SAOS had to be performed at slightly lower shear stresses than those present in the vessels under physiologic circumstances. However, vascular shear rates are not fixed, but vary dynamically in the body. A decrease in blood flow either generalized due to blood loss or due to vascular dysfunction such as in septicaemia, or more localized due to vascular occlusion following intima hyperplasia or thromboembolic events can easily lower the wall shear rates to those shear forces that we set in our experimental protocol. Since blood is a viscoelastic fluid, storage modulus (G') values were always below loss modulus (G'')-values. A percolated structure in normal plasma could only be achieved at very high unphysiologic HCT (higher than 75%). In this circumstance, a jellylike texture was the result.

2. Material and Methods

2.1. Blood samples

Whole blood of healthy man, horse, and sheep were used for this study. 4 human volunteers (2f, 2m, age: 22-35 y), 4 *Warmblut* horses (2mares, 2geldings, age: 13-26 y), and 10 female *Milchschaf* sheep (age: 1.2-3.6 y) were used. 90 mL blood was withdrawn into EDTA tubes from each individual by venous puncture (man: V. radialis, horse and sheep: V. jugularis) with a 16 G needle connected to a vacutainer system. Samples were centrifuged at 2000 rpm for 10 minutes and blood plasma was separated. New samples were reconstituted out of RBC concentrate and autologous plasma to generated whole blood samples with HCT values of 30, 40, 50, and 60%. The samples were carried in insulated bags to the laboratory and kept in the fridge prior to their measurement. All measurements were finished within 8 hours following withdrawal.

The procedure was approved by the Ethics Committee of the Medical University Vienna (1892/2013) and the Austrian Federal Ministry of Science and Research (animal license number: GZ-1744/115-97/98).

2.2. Rheological protocol

SAOS measurements were performed using the stress controlled PhysicaMCR301 rheometer (Anton Paar, Graz, Austria). 3.5 mL of each blood sample was filled into the stainless steel double gap cylinder system and was analyzed at three different temperatures (7, 22, 37°C) starting with the lowest temperature. Isothermal amplitude and frequency sweep tests were performed.

Strain dependency of blood samples was measured at fixed frequency (10 rad s^{-1}) to compare the yield stresses between the species. In order to carry out the subsequent frequency tests in linear regime, some of the human and horse blood samples were tested at frequencies between 1 and 20rad s^{-1}. After a pre-shear interval (man and sheep: 30 s rotation at 300 s^{-1} followed by a 20 s interval at 1 s^{-1}; horse: 30 s oscillation at 0.01 Pa and 10 rad s^{-1}), increasing shear stresses of 0.001 – 10 Pa were applied by a logarithmic shear stress ramp. Yield points were estimated by the Rheoplus software (version 4.2, Anton Paar, Graz, Austria).

Frequency sweep tests were performed at 0.01 Pa throughout the whole frequency range (20 - 1 rad s^{-1}). The blood samples were subjected to frequency sweep from high to low frequency only.

Descriptive statistic was performed by IBM®SPSS®Statistics (version 22).

3. Results

3.1. Amplitude sweep tests

Allowed frequencies for the frequency sweep tests were within the narrow frequency range between 3 and 15 rad s^{-1}. These frequencies correspond to the range of the physiologic heart rate of the species at rest (0.5 - 2.5 Hz).

The LVE-range could be easily defined by the software at 7°C in man and horse (as well as in sheep at high HCT: 60%), but became imprecise when the temperature increased. Especially at 37°C, the G′-values were sloping – although continuously decreasing. Occasionally, LVE at 37°C had to be determined based on G′′-values that always displayed a plateau until a certain shear stress was reached. As expected, experiments became more reproducible when HCT was high and temperature was low.

Basically, yield points and G′-values at LVE were very low in blood although traceable. G′-values increased with hematocrit and decreased with temperature. Data are provided in table 1 and figure 1. In blood samples at 40% HCT and 37°C, yield points are not provided due to increased sloping of G′, although G′′ showed a plateau value. Yield points appeared to be slightly higher in horse compared to

IOP Conf. Series: Journal of Physics: Conf. Series **790** (2017) 012035 doi:10.1088/1742-6596/790/1/012035

man. In sheep, yield points could be determined at 60% HCT only, however, the obtained values were the highest among the three species.

3.2. *Frequency sweep tests*

G′-values increased with the HCT and decreased with the temperature (see table 2 and figure 2). At 40% HCT and 37°C the interquartile distance became larger indicating greater variance among the samples. There was no G′-G′′-crossover within the allowed frequencies. Sheep with the lowest RBC aggregation showed the lowest G′-values. It is interesting to note that G′ was lower in man than in horse at 40% HCT, but higher in man than in horse at 60% HCT at each temperature. At 50% HCT the median values of man and horse were nearly identical.

Table 1. Yield point of human, equine and ovine blood. Data are presented as median values and quartiles in parentheses

$\tau_y$at 7°C in Pa	**40% HCT**	**50% HCT**	**60% HCT**
Man	0.021(0.008/0.022)	0.033(0.0127/0.046)	0.046(0.045/0.046)
Horse	0.033(0.020/0.086)	0.021(0.020/0.078)	0.072(0.013/0.185)
Sheep	-	-	0.464(0.348/0.465)
$\tau_y$at 22°C in Pa	40% HCT	50% HCT	60% HCT
Man	0.003(0.001/0.008)	0.021(0.010/0.021)	0.021(0.012/0.039)
Horse	0.027(0.009/0.046)	0.020(0.010/0)	0.060(0.020/0.185)
Sheep	-	-	0.214(0.213/0.215)
$\tau_y$at 37°C in Pa	40% HCT	50% HCT	60% HCT
Man	-	0.009(0.004/0.009)	0.009(0.004/0.018)
Horse	-	0.010(0.009/0.018)	0.054(0.004/0.189)
Sheep	-	-	0.097(0.031/0.099)

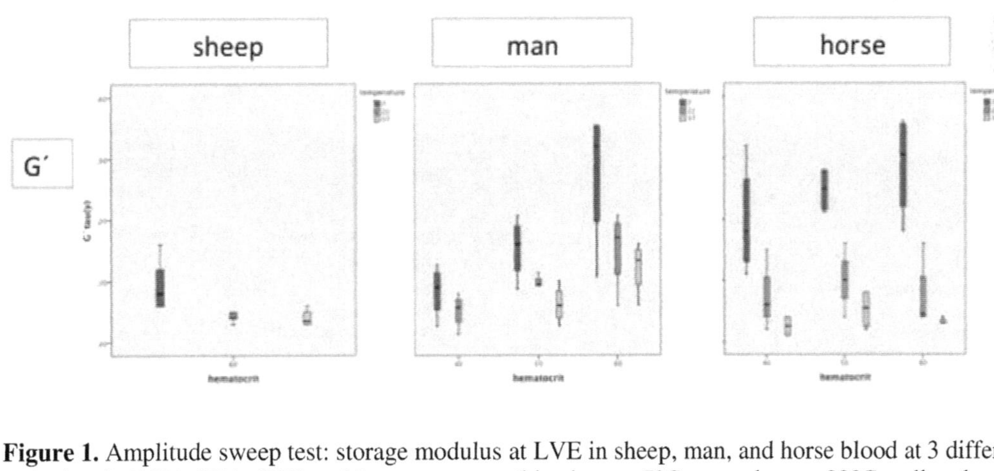

Figure 1. Amplitude sweep test: storage modulus at LVE in sheep, man, and horse blood at 3 different HCT levels (40%, 50%, 60%) and 3 temperatures (blue boxes: 7°C, green boxes: 22°C, yellow boxes: 37°C). In sheep, tests could be performed at 60% HCT only.

There were only minute changes of the tanδ-value with HCT and temperature, although the loss factor decreased with the HCT and increased with the temperature in each sample. Tanδ-values at 9.38 rad s^{-1} ranged between 2.30 (2.11/2.88) (measured in human blood at 60% HCT and 7°C) and 4.17(4.11/4.46) (measured in human blood at 40% HCT and 37°C), between 1.80 (1.58/1.82) and 1.90

(1.24/2.72) in horse, and up to 4.56 (2.98/5.38) in sheep, verifying blood as viscoelastic fluid. Data are provided in figure 3.

Table 2. G′ at 9.38 rad s^{-1} in human, equine and ovine blood samples. Data are expressed as median and quartiles in parentheses

$G'_{9rad/s}$ at 7°C in Pa	40% HCT	50% HCT	60% HCT
Man	111(92/131)	262(253/289)	441(410/534)
Horse	176(145/221)	260(238/281)	340(311/415)
Sheep	-	-	130(101/258)
$G'_{9rad/s}$ at 22°C in Pa	40% HCT	50% HCT	60% HCT
Man	51(35/67)	132(116/151)	259(225/289)
Horse	92(70/109)	132(130/139)	179(161/216)
Sheep	-	-	86(69/105)
$G'_{9rad/s}$ at 37°C in Pa	40% HCT	50% HCT	60% HCT
Man	31(19/35)	77(69/92)	171(159/181)
Horse	56(31/81)	77(66/80)	119(114/135)
Sheep	-	-	55(50/63)

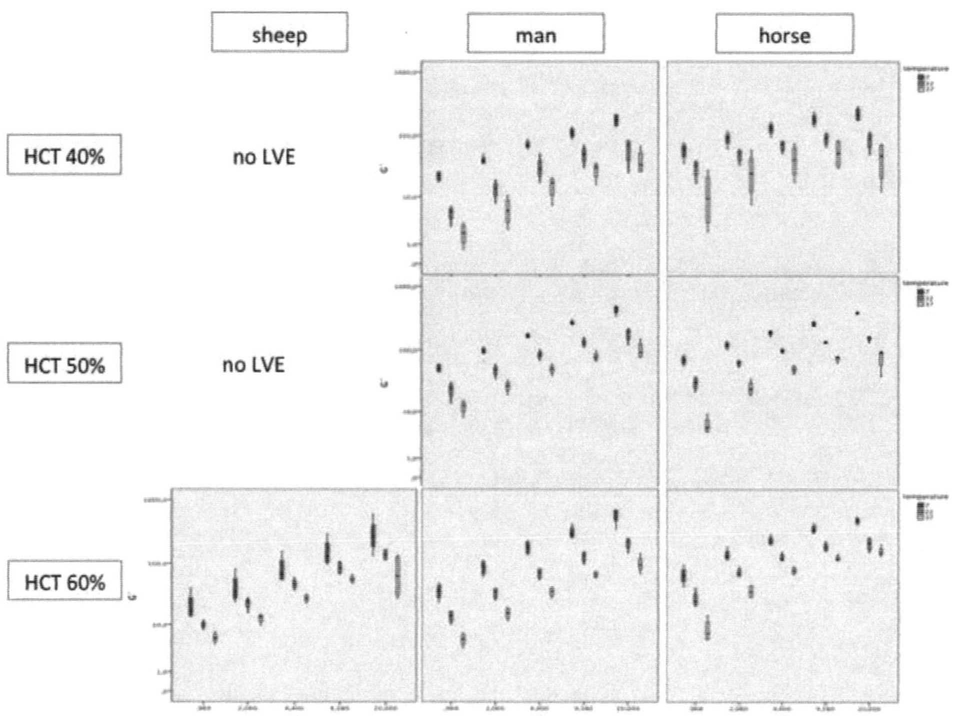

Figure 2. Frequency sweep test: storage modulus of sheep, man, and horse blood at 3 different HCT levels (40%, 50%, 60%) and 3 temperatures (blue boxes: 7°C, green boxes: 22°C, yellow boxes: 37°C).

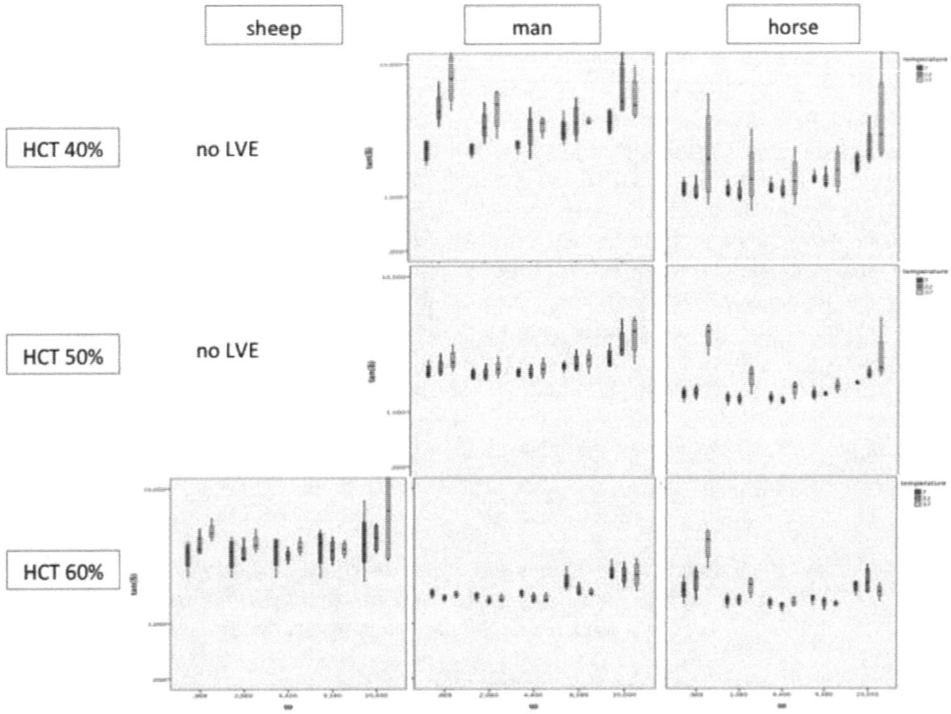

Figure 3. Frequency sweep test: loss factor (tanδ) of sheep, man, and horse blood at 3 different HCT levels (40%, 50%, 60%) and 3 temperatures (blue boxes: 7°C, green boxes: 22°C, yellow boxes: 37°C).

4. Discussion

In the present investigation we showed that whole blood can be measured by small amplitude oscillation. To study the influence of RBC properties on our protocol, we used three species: the horse representing the species with highest RBC aggregability of all, followed by man with intermediate, and sheep with nearly immeasurable RBC aggregability (16). Using autologous plasma, the samples were adjusted to several predetermined HCT values and tested at different temperatures. By this approach, we demonstrated that the shear moduli increased with the HCT and decreased with the temperature in each species, but the gradient of this increase was species-specific.

At a frequency analog to the physiologic heart rate, G′ was higher in horse blood than in human blood at the physiological HCT (40%), but lower in horse compared to man when the HCT was raised (60%). This implicates that the gradient of the G′-increase in relation to the cellular fraction in blood was lower in the horse. This could be of physiological relevance in this species. Resting horses possess a splenic reservoir for RBCs with a HCT of about80%. After an exercise-induced contraction these splenic RBCs are added to the RBCs in the circulation thereby increasing HCT to maximally 65% (35). The extra number of circulating RBCs effectively increases the blood oxygen store and the oxygen carrying capacity. However, in parallel, the blood viscosity and the elastic proportion of the blood suspension is elevated, as well. On the other hand, the low dependency of G′ on the HCT value could be a clear benefit during exercise.

In sheep, an HCT-threshold must be overcome before the desired quasi-static condition of the blood sample is achieved. Ovine RBCs are smaller than those of man and horse (36) and have a reduced deformability if exposed to shear stress (37). Based on their minute RBC aggregability (16), RBCs are thus singularly suspended in the plasma volume. Flow curves show the low shear thinning of ovine blood. Our SAOS results notice that apart from the attracting forces between the RBCs, also the forces between RBCs and the surrounding plasma must be very low.

Another interesting finding of the sheep blood concerns the yield point that could not be measured at HCT of 40% and 50%. However, at HCT of 60% the yield point is higher than in man and horse. The physiological HCT of ovine blood ranges between 30 and 38% (36). A release of splenic RBCs by a sympathetic stimulus is insignificant in sheep due to the diverse composition and function of the spleen. Sheep do not exhibit high HCT values regularly. But if they do so, the structural strength of blood that is reflected by the yield point seems to be higher than in man and horse.

In conclusion, blood is a fragile suspension that shows its best stability around the resting heart rate of the particular species. At frequencies above 15 rad $s^{-1}$and below 3 rad s^{-1}, G′-values are significantly sloping, show an irregular sequence, or are even absent. We assume that the weak structure of the bulk is disrupted at high frequencies, while at low frequencies, sedimentation of RBCs in the rheometer gap resulted in phase separation during the measurement. It is logical that the suspension stability increased with the increase of the volume fraction of cellular elements. Concerning the relevance of RBC aggregability for the suspension stability, it was shown that G′-values are higher when RBCs had the chance to form aggregates, at least at physiological circumstances.

References

[1] Waugh RE and Hochmuth RM *Mechanics and deformability of hematocytes* In: *The Biomedical Engineering Handbook* ed Bronzino JD (Boca Raton, FL: CRC) pp 474-486

[2] Branemark PI and Bagge U 1977 Intravascular rheology of erythrocytes in man *Blood Cells***3**11-24

[3] Fedosov DA, Peltomäki M, Gompper G 2014 Deformation and dynamics of red blood cells in flow through cylindrical microchannels *Soft Matter* **10** 4258-4267

[4] Gaehtgens P, Schmidt-Schönbein H 1982 Mechanism of dynamic flow adaption of mammalian erythrocytes *Naturwissenschaften* **69** 294-296

[5] Fischer TM 2004 Shape memory of human red blood cells *Biophysical Journal* **86** 3304-3313

[6] Park YK, Best CA, Auth T, Gov NS, Safran SA, Popescu G, Suresh S, Feld MS 2010 Metabolic remodeling of the human red blood cell membrane. *PNAS* **107** 1289-1294

[7] Basu H, Dharmadhikari AK, Dharmadhikari JA, Sharma S, Mathur D 2011 Tank threading of optically trapped red blood cells in shear flow *Biophysical Journal* **101** 1604-1612

[8] Dupire J, Socol M, Viallat A 2012 Full dynamics of a red blood cell in shear flow. *PNAS* **109** 20808-20813

[9] Fischer TM 2007 Tank thread frequency of the red cell membrane: dependence on the viscosity of the suspending medium *Biophysical Journal* **93** 2553-2561

[10] Tran-Son-Tray R 2006 A study of the tank thread motion of red blood cells in shear flow. PhD thesis. Washington University, St. Louis, MO, USA

[11] P Gaehtgens 1981 Microrheology of blood in capillaries *Drug Research* **31** 1995-1998

[12] Barshtein G Wajnblum D Yedgar S 2000 Kinetics of linear Rouleaux formation studied by visual monitoring of red cell dynamic organisation *Biophysical Journal* **78** 2470-2474

[13] Brust M, Aouane O, Thiebaud M, Flormann, Verdier C, Kaestner L, Laschke MW, Selmi H, Benyoussef A, Podgorski T, Coupier G, Misbah C, Wagner C 2014 The plasma protein fibrinogen stabilizes clusters of red blood cells in microcapillary flows *Scientific Reports* **4** 4348; DOI: 10.1038/srep04348

[14] Cokelet GR, Meiselman HJ 2007 *Macro- and microrheological properties of blood* In: *Handbook of Hemorheology and Hemodynamics* eds Baskurt OK, Hardeman MR, Rampling MW, Meiselman HJ (Amsterdam, Berlin, Oxford, Tokyo, Washington, DC: IOS Press) pp.45-71

[15] Johnson PC, Bishop JJ, Popel S, Intaglietta M 1999 Effects of red cell aggregation on the venous microcirculation *Biorheology* **36** 457-460

[16] Windberger U, Bartholovitsch A, Plasenzotti R, Korak HJ, Heinze G 2003 Whole blood viscosity, plasma viscosity and erythrocyte aggregation in nine mammalian species: reference values and comparison of data. *Experimental Physiology* **88.3** 431–440

[17] Windberger U, Baskurt OK 2007 *Comparative hemorheology* In: *Handbook of Hemorheology and Hemodynamics* eds Baskurt OK, Hardeman MR, Rampling MW, Meiselman HJ (Amsterdam, Berlin, Oxford, Tokyo, Washington, DC: IOS Press) pp 267-285

[18] Lowe GDO 1998 *Clinical Blood Rheology* 1-2 (Boca Raton, FL: CRC Press)

[19] Cabel M, Meiselman HJ, Popel AS, Johnson PC 1997 Contribution of red blood cell aggregation to venous vascular resistance in skeletal muscle *Am J Physiol* **272** H1020-H1032

[20] Toth K, Kesmarky G, Alexy T 2007 *Clinical significance of hemorheologic alterations* In: *Handbook of Hemorheology and Hemodynamics* eds Baskurt OK, Hardeman MR, Rampling MW, Meiselman HJ (Amsterdam, Berlin, Oxford, Tokyo, Washington, DC: IOS Press) pp 392-432

[21] Bishop JJ, Nance PR, Popel AS, Intaglietta M, Johnson PC 2001 Effect of erythrocyte aggregation on velocity profiles in venules *Am J Physiol Heart Circ Physiol* **280** H222-H236

[22] Soutani M, Suzuki Y, Tateishi N, Maeda N 1995 Quantificative evaluation of flow dynamics of erythrocytes in microvessels: influence of erythrocyte aggregation. *Am J Physiol Heart Circ Physiol* **268** H1959-H1965

[23] Charansonney O, Mouren S, Dufaux J, Duvelleroy M, Vicaut E 1993 Red blood cell aggregation and blood viscosity in an isolated heart preparation *Biorheology* **30** 75-84

[24] Pantos J, Efstathopoulos E, Katritsis G 2007 Vascular wall shear stress in clinical practice *Current Vasc Pharmacol* **5** 113-119

[25] Kim S, Popel AS, Intaglietta M, Johnson PC 2005 Aggregate formation of erythrocytes in postcapillary venules *Am J Physiol Heart Circ Physiol* **288** 584-590

[26] Bishop JJ, Popel AS, Intaglietta M, Johnson PC 2001 Effects of erythrocyte aggregation and venous network geometry on red blood cell axial migration. *Am J Physiol Heart Circ Physiol* **281** H939-H950

[27] Moncada S, Palmer RM and Higgs EA 1981 Nitric oxide: physiology, pathophysiology and pharmacology of nitric oxide *Pharmacol Reviews* **43** 109–142

[28] Malek K and Izumo S 1992 Physiological fluid shear stress causes down-regulation of endothelin-1 mRNA in bovine aortic endothelium *Am J Physiol* **263** C389–C396

[29] Ong PK, Jain S and Kim S 2012 Spatio-temporal variations in cell-free layer formation near bifurcations of small arteries *Microvasc Res* **83** 118-125

[30] Hightower CM, Vazquez BY, Park SW, et al. 2010 Integration of cardiovascular regulation by the blood/endothelium cell free layer *Wiley Interdisciplinary Reviews: Systems Biology and Medicine* **3** 458-470

[31] Yalcin O, Aydin F, Ulker P, et al. 2006 Effects of red blood cell aggregation on myocardial haematocrit gradient using two approaches to increase aggregation. *Am J Physiol Heart Circ Physiol* **290** H765-H771

[32] G Tomaliuolo 2014 Biomechanical properties of red blood cells in health and disease towards microfluidics *Biomicrofluidics* **8** 051501

[33] Baskurt OK, Meiselman H 2003 Blood rheology and hemodynamics *Semin Thromb Hemost* **29** 435-450

[34] Tomaliuolo G, Carciati A, Caserta S, Guido S 2015 Blood linear viscoelasticity by small amplitude oscillatory flow *Rheologica Acta* **5** 1-11

[35] Persson SGB 1983 *The significance of haematological data in the evaluation of soundness and fitness in the horse* In: *Equine exercise physiology* eds Snow DH, Persson SGB, Rose RJ (Cambridge, UK: Granta Editions)

[36] Kraft W, Duerr UM (eds) 2005 *Klinische Labordiagnostik in der Tiermedizin* (Stuttgart, New York: Schattauer) 6th edition pg. 59

[37] Plasenzotti R, Stoiber B, Posch M, Windberger U 2004 Red blood cell deformability and aggregation behaviour in different animal species *Clin Hemorheol Microcirc* **31** 105-111

Clay-cement suspensions - rheological and functional properties

L Wojcik, P Izak, J Mastalska-Poplawska, M Gajek

AGH University of Science and Technology, Faculty of Materials Science and Ceramics, Department of Ceramics and Refractories, Krakow, Poland

E-mail: lukasz.wojcik@agh.edu.pl

Abstract. The piping erosion in soil is highly unexpected in civil engineering. Elimination of such damages is difficult, expensive and time-consuming. One of the possibility is the grouting method. This method is still developed into direction of process automation as well as other useful properties of suspensions. Main way of modernization of the grouting method is connected it with rheology of injection and eventuality of fitting them to specific problems conditions. Very popular and useful became binders based on modified clays (clay-cement suspensions). Important principle of efficiency of the grouting method is using of time-dependent pseudothixotropic properties of the clay-cement suspensions. The pseudo-rheounstability aspect of the suspensions properties should be dedicated and fitted to dynamic changes of soil conditions destructions. Whole process of the modification of the suspension rheology is stimulated by the specific agents. This article contains a description of practical aspects of the rheological parameters managing of the clay-cement suspensions, dedicated to the building damages, hydrotechnic constructions etc.

Keywords: Pseudothixotropy, Clay-cement Suspension, Grouting, Piping Erosion

1. Introduction

Clay-cement suspensions are used in the construction and maintenance of hydrotechnic constructions such as embankments, dikes and dams, where the main factor in their suitability is high water tightness. This material, apart from the very good waterproofing properties, should be characterized by the low cost, simple fabrication and application [1-2].

During reparation of the embankments, using various drilling techniques (for example WIPS or DSM) enabling the creation of protective waterproofing barrier in the cross-section of the shaft or filling occurring cracks and voids (injection methods), binder is usually introduced as the suspension [3]. This operation aims to restore the integrity and liquidation of penetrations and waterways, whereby a major factor determining the suitability of the used material is its ability to retain water - determined by the filtration coefficient (k) [4].

Another very important factor determining suitability of the sealing material is its rheological properties. Modernization of levees, often located in inaccessible terrain, makes it necessary to transport the binder in considerable distances. The most widely used solution is to create a portable base, which is followed by the production of the binder in the form of suspension, and transport through pipelines to the place of performed work. Control and modification of the rheological properties at the place of the suspension preparation becomes one of the basic steps of determining their suitability in this type of work [5-7].

2. Materials and methods

The subject of the study were samples of the clay-cement binders prepared in laboratory using smectite-illite clay, CEM II / B-M 32.5 R cement produced by Lafarge and bonding inorganic modifier. The effect of changes in the share of individual components on the properties was investigated.

Studies have been conducted in two stages. In the first stage rheological properties of the freshly prepared suspensions were measured, whereas in the second stage mechanical strength and filtration coefficient of the binders after 90 days were checked. Studies of the hardened binders were performed by the use of Tiratest 300 apparatus and by the use of device measuring filtration coefficient by the flow pump method.

Rheological measurements were conducted by the use of two different rheometers, Brookfield R/S in the coaxial cylinders system and Anton Paar Physica MCR-301 in the parallel plates system. Flow curves, as well as G′ and G″ moduli as a function of the composition changes, were determined. The effect of changes in the cement quantities and clay substrate concentration on the rheological properties of hydrotechnic binders was identified.Two series of tests were made. In the first one, cement part at a constant concentration of the clay material was changed, whereas in the second series clay content was changed. All measurements were made on samples immediately after its preparation, at 20°C.

3. Results and discussion

3.1. Rheological measurements

Analysis of the obtained results have shown very characteristic course of the flow curves. All registered runs are accompanied by a sharp increase in the shear stress in the low shear rates until the characteristic peaks in the range of 5 to 10 s^{-1}. At higher shear rates, i.e.> 10 s^{-1} , stabilization of shear stress is obtained, what means shear thinning (pseudoplastic flow). Return flow curves corresponding to the decreasing shear rates are localized under curves recorded for increasing shear rates, indicating on thixotropic properties. In a series of samples with a variable cement content, shift of the maximum shear stress in the direction of lower shear rates with increasing cement content was registered (figure 1).

In a series with a variable clay content, shift of the shear stress maximum was not observed. But very rapid increase in the shear stress values was highlighted, especially with the participation of the two highest concentrations of clay, i.e. 29% and 32%. Registered maximum values were up to 7-fold higher than those recorded at lower concentrations (figure 2).

Oscillation tests were carried out at variable oscillation frequency, in the range of 0.05 to 500 Hz, for 120 minutes. In each case, slurry samples were examined immediately after its preparation. The measurements were performed at 20°C.

As in the case of the flow curves, tests were performed for two series - by changing respectively the content of cement and clay. Similar courses of changes in elastic modulus G' and viscous modulus G" were noticed.

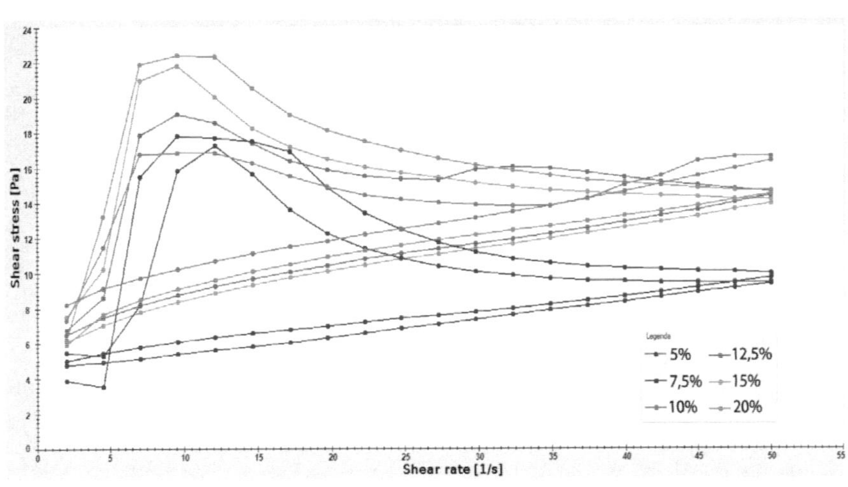

Figure 1. Influence of the cement content on clay-cement suspensions flow curves

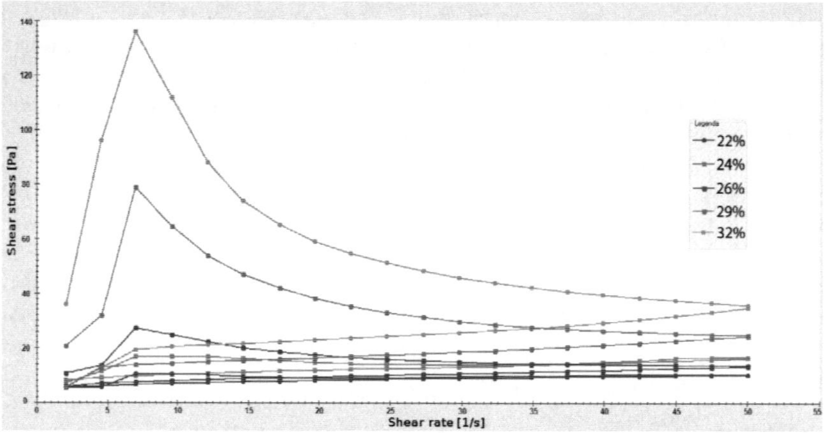

Figure 2. Influence of the clay content on clay-cement suspensions flow curves

With the passage of time and decrease of the oscillation frequency, an increase in the value of both moduli was recorded. At the beginning, values of the G" modulus were higher than the values of the G' modulus. As time passes, an increase in the value of the elasticity modulus G' was registered, and cross-over point as a function of the oscillation frequency occurred.

In a series of the samples with the increasing cement content describing phenomena also occurred (figure 3) but it is difficult to show any regularity associated with changes in the share of cement content (figure 4.).

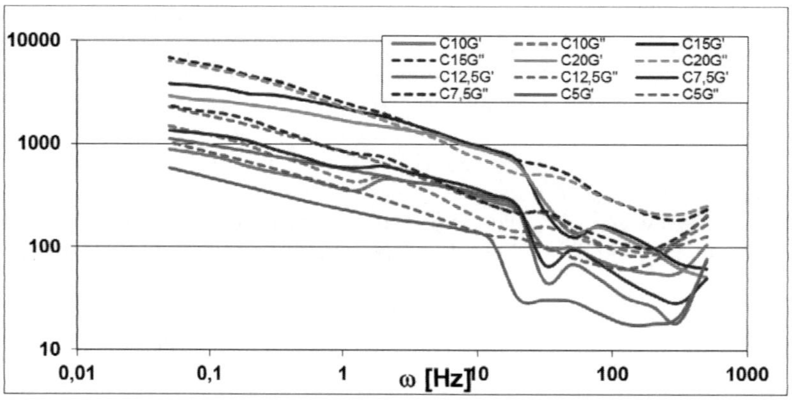

Figure 3. Influence of cement content on storage modulus G′ and loss modulus G″

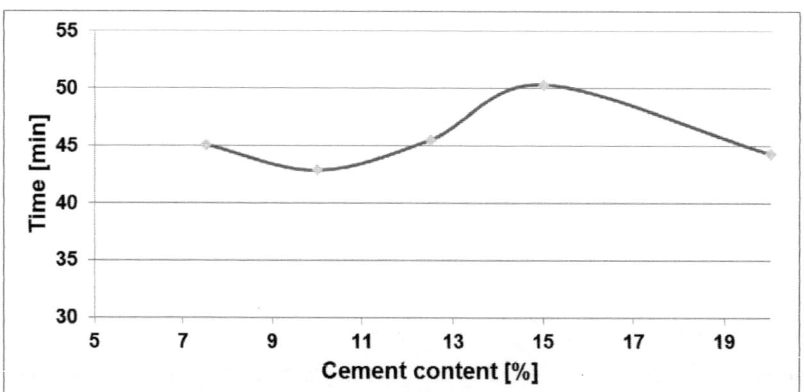

Figure 4. Influence of cement content on G′ and G″ cross-over points

In a series of samples with varying clay content (figure 5), said regularity is more visible. As for the previous series, an increase in G′ and G″ moduli values with increasing clay concentration was observed. However, the time after which "cross-over" phenomena occurs, is related to the change in composition of respondents suspensions. With the increase of the clay concentration, shortening of the cross-over occurrence follows. These values vary from 51 minutes for the lowest concentrations up to approximately 37 minutes for concentrations above 26% (figure 6). This means that the clay mineral accelerates the transition from liquid to solid state of the clay-cement suspension.

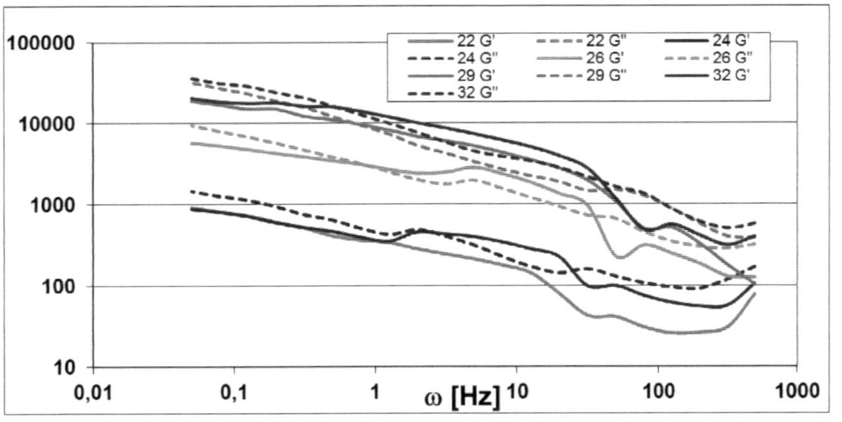

Figure 5. Influence of clay content on storage modulus G′ and loss modulus G″

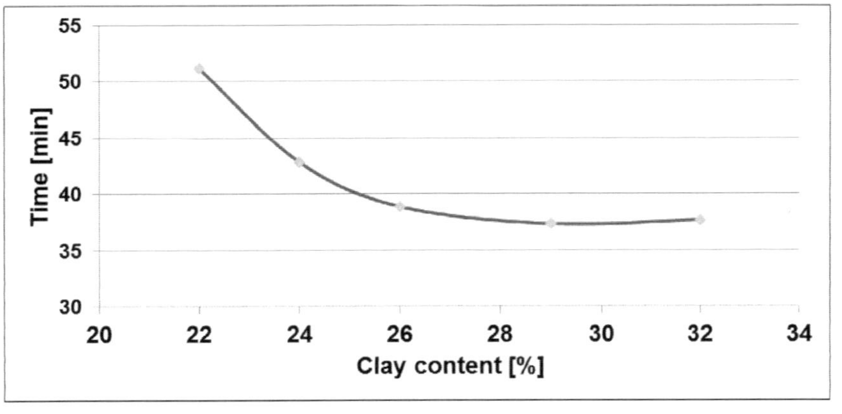

Figure 6. Influence of clay content on G′ and G″ cross-over points

3.2. Filtration coefficient and mechanical strength measurements

Filtration coefficient and mechanical strength measurements were performed on solidified binder samples after 90 days of puberty. Studies were carried out on cylindrical samples having a diameter of 35mm and 35mm height.

The results of the filtration coefficient measurements are shown in figures 7 and 8. Analyzing the obtained results, very strong relationship between the concentration of clay and the filtration coefficient of studied binders can be seen. With the increase of the clay component concentration the filtration coefficient rate decreases very rapidly. Particularly, decrease at lower concentrations of clay mineral can be seen, wherein the change in concentration of about 2% causes decrease of the coefficient rate of magnitude. For higher concentrations these drops are not so high, but trend persists. For the tested clay concentrations, values of the filtration coefficient are in the range from 10^{-8} to 10^{-11} m/s.

For a series of samples with varying cement concentrations decrease in the filtration coefficient rate with increasing concentration of cement was also observed, although these drops were not as high as in the case for a series with varying clay content. The total range of variation was approximately one order of magnitude, from 10^{-9} to 10^{-10} m/s.

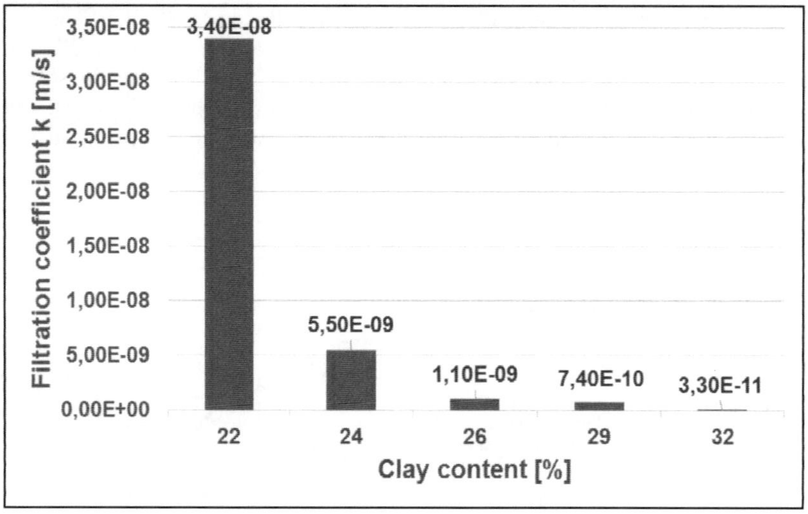

Figure 7. Influence of clay content on filtration coefficient

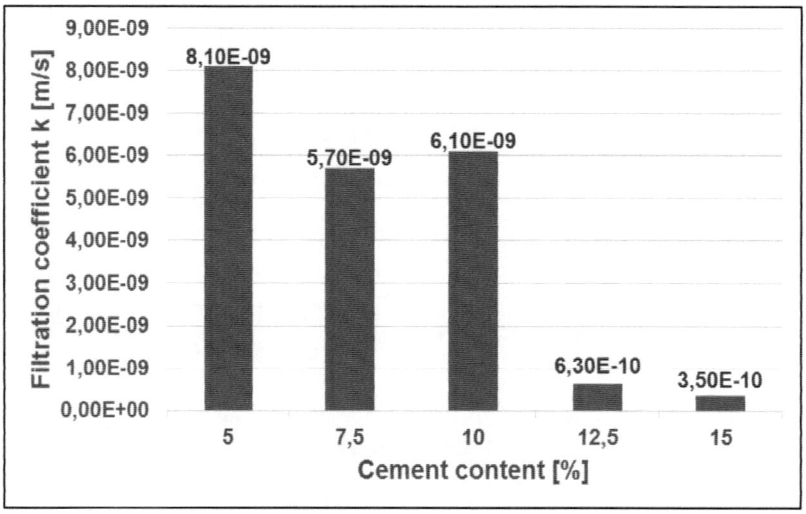

Figure 8. Influence of cement content on filtration coefficient

Mechanical strength measurements have shown that the crucial meaning has the cement content, where a steady increase in mechanical strength with an increase in the cement concentration was observed. The range of variation changed from 0.75 to 1.75 MPa. For a series with varying clay content, there were no significant influence of this parameter on the measured values. In the whole range of concentrations, values were more or less stable, approximately 1 MPa.

Results of the mechanical strength measurements are shown in figures 9 and 10.

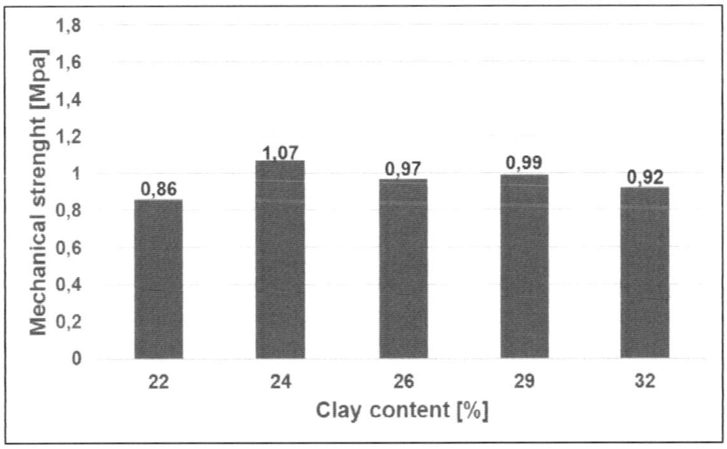

Figure 9. Influence of clay content on mechanical strength

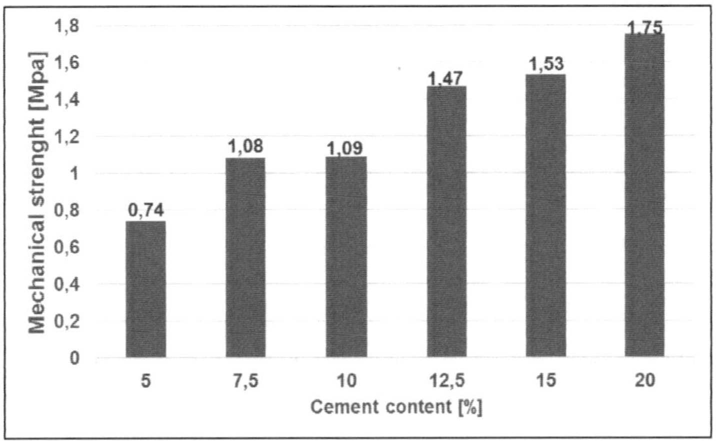

Figure 10. Influence of cement content on mechanical strength

4. Conclusions

Conducted tests showed that functional properties of clay-cement binders can be programmable in a wide range. Rheological behaviour is a result of the properties of clay materials and cement hydration products. Viscosity and shear stress could be changed in wide range by changing the composition. This enables the design of blends significantly different from each other with rheological properties, so that they can be used in different process conditions.

The oscillation moduli G′ and G″ shown that during the reaction viscoelastic properties of samples are changing. During the studies, the characteristic cross-over points, evidence of change in the properties of slurry from viscous to elastic, were observed. Time after which they appear is strongly dependent on the amount of raw mineral, i.e. illite-smectite clay. For the varying cement content there was no regularity in this case.

Mechanical strength and filtration coefficient can be modified by changing of clay or cement content. Mechanical strenght is related to cement content whereas the filtration coefficient strongly depends on the clay content.

References

[1] Kulikow J N 1984 *Comprehensive method of flood barriers sealing* (Moscow: Nedra)

[2] Dziewanski J, Malolepszy J, Pilecki Z 2005 *Optimizing the slag-alkali injection grouts compositions used in underground constructions* (Krakow: The Mineral and Energy Economy Research Institute of the Polish Academy of Sciences)

[3] Kus R 2005 *Use of materials based on polymineral clays for building waterproofing barriers and compaction of soil within the hydrotechnic objects and flood control facilities. Methods of assessment of their geotechnical and rheological properties* (Warsaw: Institute of Meteorology and Water Management)
[4] Pazdro Z, Kozerski B *1990 General Hydrogeology* (Warsaw: Geological Publishing House)
[5] Herzig J, Szczepanska J *Use of flow pump method in the measurements of filtration coefficient in poorly permeable soils* Contemporary Problems of Hydrogeology vol VII (Krakow: AGH UST Publishing House)
[6] Wojcik L, Izak P, Kus R 2008 *Ceramics* **103** 1181-88
[7] Izak P 2015 *Rheology in Ceramics* (Krakow: AGH UST Publishing House)

IOP Conf. Series: Journal of Physics: Conf. Series **790** (2017) 012037 doi:10.1088/1742-6596/790/1/012037

Effect of crosslinker length on the elastic and compression modulus of poly(acrylamide) nanocomposite hydrogels

J Zaragoza, A Chang and P Asuri

Department of Bioengineering, Santa Clara University, Santa Clara, California, United States of America

E-mail: asurip@scu.edu

Abstract. Polymer hydrogelshave shown to exhibit improved properties upon the addition of nanoparticles; however, the mechanical underpinnings behind these enhancements have not been fully elucidated. Moreover, fewer studies have focused on developing an understanding of how polymer parameters affect the nanoparticle-mediated enhancements. In this study, we investigated the elastic properties of silica nanoparticle-reinforced poly(acrylamide) hydrogels synthesized using crosslinkers of various lengths. Crosslinker length positively affected the mechanical properties of hydrogels that were synthesized with or without nanoparticles. However the degree of nanoparticle enhancement was negatively correlated to crosslinker length. Our findings enable the understanding of the respective roles of nanoparticle and polymer properties on nanoparticle-mediated enhancement of hydrogels and thereby the development of next-generation nanocomposite materials.

1. Introduction

Previous studies, both experimental and computational, have established that the addition of nanoparticles to polymer systems can have a large impact on their properties, even for very low nanoparticle weight fractions [1-6]. Despite the extensive efforts that have led to elaborate correlations among various properties of polymer systems and nanoparticles on the enhancements of the polymer properties, a complete understanding of the mechanisms in which nanoparticles act to enhance polymer properties is lacking [7-8]. While several hypotheses have been proposed to explain the mechanisms behind the enhancements, one hypothesis that has gained significant support is that strong polymer-nanoparticle interactions can facilitate the formation of additional crosslinks within the polymer network, which thereby leads to enhancements in polymer properties [9-12]. However, it is not clear how the formation of the additional 'pseudo' crosslinks affects hydrogel microarchitecture; for instance, the relative effects of nanoparticle mediated crosslinks on the ordered and disordered regions of a polymer are largely unknown.

In this study, we used chemically crosslinked polyacrylamide (pAAM) hydrogels incorporating silica nanoparticles (SiNPs) as a model system to experimentally study the role of polymer-nanoparticle interactions in enhancing the mechanical properties of hydrogels. Furthermore, we examine the effect of polymer properties, specifically by varying the crosslinker length, on the enhancements in polymer mechanical properties due to the addition of nanoparticles. We expected that determining the effect of polymer properties on the degree of nanoparticle-mediated enhancements would reveal if the nanoparticles impact both ordered and disordered regions of the hydrogel network or if they are selective to one region, thereby shedding light on the mechanisms behind the nanoparticle-mediated enhancements in hydrogel mechanical properties. Our data obtained using unconfined compression and rheological characterization of pAAm-SiNP composites showed enhancements in the elastic properties of hydrogels upon the addition of nanoparticles, consistent with previous investigations [11-14], irrespective of the choice of the crosslinker. Our data also demonstrates that the degree of nanoparticle-mediated enhancements was strongly dependent on the crosslinker length, with statistically similar modulii for neat pAAm hydrogels and pAAm-SiNP hydrogels prepared with the longest crosslinker used in this study.

2. Experimental procedure

2.1. Materials

All the materials for the polymerization reaction, acrylamide (AAm, monomer), initiator, ammonium persulfate (APS, initiator), N,N,N',N'-tetramethylethylenediamine (TEMED, catalyst), N,N'-methylenebis(acrylamide) (bisacrylamide, crosslinker) and poly(ethylene glycol) diacrylates of various chain lengths (n = 2, 4, and 17, crosslinker) were purchased from Sigma Aldrich (St. Louis, MO) and used as received. Tris-HCl buffer (pH 7.2) was obtained from Life Technologies (Carlsbad, CA) and binzil silica nanoparticle colloid solution with mean particle size of 4 nm was obtained as a gift from AkzoNobel Pulp and Performance Chemicals Inc. (Marietta, GA).

2.2. Polymerization reaction

Chemically crosslinked poly(acrylamide) hydrogels were prepared as previously reported [14]. Briefly, AAm and crosslinker stocks were diluted to their desired concentrations in pH 7.2, 250 mMTris-HCl buffer, followed by the addition of TEMED (0.1% of the final reaction volume) and 10% w/v APS solution (1% of the final reaction volume). For nanocomposite hydrogels, various amounts of silica nanoparticles were added to the reaction mixture prior to the addition of APS and TEMED. Polymerization reactions were performed at 25 °C between parallel plates (of a mold or the rheometer cell) to minimize exposure to air as oxygen inhibits the free radical polymerization reaction.

2.3. Compressive modulus measurements

Compressive modulus measurements for the pAAm hydrogels were performed as previously reported (Zaragoza, 2015). Briefly, we first prepared pAAm hydrogel disks using an acrylic mold (1.6 mm height and 6.5 mm in radius) by pipetting 210 μL of a well-mixed reaction mixture into individual molds. After complete gelation (gelation usually occurs within 20 minutes), pAAm hydrogel discs were removed from the mold, wiped gently with tissue paper to remove any excess water, before adding the samples to the Mach-1 mechanical testing system (Biomomentum, Canada). The pAAm gel disks were compressed under unconfined conditions and at room temperature at 0.1 mm/s to 50% of sample thickness, and the compressive modulus was determined by calculating the slope of the linear region of the stress-strain curves (typically between 10–15% strain). The compressive moduli are reported as average of three independent measurements.

2.4. Rheological characterization

Rheological measurements of pAAm hydrogels were carried out, as previously described, using the MCR302 rotational rheometer (AntorPaar, Austria). Briefly, 500 μL of a well-mixed reaction mixture was pipetted onto the lower plate of the rheometer and the upper plate was lowered until the desired gap distance (1 mm) was achieved. Amplitude sweeps at a constant frequency of 1Hz were then carried out to ensure measurements were carried out in the linear viscoelastic regime of the acrylamide hydrogels. Next, dynamic sweep tests over frequencies ranging from 0.1-100 Hz. Final hydrogel parameters were determined by following the gelation for 90 minutes at 1 Hz and 1% strain for all samples. The elastic moduli are reported as average of three independent measurements.

3. Results and discussion

3.1. Effect of nanoparticles on hydrogel elastic modulus

Poly(acrylamide) (pAAm), synthesized using 10% w/v acrylamide and 0.5 w/v bisacrylamide,wasused as the model because of its widespread application and availability, ease of synthesis, and previous use in experimental analyses. Commercially available silica nanoparticles (4 nm diameter) were used due to their reported ability to strongly interact with pAAmchains and thereby influence various pAAmhydrogel properties [14]. Elastic moduli of the hydrogels prepared with and without SiNPs were tested using rotational rheology, which indicated that bisacrylamide-crosslinked pAAm hydrogels exhibit enhancements in mechanical properties upon the addition of SiNPs at weight fractions as low as 1% w/v [Figure 1]. The SiNP-mediated enhancements increase linearly with nanoparticle concentration and plateau at 4% w/vSiNPs. The observed results are attributed to the formation of "pseudo-crosslinks" facilitated by strong nanoparticle-polymer interactions and eventual saturation in crosslinker density [12, 14, 15].

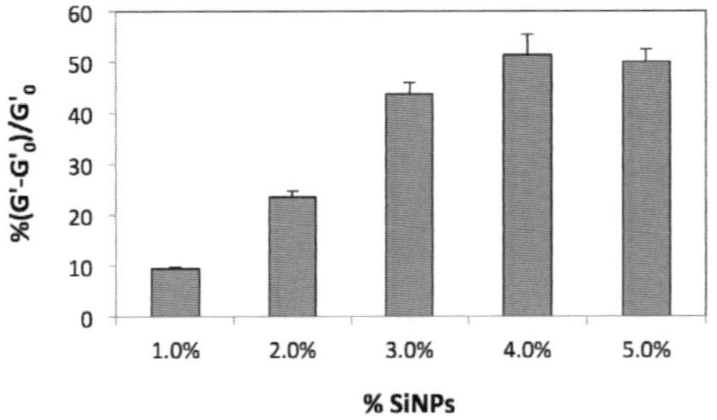

Figure 1. Percent relative elastic modulus of pAAm hydrogels as a function of 4 nm SiNP concentration. The values for relative elastic modulus were calculated by normalizing the modulus for pAAm-SiNP hydrogels (G') to the modulus for control pAAm gels (G'_0). Error bars indicate standard deviation of triplicate measurements.

3.2. Effect of crosslinker length on hydrogel mechanical properties

Next, we used PEG diacrylates as crosslinkers instead of bisacrylamide since PEG diacrylates are commercially available in various chain lengths and therefore allow assessment of crosslinker length on hydrogel properties. Mechanical testing revealed a length dependent increase in mechanical properties. Specifically, rheological tests indicated that the longer the PEG chain length, the higher the viscoelastic modulus of the resulting hydrogel (Figure2), with the dimer crosslinker leading to the lowest value in elastic modulus. These findings are corroborated by unconfined compression tests which showed that pAAm hydrogels crosslinked with 17-mer PEG exhibited highest modulus followed by those crosslinked using tetramer and dimer crosslinkers (Figure 2). To account for these observations, we hypothesize that the longer crosslinkers lead to increases in the ordered structure within the hydrogel network, which in turn leads to increases in hydrogel mechanical properties. Our findings and hypothesis are consistent with previous studies that indicate that hydrogel modulus can be controlled by varying crosslinker length [16, 17]. Further experiments including X-ray diffraction are necessary to confirm mechanisms by which crosslinker length affects hydrogel properties.

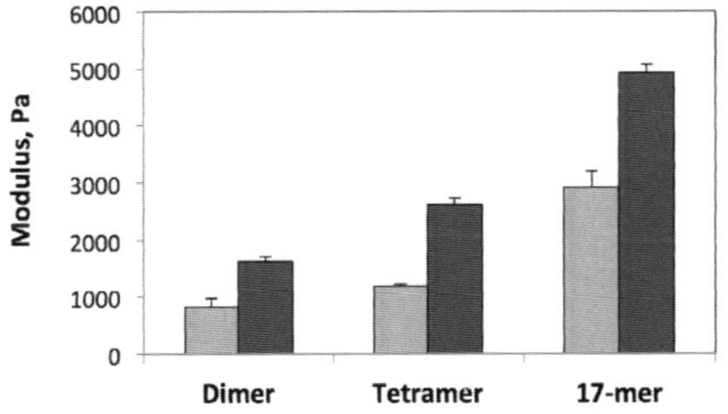

Figure 2. Variation in elastic (light grey) and compressive (dark grey) modulus of pAAm hydrogels crosslinked with PEG diacrylate crosslinkers of varying chain length. Error bars indicate standard deviation of triplicate measurements.

3.3. Effect of crosslinker length on nanoparticle-mediated enhancements in hydrogel properties

Finally, we explored the effect of adding nanoparticles to hydrogels synthesized using crosslinkers of various lengths. We selected 3% (w/v) SiNPsas it led to the highest level of enhancements within the linear regime for Bis-crosslinked pAAM hydrogels (Figure 1). Consistent with previous results, addition of nanoparticles enhanced the elastic modulus of pAAm hydrogels crosslinked with the various PEG diacrylates. However, the degree of enhancement was inversely correlated to crosslinker length (Figure 3a), with statistically similar values of elastic modulus for hydrogels crosslinked using the longest crosslinker (i.e.,17-mer PEGdiacrylate) prepared with or without nanoparticles (Figure2). In contrast, we observed >50% enhancement in elastic modulus for pAAm hydrogels crosslinked using the PEG dimerupon the addition of SiNPs(Figure3). Similar trends, i.e., decrease in the degree of SiNP-mediated enhancement with increasing crosslinker length, were also observed for unconfined compression (Figure4). The observed saturation in SiNP-mediated increases in modulus suggests that modifying either parameter, i.e. increasing nanoparticle concentration or crosslinker chain length may have a similar effect on hydrogel mechanical properties and can be attributed to the hypothesis that both chemical crosslinkers and nanoparticle-based pseudo crosslinkers lead to enhancements in hydrogel modulus by increasing the ordered structure of the polymer network [12, 14, 18].

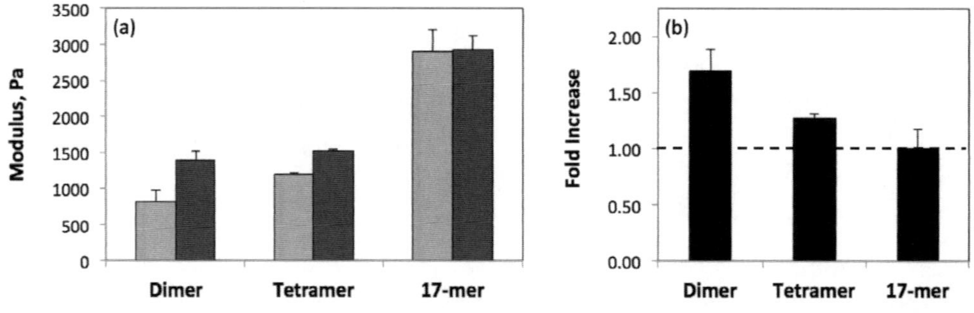

Figure 3. (a)Elastic modulus of pAAm hydrogels prepared with or without 4 nm SiNPs and PEG diacrylate crosslinkers of varying chain length – control pAAm hydrogels (light grey) and pAAm-SiNP hydrogels (dark grey). (b) Fold increase in the modulus of pAAm hydrogels due to the addition of 4 nm SiNPs. Error bars indicate standard deviation of triplicate measurements

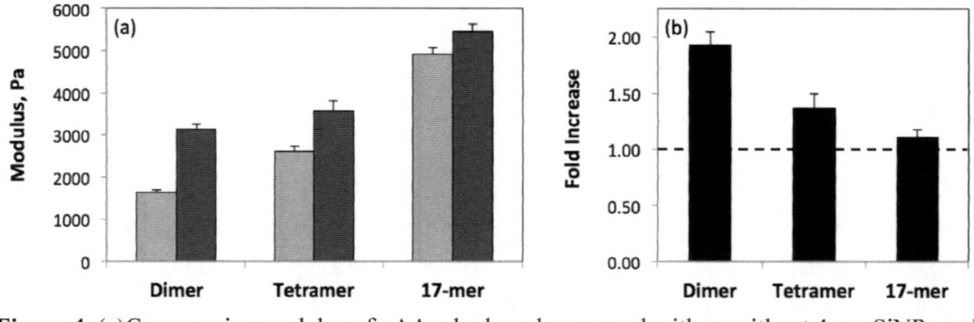

Figure 4. (a)Compressive modulus of pAAm hydrogels prepared with or without 4 nm SiNPs and PEG diacrylate crosslinkers of varying chain length – control pAAm hydrogels (light grey) and pAAm-SiNP hydrogels (dark grey). (b) Fold increase in the modulus of pAAm hydrogels due to the addition of 4 nm SiNPs. Error bars indicate standard deviation of triplicate measurements

4. Conclusions

In this work, we performed experiments that improve our understanding of the role of nanoparticles on the mechanical properties of hydrogels. Using poly(acrylamide) hydrogels incorporating silica nanoparticles as the model system, we first demonstrated that the addition of nanoparticles led to significant enhancement in hydrogel mechanical properties, possibly due to the formation of additional pseudo-crosslinks in the hydrogel network. Next, using PEG diacrylate of various chain lengths we show that pAAM hydrogels crosslinked with longer chain crosslinkers presented higher elastic and compressive moduli, relative to those crosslinked using shorter crosslinkers. We hypothesize that longer crosslinks lead to an increase in the ordered structure of the hydrogel network, although further experiments using X-ray diffraction are necessary to confirm the hypothesis. Finally, our results also demonstrate that the degree of nanoparticle-mediated enhancements was inversely correlated to the length of the crosslinker. In combination, our data indicates that nanoparticles and crosslinkers act similarly in that they both lead to an increase in the ordered structure of the polymer network, thereby leading to saturation in the enhancementsin hydrogel mechanical properties afforded by (covalent or pseudo) crosslinking. Therefore, for the pAAm hydrogels crosslinked with the 17-merPEGdiacrylatecrosslinkers, we observe statistically similar elastic and compression modulus values for hydrogels prepared with or without nanoparticles. It is however important to note that nanoparticles contribute to additional crosslinking only for systems characterized by strong polymer-nanoparticle interactions. The implications of these results are two-fold: first, the data demonstrates that nanoparticle-mediated enhancements in polymer mechanical properties are strongly dependent on the crosslinker length and not simplyits interactions with the polymer backbone. Second, since previous studies have indicated that chemical crosslinkers improve polymer mechanical properties by increasing the ordered structure within the polymer network, saturation of nanoparticle enhancements at longer crosslinker length indicates that nanoparticles behave similarly to chemical crosslinkers, i.e., they impact the ordered regions of the hydrogel network. Overall, these results suggest a new understanding of the roles of chemical crosslinkers and nanoparticles in mechanical reinforcement of polymer nanocomposites.

Acknowledgments

This work was supported by the School of Engineering at Santa Clara University. Silica nanoparticles were a kind gift from Akzo Nobel Pulp and Performance Chemicals Inc.

References

[1] Jordan J, Jacob KI, Tannenbaum R, Sharaf MA and Jasiuk I 2005 Experimental trends in polymer nanocomposites – a review *Materials Science and Engineering A* **393**1–11

[2] Hussain F, Hojjati M, Okamoto M and Gorga R 2006 Polymer-matrix nanocomposites, processing, manufacturing, and application*Journal of Composite Materials***40** 1511-75

[3] Tjong SC 2006Structural and mechanical properties of polymer nanocomposites *Materials Science and Engineering R***53**73–197

[4] Münstedt H and Triebel C 2009 Elastic properties of polymer melts filled with nanoparticles*AIP Conference Proceedings***1375**201–7

[5] Sen S, Thomin JD, Kumar SK and Keblinski P 2007 Molecular underpinnings of the mechanical reinforcement in polymer nanocomposites *Macromolecules***40(11)**4059–67

[6] Xiao J, Huang Y and Manke CW 2010 Computational design of polymer nanocomposite coatings: a multiscale hierarchical approach for barrier property prediction*Industrial & Engineering Chemistry Research***49**7718–27

[7] Yan L and Xie X 2013 Computational modeling and simulation of nanoparticle self-assembly in polymeric systems: Structures, properties and external field effects *Progress in Polymer Science***38**369–405

[8] Balazs AC, Emrick T and Russell T 2006 Nanoparticle polymer composites: where two small worlds meet *Science***314** 1107–10

[9] Jancar J, Douglas JF, Starr FW, Kumar SK, Cassagnau P, Lesser AJ, Sternsteinh SS and Buehlerb MJ 2010 Current issues in research on structure-property relationships in polymer nanocomposites *Polymer***51** 3321–43

[10] Kutvonen A, Rossi G, Puisto SR, Rostedt NKJ and Ala-Nissila T 2012 Influence of nanoparticle size, loading, and shape on the mechanical properties of polymer nanocomposites *The Journal of Chemical Physics***137** 214901

[11] Wu CJ, Wilker JJ and Schmidt G2013 Robust and adhesive hydrogels from cross-linked poly(ethylene glycol) and silicate for biomedical use *Macromol Bioscience* **13**59–66

[12] Prado-Gotor R, Lopez-Perez G, Martin MJ, Cabrera-Escribano F and Franconetti A2014 Use of gold nanoparticles as crosslink agent to form chitosan nanocapsules: Study of the direct interaction in aqueous solutions*Journal of Inorganic Biochemistry***135**77–85

[13] Xia LW, Xie R, Ju XJ, Wang W, Chen QM and Chu LY 2013Nano-structured smart hydrogels with rapid response and high elasticity*Nature Communications* **4**2226

[14] Zaragoza J, Babhadiashar N, O'Brien V, Chang A, Blanco M, Zabalegui A, Lee H and Asuri P 2015 experimental investigation of mechanical and thermal properties of silica nanoparticle-reinforced poly(acrylamide) nanocomposite hydrogels *PLoS ONE***10** e0136293

[15] Fu SY, Feng XQ, Lauke B and Mai YW 2008 Effects of particle size, particle/matrix interface adhesion and particle loading on mechanical properties of particulate-polymer composites *Composites Part B: Engineering***39** 933–61

[16] Syverud K, Pettersen S, Draget K and Chinga-Carrasco G 2014 Controlling the elastic modulus of cellulose nanofibril hydrogels – scaffolds with potential in tissue engineering *Cellulose***22** 473–81

[17] Wagner B, Tharmann R, Haase I, Fischer M, Bausch AR 2006 Cytoskeletal polymer networks: the molecular structure of cross-linkers determines macroscopic properties *Proceedings of the National Academy of Sciences***103**13974–8

[18] Ramesan M 2013 Fabrication, characterization, and properties of poly(ethylene-co-vinyl acetate)/magnetite nanocomposites *Journal of Applied Polymer Science***131**40116

Institute of Physics
Dirac House, Temple Back
Bristol BS1 6BE UK

ISSN: 1742-6588
ISBN 978-1-5108-3678-5

CURRAN ASSOCIATES INC.
proceedings.com

International Conference on Environmental, Industrial and Energy Engineering (EI2E 2017)

IOP Conference Series: Earth and Environmental Science
Volume 104

Shanghai, China
19 – 22 October 2017

IOP
Institute of Physics